T0211112

Communications
in Computer and Information Science   1450

Ladjel Bellatreche · Marlon Dumas ·
Panagiotis Karras · Raimundas Matulevičius ·
Ahmed Awad · Matthias Weidlich ·
Mirjana Ivanović · Olaf Hartig (Eds.)

New Trends in Database and Information Systems

ADBIS 2021 Short Papers, Doctoral Consortium and Workshops:
DOING, SIMPDA, MADEISD, MegaData, CAoNS
Tartu, Estonia, August 24–26, 2021
Proceedings

 Springer

Editors
Ladjel Bellatreche (iD)
LIAS/ISAE-ENSMA
Chasseneuil-du-Poitou, France

Panagiotis Karras (iD)
Aarhus University
Aarhus, Denmark

Ahmed Awad (iD)
University of Tartu
Tartu, Estonia

Mirjana Ivanović (iD)
University of Novi Sad
Novi Sad, Serbia

Marlon Dumas (iD)
University of Tartu
Tartu, Estonia

Raimundas Matulevičius (iD)
University of Tartu
Tartu, Estonia

Matthias Weidlich (iD)
Humboldt-Universität zu Berlin
Berlin, Germany

Olaf Hartig (iD)
Linköping University
Linköping, Sweden

ISSN 1865-0929   ISSN 1865-0937 (electronic)
Communications in Computer and Information Science
ISBN 978-3-030-85081-4   ISBN 978-3-030-85082-1 (eBook)
https://doi.org/10.1007/978-3-030-85082-1

This Springer imprint is published by the registered company Springer Nature Switzerland AG
The registered company address is: Gewerbestrasse 11, 6330 Cham, Switzerland

Preface

This volume contains a selection of the papers presented at the 25th European Conference on Advances in Databases and Information Systems (ADBIS 2021), held during August 24–26, 2021, at Tartu, Estonia.

The ADBIS series of conferences aims at providing a forum for the presentation and dissemination of research on database and information systems, the development of advanced data storage and processing technologies, and the design of data-enabled systems/software/applications. ADBIS 2021 in Tartu continues after St. Petersburg (1997), Poznań (1998), Maribor (1999), Prague (2000), Vilnius (2001), Bratislava (2002), Dresden (2003), Budapest (2004), Tallinn (2005), Thessaloniki (2006), Varna (2007), Pori (2008), Riga (2009), Novi Sad (2010), Vienna (2011), Poznań (2012), Genoa (2013), Ohrid (2014), Poitiers (2015), Prague (2016), Nicosia (2017), Budapest (2018), Bled (2019), and Lyon (2020). This edition has been totally managed during the COVID-19 pandemic.

The program of ADBIS 2021 included keynotes, research papers, thematic workshops, and a doctoral consortium. The conference attracted 70 paper submissions from 261 authors in 39 countries and from all continents. After rigorous reviewing by the Program Committee, the 18 papers selected as full contributions are included in the LNCS proceedings volume.

Furthermore, the Program Committee selected 8 more papers as short contributions and 21 papers from the workshops and doctoral consortium which are published in this CCIS volume on New Trends in Databases and Information Systems. All papers were evaluated by at least three reviewers and some by four reviewers. The selected papers span a wide spectrum of topics in databases and related technologies, tackling challenging problems and presenting inventive and efficient solutions.

ADBIS 2021 strived to create conditions for more experienced researchers to share their knowledge and expertise with young researchers. In addition, the following five workshops and doctoral consortium associated with ADBIS were co-allocated with the main conference:

- Intelligent Data - From Data to Knowledge (DOING 2021), organized by Mírian Halfeld Ferrari (Université d' Orleans, INSA CVL, LIFO EA, France) and Carmem S. Hara (Universidade Federal do Paraná, Brazil).
- Data-Driven Process Discovery and Analysis (SIMPDA 2021), organized by Paolo Ceravolo (Università degli Studi di Milano, Italy), Maurice van Keulen (University of Twente, The Netherlands), and Maria Teresa Gomez Lopez (University of Seville, Spain),
- Modern Approaches in Data Engineering and Information System Design (MADEISD 2021), organized by Ivan Luković (University of Belgrade, Serbia), Slavica Kordić (University of Novi Sad, Serbia), and Sonja Ristić (University of Novi Sad, Serbia).

- Advances in Data Systems Management, Engineering, and Analytics (MegaData 2021), organized by Yaser Jararweh (Duquesne University, USA), Tomás F. Pena (University of Santiago de Compostela, Spain), and Feras M. Awaysheh (University of Tartu, Estonia).
- Computational Aspects of Network Science (CAoNS 2021), organized by Dimitrios Katsaros (University of Thessaly, Greece) and Yannis Manolopoulos (Open University of Cyprus and Aristotle University of Thessaloniki, Greece).
- Doctoral Consortium (DC), co-chaired by Mirjana Ivanović (University of Novi Sad, Serbia) and Olaf Hartig (Linköping University, Sweden).

We would like to express our gratitude to every individual who contributed to the success of ADBIS 2021. First, we thank all authors for submitting their research papers to the conference. We are also indebted to the members of the community who offered their precious time and expertise in performing various roles ranging from organizational to reviewing - their efforts, energy, and degree of professionalism deserve the highest commendations. Special thanks to the Program Committee members and the external reviewers for evaluating papers submitted to ADBIS 2021, ensuring the quality of the scientific program, despite the COVID-19 pandemic. Thanks also to all the colleagues, secretaries, and engineers involved in the conference organization, as well as the workshop organizers. Special thanks are due to the members of the Steering Committee, in particular, its chair, Yannis Manolopoulos, for all their help and guidance. A particular thank you to the University of Tartu's Institute of Computer Science for hosting and supporting the conference despite the uncertainties created by the ongoing pandemic.

Finally, we thank Springer for publishing the proceedings containing invited papers and research papers in the LNCS and CCIS series. The Program Committee work relied on EasyChair, and we thank its development team for creating and maintaining it; it offered great support throughout the different phases of the reviewing process.

Last but not least, we thank the participants of ADBIS 2021 for sharing their work and presenting their achievements, thus providing a lively, fruitful, and constructive forum, and giving us the pleasure of knowing that our work was purposeful.

June 2021

Ladjel Bellatreche
Marlon Dumas
Panagiotis Karras
Raimundas Matulevičius

Preface to ADBIS 2021 International Workshops and Doctoral Consortium

This volume contains a selection of the papers presented at the workshops and the doctoral consortium of the 25th European Conference on Advances in Databases and Information Systems (ADBIS 2021), held during August 24–26, 2021, in Tartu, Estonia. This year, ADBIS attracted five workshop proposals. After evaluation of these proposals by the workshop chairs, four proposals were selected directly and one was accepted after a revision with the organizers. Each team of organizers established their International Program Committee, announced the workshop, organized the review process, and assembled the scientific program. All five workshops were eventually co-located with ADBIS 2021.

This volume contains 16 long papers and 5 short papers selected to be presented in these workshops and the doctoral consortium (DC). The selected papers span a wide spectrum of topics, ranging from information extraction, through data-driven process analysis and architectures for data processing infrastructures, to networked systems. Moreover, there is also a notable diversity in the adopted research methodologies, with the results including novel formal methods, contributions related to system design, and empirical studies. Further details about the scope of individual satellite events, as well as the selected papers, are given separately by the organizers as part of this preface.

We would like to express our gratitude to all people who contributed to the success of the ADBIS workshops and DC. We thank the authors for submitting their research work and are indebted to the members of the community that supported the selection of papers by providing insightful reviews. Moreover, we are particularly grateful for the efforts of the organizers of the workshops and the DC to promote and coordinate the respective events, and for setting up a high-quality program. The support of the ADBIS 2021 general chairs, Marlon Dumas and Raimundas Matulevičius, was also much appreciated. Moreover, we thank Springer, for publishing the proceedings as part of their CCIS series, and the team behind EasyChair as the tool was of great help throughout the different phases of the reviewing process. Last but not least, we are grateful to the participants of the ADBIS 2021 workshops and DC, for their contributions to making the satellite events a forum for lively, fruitful, and constructive discussions.

June 2021

Ahmed Awad
Matthias Weidlich

Organization

General Chairs

Marlon Dumas   University of Tartu, Estonia
Raimundas Matulevičius   University of Tartu, Estonia

Program Committee Co-chairs

Ladjel Bellatreche   ISAE-ENSMA Poitiers, France
Panagiotis Karras   Aarhus University, Denmark

Workshop Co-chairs

Ahmed Awad   University of Tartu, Estonia
Matthias Weidlich   Humboldt University of Berlin, Germany

Doctoral Consortium Co-chairs

Mirjana Ivanović   University of Novi Sad, Serbia
Olaf Hartig   Linköping University, Sweden

Webmaster

Mubashar Iqbal   University of Tartu, Estonia

Proceedings Technical Editor

Abasi-Amefon Obot Affia   University of Tartu, Estonia

Technical Arrangements

Orlenys Lopez-Pintado   University of Tartu, Estonia

Financial and Local Arrangements

Anneli Vainumae   University of Tartu, Estonia

Steering Committee

Yannis Manolopoulos   Open University of Cyprus, Cyprus
  (Chair)
Ladjel Bellatreche   ISAE-ENSMA Poitiers, France

Maria Bielikova	Kempelen Institute of Intelligent Technologies, Slovakia
Barbara Catania	University of Genoa, Italy
Jérôme Darmont	University of Lyon 2, France
Johann Eder	Alpen Adria Universität Klagenfurt, Austria
Tomáš Horváth	Eötvös Loránd University, Hungary
Mirjana Ivanović	University of Novi Sad, Serbia
Marite Kirikova	Riga Technical University, Latvia
Rainer Manthey	University of Bonn, Germany
Manuk Manukyan	Yerevan State University, Armenia
Tadeusz Morzy	Poznan University of Technology, Poland
Kjetil Nørvåg	Norwegian University of Science and Technology, Norway
Boris Novikov	National Research University, Higher School of Economics, Saint Petersburg, Russia
George Papadopoulos	University of Cyprus, Cyprus
Jaroslav Pokorny	Charles University in Prague, Czech Republic
Bernhard Thalheim	Christian-Albrechts University of Kiel, Germany
Goce Trajcevski	Iowa State University, USA
Valentino Vranić	Slovak University of Technology in Bratislava, Slovakia
Tatjana Welzer	University of Maribor, Slovenia
Robert Wrembel	Poznan University of Technology, Poland
Ester Zumpano	University of Calabria, Italy

Program Committee

Alberto Abello	Universitat Politècnica de Catalunya, Spain
Reza Akbarinia	Inria, France
Bernd Amann	Sorbonne Université, France
Hassan Badir	ENSA Tangier, Morocco
Amin Beheshti	Macquarie University, Australia
Andreas Behrend	Technical University of Cologne, Germany
Sadok Ben Yahia	Tallinn University of Technology, Estonia
Soumia Benkrid	ESI Algiers, Algeria
Djamal Benslimane	University of Lyon 1, France
Fadila Bentayeb	University of Lyon 2, France
Miklos Biro	Software Competence Center Hagenberg, Austria
Kamel Boukhafa	USTHB, Algeria
Barbara Catania	University of Genoa, Italy
Tania Cerquitelli	Politecnico di Torino, Italy
Richard Chbeir	University of Pau and Pays de l'Adour, France
Isabelle Comyn-Wattiau	ESSEC Business School, France
Ajantha Dahanayake	Lappeenranta University of Technology, Finland
Jérôme Darmont	University of Lyon 2, France
Christos Doulkeridis	University of Piraeus, Greece

Cedric Du Mouza	CNAM, France
Markus Endres	University of Passau, Germany
Philippe Fournier-Viger	Harbin Institute of Technology, Shenzhen, China
Johann Gamper	Free University of Bozen-Bolzano, Italy
Gabriel Ghinita	University of Massachusetts at Boston, USA
Olga Gkountouna	George Mason University, USA
Giancarlo Guizzardi	Free University of Bozen-Bolzano, Italy, and University of Twente, The Netherlands
Hele-Mai Haav	Tallinn University of Technology, Estonia
Dirk Habich	TU Dresden, Germany
Mirian Halfeld-Ferrari	Université d'Orléans, France
Tomáš Horváth	Eötvös Loránd University, Hungary
Mirjana Ivanović	University of Novi Sad, Serbia
Stefan Jablonski	University of Bayreuth, Germany
Stéphane Jean	LIAS, Poitiers University, France
Zoubida Kedad	University of Versailles, France
Marite Kirikova	Riga Technical University, Latvia
Attila Kiss	Eötvös Loránd University, Hungary
Sergio Lifschitz	Pontifícia Universidade Católica do Rio de Janeiro, Brazil
Sebastian Link	University of Auckland, New Zealand
Ivan Luković	University of Belgrade, Serbia
Zakaria Maamar	Zayed University, United Arab Emirates
Wojciech Macyna	Wroclaw University of Technology, Poland
Federica Mandreoli	University of Modena, Italy
Yannis Manolopoulos	Open University of Cyprus, Cyprus
Patrick Marcel	Université de Tours, France
Pascal Molli	University of Nantes, France
Anirban Mondal	Ashaka University, India
Tadeusz Morzy	Poznan University of Technology, Poland
Kjetil Nørvåg	Norwegian University of Science and Technology, Norway
Boris Novikov	National Research University, Higher School of Economics, Saint Petersburg, Russia
Andreas Oberweis	Karlsruhe Institute of Technology, Germany
Carlos Ordonez	University of Houston, USA
Oscar Pastor	Universidad Politécnica de Valencia, Spain
Jaroslav Pokorný	Charles University in Prague, Czech Republic
Franck Ravat	IRIT, Université Toulouse Capitole, France
Stefano Rizzi	University of Bologna, Italy
Oscar Romero	Universitat Politècnica de Catalunya, Spain
Carmem S. Hara	Universidade Federal do Paraná, Brazil
Gunter Saake	University of Magdeburg, Germany
Kai-Uwe Sattler	Technical University Ilmenau, Germany
Milos Savic	University of Novi Sad, Serbia
Kostas Stefanidis	Tampere University, Finland

Sergey Stupnikov   Russian Academy of Sciences, Russia
Olivier Teste   Université de Toulouse, France
Maik Thiele   TU Dresden, Germany
Goce Trajcevski   Iowa State University, USA
Anton Tsitsulin   University of Bonn, Germany
Panos Vassiliadis   University of Ioannina, Greece
Thanasis Vergoulis   "Athena" Research Center, Greece
Tatjana Welzer   University of Maribor, Slovenia
Marek Wojciechowski   Poznan University of Technology, Poland
Robert Wrembel   Poznan University of Technology, Poland
Vladimir Zadorozhny   University of Pittsburgh, USA

Additional Reviewers

Petar Jovanovic   Universitat Politècnica de Catalunya, Spain
Elio Mansour   University of Pau and Pays de l'Adour, France
Sergi Nadal   Universitat Politècnica de Catalunya, Spain
Rediana Koçi   Universitat Politècnica de Catalunya, Spain
Nadia Yacoubi   Ayadi Institut Supérieur de Gestion de Tunis, Tunisia
Serafeim Chatzopoulos   "Athena" Research Center, Greece

Contents

ADBIS 2021 Short Papers

Learned What-If Cost Models for Autonomous Clustering 3
   Daniel Lindner, Alexander Löser, and Jan Kossmann

Cost-Sensitive Predictive Business Process Monitoring 14
   Martin Käppel, Stefan Jablonski, and Stefan Schönig

A Hybrid Data Model and Flexible Indexing for Interactive Exploration
of Large-Scale Bio-science Data 27
   Gajendra Doniparthi, Timo Mühlhaus, and Stefan Deßloch

Relational Conditional Set Operations 38
   Alexis I. Aspauza Lescano and Robson L. F. Cordeiro

GASP: Graph-Based Approximate Sequential Pattern Mining for Electronic
Health Records 50
   *Wenqin Dong, Eric W. Lee, Vicki Stover Hertzberg, Roy L. Simpson,
   and Joyce C. Ho*

Semi-synthetic Data and Testbed for Long-Distance E-Vehicle Routing 61
   *Andrius Barauskas, Agnė Brilingaitė, Linas Bukauskas, Vaida Čeikutė,
   Alminas Čivilis, and Simonas Šaltenis*

Looking for COVID-19 Misinformation in Multilingual Social
Media Texts 72
   *Raj Ratn Pranesh, Mehrdad Farokhnejad, Ambesh Shekhar,
   and Genoveva Vargas-Solar*

Semantic Discovery from Sensors and Image Data for Real-Time
Spatio-Temporal Emergency Monitoring 82
   Ilia Triapitcin, Ajantha Dahanayake, and Bernhard Thalheim

**ADBIS 2021 Workshop: Intelligent Data – From Data
to Knowledge – DOING**

Standard Matching-Choice Expressions for Defining Path Queries
in Graph Databases 97
   Ciro Medeiros, Umberto Costa, and Martin Musicante

COVID-19 Portal: Machine Learning Techniques Applied to the Analysis
of Judicial Processes Related to the Pandemic.   109
   Ana Sodré, Dimmy Magalhães, Luis Floriano, Aurora Pozo,
   Carmem Hara, and Sidnei Machado

LACLICHEV: Exploring the History of Climate Change in Latin America
Within Newspapers Digital Collections   121
   Genoveva Vargas-Solar, José-Luis Zechinelli-Martini,
   Javier A. Espinosa-Oviedo, and Luis M. Vilches-Blázquez

Public Health Units - Exploratory Analysis for Decision Support   133
   Tatiane Lautert, Nádia P. Kozievitch, Ismael Villanueva-Miranda,
   and Monika Akbar

The Formal-Language-Constrained Graph Minimization Problem   139
   Ciro Medeiros, Martin Musicante, and Mirian Halfeld-Ferrari

Interpreting Decision-Making Process for Multivariate Time Series
Classification   146
   Rufat Babayev and Lena Wiese

**ADBIS 2021 Workshop: Data-Driven Process Discovery
and Analysis – SIMPDA**

Process Mining Encoding via Meta-learning for an Enhanced
Anomaly Detection   157
   Gabriel Marques Tavares and Sylvio Barbon Junior

OCEL: A Standard for Object-Centric Event Logs   169
   Anahita Farhang Ghahfarokhi, Gyunam Park, Alessandro Berti,
   and Wil M. P. van der Aalst

**ADBIS 2021 Workshop: Modern Approaches in Data Engineering
and Information System Design – MADEISD**

Natural Semantics for Domain-Specific Language   181
   William Steingartner and Valerie Novitzká

A RESTful Privacy-Aware and Mutable Decentralized Ledger   193
   Sidra Aslam and Michael Mrissa

Segmentation Quality Refinement in Large-Scale Medical Image Dataset
with Crowd-Sourced Annotations   205
   Jan Cychnerski and Tomasz Dziubich

Process of Medical Dataset Construction for Machine Learning - Multifield
Study and Guidelines.   217
    Jan Cychnerski and Tomasz Dziubich

**ADBIS 2021 Workshop: Advances in Data Systems Management,
Engineering, and Analytics – MegaData**

A Federated Interactive Learning IoT-Based Health Monitoring Platform   235
    *Sadi Alawadi, Victor R. Kebande, Yuji Dong, Joseph Bugeja,
    Jan A. Persson, and Carl Magnus Olsson*

Augmenting SQLite for Local-First Software   247
    Iver Toft Tomter and Weihai Yu

**ADBIS 2021 Workshop: Computational Aspects of Network
Science – CAoNS**

Scalable and Explainable User Role Detection in Social Media.   263
    Johannes Kastner and Peter M. Fischer

Multi-dimensional Ranking via Majorization.   276
    Georgios Stoupas and Antonis Sidiropoulos

Inferring Missing Retweets in Twitter Information Cascades.   287
    Jennifer Neumann and Peter M. Fischer

ADBIS 2021 Doctoral Consortium

Semantically Diverse Constrained Queries   297
    Xu Teng and Goce Trajcevski

An Approach to Validation of Business-Oriented Smart Contracts Based
on Process Mining   303
    Vladimir Ivković and Ivan Luković

Challenges in Lifelong Pathways Recommendation   310
    Nicolas Ringuet

The Descriptiveness of Feature Descriptors with Reduced Dimensionality . . .   317
    Dániel Varga, János Márk Szalai-Gindl, and Sándor Laki

Author Index   323

ADBIS 2021 Short Papers

Learned What-If Cost Models
for Autonomous Clustering

Daniel Lindner, Alexander Löser, and Jan Kossmann[(✉)]

Hasso Plattner Institute, University of Potsdam, Potsdam, Germany
{daniel.lindner,alexander.loeser}@student.hpi.de, jan.kossmann@hpi.de

Abstract. Clustering database tables, i.e., storing similar values in physical proximity, can enhance the performance of database systems. Optimizing clustering configurations manually is cumbersome because the solution space is large and the performance impact depends on the system's workload. Autonomous clustering approaches can support database administrators in this task. However, determining the optimal clustering by physically applying all possible clusterings is prohibitively expensive. Our learned cost model *simulates* the effects of a hypothetical clustering and *estimates* the resulting workload latency based on previously trained cost models. The model accurately estimates the latencies of TPC-H workloads (scale factor 1) for different clusterings with a relative error of at most 5%, thereby enabling the determination of optimal clustering configurations.

Keywords: Physical database design · Database clustering · Autonomous database · Learned cost models

1 Introduction

The physical data layout of a database table has a large impact on database performance [6]. Database clustering is the process of reorganizing a table's physical layout such that tuples with similar values in certain columns are grouped and stored together. Thereby, clustering increases data locality and may enable further optimizations, such as partition pruning [8]. Database vendors such as Oracle [23], IBM [13], Snowflake [19], and Microsoft [1] have acknowledged the importance of clustering, and provide tools to cluster the stored data by multiple columns. However, real-world databases often contain wide tables with numerous columns [2,12], e.g., enterprise resource planning systems consist of tables with hundreds of columns. In such scenarios, a tremendous number of clustering options exists, which makes it challenging to determine the clustering that results in the best performance. Traditionally, this choice is made by database administrators [9] who must have detailed knowledge of the database and its workload.

Autonomous database management systems (DBMS) promise to simplify the work of database administrators [17]. Not only does the DBMS *itself* decide on its

© Springer Nature Switzerland AG 2021
L. Bellatreche et al. (Eds.): ADBIS 2021 Short Papers, Workshops and Doctoral Consortium,
CCIS 1450, pp. 3–13, 2021.
https://doi.org/10.1007/978-3-030-85082-1_1

configuration, but it may even identify settings superior to human decisions: The database has the most profound knowledge of workload and data and thus has also the best rationale for decisions [22]. For determining optimal configurations, various configurations are simulated and their performance impact is predicted by learned cost models which estimate the runtime of database operations based on a set of *features*. The accuracy of these estimations is crucial for determining optimal configurations. Learned cost models have been applied to, e.g., query optimization and index selection [22].

Contributions. In this work, we use a learned what-if[1] cost model to identify how a clustering impacts the database's performance. First, we perform a rule-based clustering simulation that predicts the effect of a clustering on database operations (e.g., whether the order of table scans changes). Second, a learned cost model estimates the workload's latency given the simulated clustering.

The remainder of this paper is structured as follows: Sect. 2 introduces basic concepts of Hyrise, the DBMS we used for our investigations. In Sect. 3, we present our approach on what-if-based latency estimation. After describing how we obtain training data in Sect. 3.3, Sect. 4 evaluates both steps of our model, the clustering simulation in Sect. 4.1 and the learned cost model in Sect. 4.2. Section 5 discusses related work before Sect. 6 concludes the paper and provides ideas for future work.

2   The Hyrise Database System

This work was developed for the research database Hyrise[2] [5]. This section provides an overview of the parts and concepts of the Hyrise database system that are necessary for understanding our approach presented in Sect. 3.

Storage Layout. Hyrise is a relational, memory-resident research database. Tables are stored in a column-oriented fashion but are partitioned into *chunks* with a fixed number of rows, 65535 by default. Thus, each chunk contains fractions of all columns, which are called *segments*. For each segment, Hyrise stores aggregated segment information [16], such as the minimum and maximum value present in the segment. During query optimization, such values may be used to prune chunks, i.e., to exclude chunks from the query execution without scanning their actual segment data. Hyrise supports various encoding types. All measurements in this work were performed with dictionary-encoded segments. Please refer to [5, Fig. 2] for more details on encodings and the storage layout.

SQL Pipeline. Hyrise can be queried via SQL. When Hyrise receives an SQL query, the *SQL pipeline* is triggered: First, the SQL query is translated into a logical query plan, called LQP. An LQP is a directed acyclic graph (DAG) whose

---

[1] We consider *what-if* as an analogy to *what-if optimization*, which was initially presented by Chaudhuri et al. [4] in the area of index selection.

[2] https://github.com/hyrise/hyrise.

nodes represent operations of the relational algebra, such as joins or aggregates. The LQP is then optimized by the query optimizer and finally translated into a physical query plan, called PQP. A PQP is a DAG whose nodes are specific implementations of database operations. For example, a join in the LQP may be translated to a hash join, a sort-merge join, or a nested loop join in the PQP. In Hyrise, those specific implementations are called *operators*. Hyrise maintains a cache of recently executed query plans, called *PQP cache*.

Reference Segments. In Hyrise, the output of an operator is a so-called *reference table*, which contains *reference segments*. These segments do not store the data, but only point to where the data is located in the original table. The process of accessing the actual values behind a reference segment is called *materialization*.

Operator Performance Data. In Hyrise, operators collect statistics about their execution. These statistics include, e.g., input and output row counts, the input's sort order, the execution time, and the accessed columns. These statistics are available via the *PQP cache*. In our clustering model, we use the entirety of all executed operators and their performance data to represent workloads.

3   What-If Latency Estimation for Autonomous Clustering

This section presents our approach on what-if-based latency estimation for autonomous clustering. While we provide a brief overview of the entire clustering model in Sect. 3.1, this work focuses on our latency estimation technique and its training data, which we describe in Sect. 3.2 and Sect. 3.3, respectively.

3.1   Clustering Model

We use the clustering model that was developed during the master's thesis of Löser [14]. Given the current workload, the model aims to find a clustering that minimizes the workload's latency, i.e., the sum of all individual query latencies.

The model operates in two steps: First, the *candidate creator* analyzes the current workload and determines a list of clustering candidates. Second, for each candidate, the *latency estimator* is used to estimate the impact of such a clustering on the current workload's performance. The database can then weight the expected performance gain against the expensive process of reclustering.

Candidate Creator. The first step to create clustering candidates is to identify *interesting columns*. A column is considered interesting if a scan or a join predicate operates on it. Table scans on a clustered column can benefit, e.g., from pruning [6]; hash-based joins may benefit from better cache hit rates.

An arbitrary number of interesting columns can be chosen for clustering, thereby allowing for n-dimensional clusterings. However, in this work, we only consider one-dimensional clusterings, i.e., with exactly one clustering column.

Latency Estimator. The most precise way to determine a clustering's performance impact is to physically implement the clustering, and measure the resulting latency. Due to the high costs for adjusting the storage layout, we consider the aforementioned approach to be unfeasible, in particular, if numerous clustering configurations have to be evaluated. Consequently, the latency estimator works on a what-if basis: Instead of physically applying a clustering, the estimator only assumes the clustering was implemented and provides a latency estimate.

Löser [14] uses a set of handcrafted rules to estimate the latency of table scans and hash joins. In this work, we replace the handcrafted rules with a learned cost model that predicts latencies for table scans, hash joins, and hash aggregates.

3.2   Latency Estimation

Determining how a certain clustering will affect the current workload's latency is a crucial part of our clustering model. The latency estimation is conducted per operator for aggregates, scans, and joins. For all other operators, we assume that their latency is unaffected by the clustering. This simplifying assumption is justified by the fact that the aforementioned three operators are responsible for more than 90% of the execution time of the TPC-H benchmark [6].

Our latency estimation approach consists of two steps: First, we perform a *clustering simulation.* For this simulation, we apply rule-based transformations to the operators. The aim is to predict how a clustering would affect the operators' clustering-dependent features, e.g., the number of input rows. For instance, when estimating the latency for clustering by column X, the simulation considers an input reduction through pruning for filter predicates on column X.

In the second step, we perform the latency estimation using learned physical cost models. The cost models receive the simulation's output as input and yield latency estimates as output. Figure 1 displays the latency estimation process.

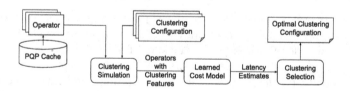

Fig. 1. Architecture of the latency estimation.

Clustering Simulation. In Hyrise, operators track certain performance data, see Sect. 2. Our clustering simulation[3] applies a predefined set of rules to identify the workload's operators impacted by the clustering and estimates how their

---

[3] https://github.com/aloeser/hyrise/blob/whatif/python/clustering/what_if_model.py.

input and output rows, sortedness, and operator-specific features will change. We identified that PQP awareness and predicate reordering need to be considered for a precise clustering simulation.

PQP Awareness. It is crucial to be aware of an operator's position in the physical query plan, and, thereby, its surrounding operators to determine whether the clustering will influence its performance. Consider the excerpt from Hyrise's query plan for query 20 of the TPC-H benchmark [3,21] depicted in Fig. 2: Hyrise first performs semi joins on the l_partkey and l_suppkey columns, before performing a scan on the l_shipdate column. By default, the lineitem table is sorted by the l_orderkey column. If lineitem was clustered by l_shipdate instead, the scan on l_shipdate will benefit from pruning. The semi joins are executed before the scan; however, pruning is applied before any operators are executed. Thus, despite the semi joins neither operate on l_orderkey nor on l_shipdate, clustering by l_shipdate has a significant impact on their input size, reducing it by factor 6. Above's examples demonstrates the importance of considering an operator's context, i.e., the surrounding PQP, to obtain precise information about runtime-critical properties.

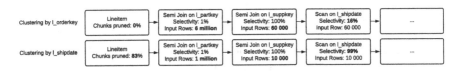

Fig. 2. Partial PQP of TPC-H Query 20 at scale factor 1, with l_orderkey and l_shipdate clustering. Reprinted from *Automatic Clustering in Hyrise* [14].

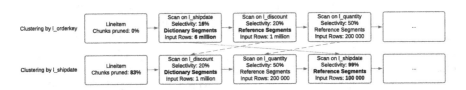

Fig. 3. Partial PQP of TPC-H Query 6 at scale factor 1, with l_orderkey and l_shipdate clustering. Reprinted from *Automatic Clustering in Hyrise* [14].

Predicate Reordering. In Hyrise, the most important types of *predicates* are table scans and semi joins. Predicate reordering effects are discussed in the following.

Hyrise's query optimizer re-arranges predicates so that the predicate with the highest selectivity (ratio of output and input rows) is executed last. On clustered columns, large parts of the non-matching values may have been pruned before the predicate's operator is executed. Consequently, the scan will have a selectivity close to 1 and will be executed last, i.e., its input size will be further reduced by previous predicates. Additionally, the reordering may cause another predicate to operate on data segments instead of reference segments and vice versa, which should be considered in the clustering simulation, too. Figure 3 visualizes both effects on the basis of query 6 of the TPC-H benchmark.

Learned Cost Model. We perform the latency estimation with a machine learning model that predicts operator runtimes. We use a gradient boosting regressor [7] with a Huber loss function [10] provided by scikit-learn [18]. Boosting allows for different, non-linear types of dependencies within variables, without explicitly modeling these types [11, p. 314]. Linear, ridge, or lasso regression did not yield accurate predictions in our case. We assume that the non-linear relationships of features like a column's data type harm their performance.

In Hyrise, operators can have multiple implementations, cf. Sect. 2. We train one model for each of these implementations. Furthermore, we observed that the input type (data or reference) has a strong influence on the performance of hash joins. Thus, we add models for each combination of input types of the build and probe input table. A complete overview of the features used to predict the operator latency can be found on GitHub[4].

3.3   Obtaining Data for Training

As training data, we use a combination of information on the clustering configuration and the operator performance data, cf. Sect. 2. This training data is obtained from three sources: (i) the execution of the TPC-H on the unmodified (lineitem sorted by l_orderkey) TPC-H dataset, (ii) the execution of the TPC-H after shuffling l_lineitem, and (iii) dedicated calibration data covering cases that are not handled by the TPC-H workloads described above.

(i) enables efficient training for operators that are not affected by changes to the clustering configuration. (ii) generates training data that represents the processing of unsorted data for operators that were initially executed on sorted data under (i), but that would be executed on such unsorted data for other clustering configurations. Further, we have seen that clustering may affect various operators, e.g., the input sizes and table types of a table scan may change due to the clustering, cf. Fig. 3. Such changes are not covered sufficiently in (i) or (ii). For this reason, (iii) adds information of synthetic PQPs, covering variations of the cost model's input features. For example, we add operators that are executed on different input sizes, i.e., table scans with a variable amount of pruned chunks, to our training data. Overall, we include the execution information of more than 90000 dedicated PQPs in the training data.

3.4   Limitations

Our latency estimation process has two limitations. First, we currently consider only one-dimensional clusterings. Second, due to the chosen machine learning

---

[4] https://github.com/aloeser/hyrise/blob/whatif/cost_model_features.md

model type, our learned cost model can hardly interpolate observations. Thus, our model does not yield reasonable estimates for previously unseen operators.

4   Results and Discussion

In this section, we evaluate our latency estimation approach. For that purpose, we use our model to predict the latencies of all TPC-H benchmark queries, assuming a different clustering than the default (l_orderkey). We operate on lineitem because it is the largest of the TPC-H tables, and responsible for the major share of execution time. For the alternative clustering columns, we consider a column mostly used in filter predicates, l_shipdate, and a column mostly used for joins, l_partkey.

Our open-source[5] model was developed based on TPC-H data with a scale factor of 1. Hence, the evaluation is focused on the same data, but we also evaluate scale factor 10 to determine how well our model scales. Like the approach itself, our evaluation is divided into two steps: We evaluate the clustering simulation's precision in Sect. 4.1 and the latency estimates' accuracy in Sect. 4.2.

4.1   Clustering Simulation

For scans and joins, we evaluate our predictions of input rows, output rows, and whether data or reference segments are processed. For the input and output rows, we consider relative errors, i.e., the ratio between predicted and actual size; for the segment type, we consider the ratio of correctly predicted types.

Table Scans. By default, Hyrise stores data in dictionary encoded segments. The dictionaries of those segments contain all of the segment's unique values, which enables an early-out optimization for scans: For range predicates, e.g., $A < 4$, the dictionaries can be used to determine whether none or all of the segment's values will match: if, for instance, the maximum of the dictionary is smaller than the filter value, all values in the segment will match. Consequently, Hyrise only has to write the output, without actually scanning the segment. In addition to input rows, output rows, and the segment type, our clustering simulation also predicts whether the early-out optimization is applicable. Table 1 shows the precision of our clustering simulation for scans that operate on lineitem. First of all, we observe that the simulation yields precise results for most of the table scans. Further, the simulation is more precise for the l_partkey clustering, which is expected, as neither clustering by l_partkey nor by l_orderkey allows any pruning or sorted searches; i.e., the features of the table scans remain unaffected.

The simulation achieves similar performance on data with a scale factor of 10. We conclude that our clustering simulation is sufficiently precise for table scans, even for higher scale factors.

---

[5] Implementation: https://github.com/aloeser/hyrise/tree/db73923.

Table 1. Aggregated statistics for the clustering simulation of scans on `lineitem`. For the input and output rows, the share of operators for which the prediction is within 10% (or 20%, respectively) of the actual value is listed. For the segment type and early-out optimization (binary values), the share of correctly predicted values is displayed.

Clustering	Input rows		Output rows		Segment type	Early-out applicable
	$\leq$10%	$\leq$20%	$\leq$10%	$\leq$20%		
`l_shipdate`	75%	86%	83%	93%	97%	90%
`l_partkey`	99%	99%	99%	99%	100%	100%

Hash Joins. In addition to the already mentioned features, we predict (and evaluate) the number of chunks of the original table that a join's input will refer to. Intuitively, a join which receives 1000 input rows that refer to 1000 different chunks should be slower than a join whose 1000 input rows refer to two chunks, as we need to access fewer dictionaries when materializing the reference segments. Table 2 shows the precision of our clustering simulation for joins on `lineitem`. We observe that the clustering simulations yield precise results for most of the hash joins. Analogous to the table scans, changes in the input size only occur through pruning; which explains why the simulation performs better for the `l_partkey` clustering. We achieve similar performance for scale factor 10 and conclude that our clustering simulation yields precise results for hash joins.

Table 2. Aggregated statistics for the clustering simulation of hash joins on `lineitem`. For the input rows, output rows, and the number of referenced chunks, the share of operators for which the prediction is within 10% (or 20%, respectively) of the actual value is listed. For the segment type the share of correctly predicted values is listed.

Clustering	Input rows		Output rows		Segment type	Referenced chunks	
	$\leq$10%	$\leq$20%	$\leq$10%	$\leq$20%		$\leq$10%	$\leq$20%
`l_shipdate`	96%	96%	99%	99%	100%	92%	92%
`l_partkey`	100%	100%	100%	100%	100%	100%	100%

Sortedness. An essential part of the clustering simulation is to determine whether the original sort order of a table is maintained by subsequent operators. Hash-based aggregates and joins are impacted by the input data's sort order. However, some operators, such as hash joins, do not forward information about sortedness, even if they do maintain the original sort order. As a consequence, we cannot provide a meaningful evaluation for our sort order predictions.

4.2   Learned Cost Model

In the following, we investigate the model's runtime estimates. To identify the most suitable clustering regarding a workload scenario, we focus on the combined latencies of table scans, joins, and aggregates that operate on the lineitem table for each clustering candidate. Training and test data are not identical, however, some operators are not affected by clustering and thus resemble the ones in the l_orderkey training dataset. The accumulated runtimes for the three operators are displayed in Fig. 4.

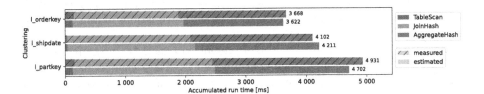

Fig. 4. Accumulated runtimes of selected operators on lineitem, estimated vs. measured, for three clusterings.

For all three candidates, the learned cost model estimates the overall latencies with a relative error <5%. Though the joins are consistently over-estimated, the aggregates' runtimes are underestimated in two cases. This observation also applies to the l_orderkey clustering that was part of the training data. Despite the misestimations, the models are accurate enough to convey the most important information. The optimal one-dimensional clustering is suggested because the predicted runtime-based order matches the order based on measurements.

Table 3 shows the mean squared error (MSE) of the accumulated latencies for each clustering and operator. Though these MSEs are high compared to the average runtimes of the operators, we have already seen that the accumulated latencies are accurate in Fig. 4.

Table 3. Prediction metrics for scans, joins, and aggregates for different clusterings.

Clustering	Mean squared error (ms^2)			∅ Runtime (ms)		
	Table scan	Join	Aggregate	Table scan	Join	Aggregate
l_orderkey	11	39	1085	9	40	94
l_shipdate	1	346	1574	2	48	106
l_partkey	14	167	2258	10	54	131

For scale factor 10, the results are comparable: The estimations of the accumulated runtimes never exceed 7%. Consistent under-estimations of aggregates and over-estimations of joins result in small relative errors. Given that the calibration data was designed for scale factor 1, this is a satisfying result.

Overall, our model predicts the latency of the workload for given clustering scenarios with an adequate accuracy: We obtain the correct order for clustering candidates and the estimations are precise enough to make a deliberate decision regarding the application of a clustering.

5   Related Work

Recently, there have been advancements in both research areas, learned cost models and database clustering.

Zhou et al. [22] survey current applications of learned cost models in databases. These cost models are used to tackle various problems, e.g., query optimization, index selection, or transaction management. However, none of the cited authors use machine learning techniques for latency estimations of different clusterings.

Marcus et al. [15] and Sun et al. [20] present cost models to estimate query latencies with tree-structured deep neural networks. These models have a high precision, but require a lot of training time, and they infer knowledge about the current data layout, which hinders the simulation of hypothetical data layouts.

Hilprecht et al. [9] present a clustering model for cloud databases. They argue that obtaining exact latency estimates is challenging, but essential for identifying an optimal clustering. As a consequence, they suggest a model based on deep reinforcement learning: the model learns the latencies of different clusterings by implementing a clustering and measuring its latency. However, when considering a large number of clusterings, the training process becomes very expensive. To speed up the training, Hilprecht et al. argue that – being in a cloud environment – the latency of transferring data between different nodes accounts for the dominating part of the query latencies. Thus, for approximate latency estimates, it is sufficient to estimate only the amount of data that needs to be transferred. However, this cost model cannot be applied to non-distributed database systems like Hyrise, as no data is transferred over the network.

6   Conclusion and Future Work

We have presented a what-if model for clustering-specific latency estimation based on a clustering simulation and a learned cost model. The results of the rule-based clustering simulation are the input of a gradient boosting regressor to estimate the latencies of a given workload. These latencies are estimated with a relative error of at most 5% for the evaluated TPC-H (scale factor 1) workloads; our model correctly identifies the optimal one-dimensional clustering. Compared to rule-based approaches, the learned cost model is more flexible and can be retrained if the implementation or underlying hardware changes.

There are interesting topics for future work. The model could consider multidimensional clusterings. Also, additional, more complex workloads and calibration data covering a more extensive range of characteristics could generalize the stated observations. In addition, other types of learned cost models, which support interpolation and fine-tuning in a more sophisticated manner, could further improve latency estimation.

References

1. Agrawal, S., Narasayya, V.R., Yang, B.: Integrating vertical and horizontal partitioning into automated physical database design. In: SIGMOD, pp. 359–370 (2004)

2. Bian, H., et al.: Wide table layout optimization based on column ordering and duplication. In: SIGMOD, pp. 299–314 (2017)
3. Boncz, P.A., et al.: TPC-H analyzed: hidden messages and lessons learned from an influential benchmark. In: Performance Characterization and Benchmarking - TPC Technology Conference, vol. 8391, pp. 61–76 (2013)
4. Chaudhuri, S., Narasayya, V.R.: An efficient cost-driven index selection tool for microsoft SQL server. In: VLDB, pp. 146–155 (1997)
5. Dreseler, M., et al.: Hyrise re-engineered: an extensible database system for research in relational in-memory data management. In: EDBT, pp. 313–324 (2019)
6. Dreseler, M., et al.: Quantifying TPC-H choke points and their optimizations. PVLDB **13**(8), 1206–1220 (2020)
7. Friedman, J.H.: Greedy function approximation: a gradient boosting machine. Ann. Stat. **29**(5), 1189–1232 (2001)
8. Herodotou, H., et al.: Query optimization techniques for partitioned tables. In: SIGMOD, pp. 49–60 (2011)
9. Hilprecht, B., et al.: Learning a partitioning advisor for cloud databases. In: SIGMOD, pp. 143–157 (2020)
10. Huber, P.J.: Robust estimation of a location parameter. Ann. Math. Stat. **35**(1), 73–101 (1964)
11. James, G., et al.: An Introduction to Statistical Learning with Applications in R. Springer, Heidelberg (2013)
12. Krüger, J., et al.: Fast updates on read-optimized databases using multi-core CPUs. PVLDB **5**(1), 61–72 (2011)
13. Lightstone, S., Bhattacharjee, B.: Automated design of multidimensional clustering tables for relational databases. In: VLDB, pp. 1170–1181 (2004)
14. Löser, A.: Automatic clustering in hyrise. Master's thesis, Hasso-Plattner-Institute, University of Potsdam (2020). https://arxiv.org/abs/2103.15509
15. Marcus, R.C., Papaemmanouil, O.: Plan-structured deep neural network models for query performance prediction. PVLDB **12**(11), 1733–1746 (2019)
16. Moerkotte, G.: Small materialized aggregates: a light weight index structure for data warehousing. In: VLDB, pp. 476–487 (1998)
17. Pavlo, A., et al.: Self-driving database management systems. In: 8th Biennial Conference on Innovative Data Systems Research, CIDR Online Proceedings (2017)
18. Pedregosa, F., et al.: Scikit-learn: machine learning in Python. J. Mach. Learn. Res. **12**, 2825–2830 (2011)
19. Snowflake: Clustering Keys & Clustered Tables (2020). https://docs.snowflake.com/en/user-guide/tables-clustering-keys.html
20. Sun, J., Li, G.: An end-to-end learning-based cost estimator. PVLDB **13**(3), 307–319 (2019)
21. Transaction Processing Performance Council: TPC-H Specification (2014). http://www.tpc.org/tpc_documents_current_versions/pdf/tpc-h_v2.17.1.pdf
22. Zhou, X., et al.: Database meets artificial intelligence: a survey. TKDE (2020)
23. Ziauddin, M., et al.: Dimensions based data clustering and zone maps. PVLDB **10**(12), 1622–1633 (2017)

Cost-Sensitive Predictive Business Process Monitoring

Martin Käppel[1(✉)], Stefan Jablonski[1], and Stefan Schönig[2]🔘

[1] Institute for Computer Science, University of Bayreuth, Bayreuth, Germany
{martin.kaeppel,stefan.jablonski}@uni-bayreuth.de
[2] University of Regensburg, Regensburg, Germany
stefan.schoenig@ur.de

Abstract. In predictive business process monitoring current and historical process data from event logs is used to predict the evolvement of running process instances. A wide number of machine learning approaches, especially different types of artificial neural networks, are successfully applied for this task. Nevertheless, experimental studies revealed that the resulting predictive models are not able to properly predict non-frequent activities. In this paper we investigate the usefulness of the concept of cost-sensitive learning, which introduces a cost model for different activities to better represent them in the training phase. An evaluation of this concept applied to common predictive monitoring approaches on various real life event logs shows encouraging results.

Keywords: Process mining · Predictive business process monitoring · Cost-sensitive learning · Process prediction

1 Introduction

Predictive business process monitoring methods support participants performing a running process instance of a business process [1]. These methods use current and historical process data from event logs to make predictions how a running process instance will evolve up to its completion. These predictions encompass, among other things, performance predictions, predictions regarding the outcome of a process [20], and predictions about upcoming events, including information about activities performed next [4,5,7,16] and by whom [4,17].

Usually, activities (or in general values of event attributes) occur in an event log in a different number. Some activities are executed almost in every process instance, while other activities only occur in a handful of instances. Often, however, such rare activities are of particular importance, since they are only executed in case of handling special or error cases. Usually, such cases bear some challenges (e.g., high costs, lack of experience, increased risk), why it is important to support involved process participants with precise predictions.

In the last years research has shown that deep learning approaches are highly customizable and precise for prediction tasks. However, deeper investigations

© Springer Nature Switzerland AG 2021
L. Bellatreche et al. (Eds.): ADBIS 2021 Short Papers, Workshops and Doctoral Consortium,
CCIS 1450, pp. 14–26, 2021.
https://doi.org/10.1007/978-3-030-85082-1_2

reveal that the imbalance of activities (or other event attributes) prevent a precise prediction of non-frequent activities [12,16]. As a result, frequent trace variants (i.e., standard cases) are predicted very well, while rare trace variants are barely predicted correctly, since they are treated in the prediction as a standard case.

Conventional machine learning (ML) algorithms assume an approximately uniform distribution of examples and assume that all prediction errors made by the predictive model are equally weighted [8]. Real-world applications typically have different interpretations for each error. For example, in a helpdesk process with activities for closing, reviewing, and rejecting a ticket, predicting a review of the ticket instead of closing the ticket is less problematic than predicting a rejection of the ticket. Similar problems occur, for instance, in medical treatment processes or processes for claim settlement.

We address the issue of predicting non-frequent trace variants by applying the concept of cost-sensitive learning. Therefore, each class is associated with a given misclassification cost and the used ML algorithm is modified to minimize the total classification costs instead of the number of incorrect predictions.

Next, we provide a introduction in basic terminology and give a formal problem definition. Sect. 3 introduces our approach, while Sect. 5 presents its evaluation. In Sect. 4 we discuss related work. Finally, Sect. 6 outlines future work.

2 Basic Terminology and Problem Statement

2.1 Process Mining and Predictive Business Process Monitoring

The main input of process mining techniques is a *(process) event log*. An event log is a set of traces that are related to a certain business process. A *trace* (also called *case*) is a temporarily ordered sequence of events representing the execution of a process instance. An *event* capsulates information about the execution of an activity (i.e., a step in a business process) within a process instance. These information are characterized by various *event attributes* such as the case id $(\mathcal{C})$, date and time of occurrence $(\mathcal{T})$, the name of the corresponding activity $(\mathcal{A})$, the process participants or systems involved in executing the activity $(\mathcal{L})$, and further data payload $(\mathcal{D})$. Note, that only case identifier, timestamp, and corresponding activity are mandatory, all other event attributes are optional.

Definition 1. *Let $\mathcal{E}$ be the set of all possible event identifiers, $\mathcal{P}$ the set of event attributes, and ε the empty element. For each $p \in \mathcal{P}$, we define a function $\pi_p : \mathcal{E} \rightarrow dom\,(p) \cup \{\varepsilon\}$ that assigns a value of the domain of p to an event. However, the empty element can only be attached to optional event attributes.*

Let us consider the fragment of an event log reported in Table 1. This event log provides the following event attributes: a *case identifier*, the name of the executed *activity*, the *timestamp* of execution, the involved *resource*, and further information (*amount, key*) in form of data payload. For example, for event e_{12}, holds $\pi_{\mathcal{A}}(e_{12})$ = "T", $\pi_{\mathcal{C}}(e_{12})$ = "Case1", $\pi_{\mathcal{L}}(e_{12})$ = "SL", $\pi_{\mathcal{T}}(e_{12})$ = "2020-10-09T14:51:01", $\pi_{\mathcal{D}_{amount}}(e_{12})$ = ε, and $\pi_{\mathcal{D}_{key}}(e_{12})$ = "HG-4".

Table 1. Excerpt of a process event log

Case ID	Event ID	Activity	Timestamp	Resource	Amount	Key
C1	e_{11}	A	2020-10-09T14:50:17	MF		SD-1
C1	e_{12}	T	2020-10-09T14:51:01	SL		HG-4
C1	e_{13}	W	2020-11-09T12:54:39	KH		HZ-2
C2	e_{21}	A	2019-04-03T08:55:38	MF		SD-2
C2	e_{22}	T	2019-04-03T08:55:53	SL	340	HK-7
C2	e_{23}	C	2019-05-19T09:00:28	KH		SGH-3
...	...	...	...	...	...	...

Definition 2. *Let S be the universe of all traces. A **trace** $\sigma \in S$ is a finite non-empty sequence of events $\sigma = \langle e_1, ..., e_n \rangle$ such that for $1 \leq i < j \leq n : e_i, e_j \in \mathcal{E} \wedge \pi_C(e_i) = \pi_C(e_j) \wedge \pi_T(e_i) \leq \pi_T(e_j)$, where $|\sigma| = n$ denotes the **length** of σ. A trace σ is called **completed** if there is no $e' \in \mathcal{E}$ such that $\pi_C(e') = \pi_C(e)$ with $e' \notin \sigma$ and $e \in \sigma$. An **event log** L is a set $L = \{\sigma_1, ..., \sigma_l\}$ of completed traces.*

Predictive business process monitoring approaches often partition traces for training in sets of prefixes and suffixes to represent uncompleted traces.

Definition 3. *Let $\sigma = \langle e_1, ..., e_n \rangle \in S$ be a trace. We define two functions hd and tl, which return the **event prefix** of length r (i.e., the first r elements of a trace), respectively the **event suffix** of length r (i.e., the last r elements of a trace) of σ as follows:*

$$hd : S \times \mathbb{N}_{\geq 0} \to S, \qquad (\sigma, r) \mapsto \begin{cases} \langle \rangle & \text{if } r = 0 \\ \langle e_1, e_2, ..., e_r \rangle & \text{if } 0 < r \leq |\sigma| \\ \sigma & \text{if } r > |\sigma| \end{cases}$$

$$tl : S \times \mathbb{N}_{\geq 0} \to S, \qquad (\sigma, r) \mapsto \begin{cases} \langle \rangle & \text{if } r = 0 \\ \langle e_{r+1}, ..., e_n \rangle & \text{if } 0 < r \leq |\sigma| \\ \sigma & \text{if } r > |\sigma| \end{cases}$$

where $\langle \rangle$ represents the empty trace.

Depending on the prediction target (i.e., event attribute), we can define different prediction functions:

Definition 4. *Let $\sigma = \langle e_1, ..., e_n \rangle \in S$ be an uncompleted trace, $\mathcal{P}$ the set of recorded event attributes, $e' \in \mathcal{E}$ a predicted event, and $r \in \mathbb{N}$, then for each $p \in \mathcal{P}$ the **prediction problem** is defined as $\Omega_p(hd(\sigma, r)) = \pi_p(e'_{r+1})$.*

For example, the prediction of the next activity is given by $\Omega_A(hd(\sigma, r)) = \pi_A(e_{r+1})$. If we not only want to predict the next event, rather than a whole suffix, we can apply Ω_p recursively over the predictions.

2.2 Formal Problem Definition

According to Definition 4 the prediction of a categorical event attribute $p \in \mathcal{P}$ (e.g., the next activity) is a classification problem γ_k^p with k classes. We assign a class label (i.e., a value of the domain of p) in dependency of the prediction target to a running process instance. We denote the set of class labels for p as $\mathcal{F}_p$. If we consider the class distribution of event attributes in different event logs (or subsets used for training), we always observe an imbalanced class distribution: For some classes (*majority classes*) there are significantly more examples than for other classes (*minority classes*). This observation is typically for the process domain. We denote with $\eta = (\eta_1, \cdots, \eta_k) \in [0,1]^k$ the corresponding probability distribution, where $\eta_i = p(c_i)$ stands for the probability of class c_i. Based on this we define majority and minority classes formally as follows:

Definition 5. *Let γ_k be a classification problem with k classes. A class c_i is called **minority class**, if $\eta_i = p(c_i) < \frac{1}{k}$. Otherwise, we call c_i **majority class**.*

Hence, we distinguish as proposed in [15] three types of classification problems:

Definition 6. *Let γ_k be a classification problem with k classes and $\mathbf{1}(x)$ the indicator function (takes the value 1, if x is true, otherwise 0).*

1. *γ_k is called **balanced**, if η describes a uniform distribution.*
2. *γ_k is called **multi-majority**, if $\sum_{i=1}^{k} \mathbf{1}\left(\eta_i \geq \frac{1}{k}\right) \geq \frac{k}{2}$.*
3. *γ_k is called **multi-minority**, if $\sum_{i=1}^{k} \mathbf{1}\left(\eta_i < \frac{1}{k}\right) > \frac{k}{2}$.*

We estimate η by the empirical probability ζ from the existing dataset $\mathcal{D} = \{(x^{(n)}, c^{(n)})\}_{n=1}^{l}$, i.e.,

$$\eta_i = \zeta_i = \frac{1}{l} \sum_{n=1}^{l} \mathbf{1}\left(c^{(n)} = c_i\right).$$

Problem types 2 and 3 are known as *class imbalance problem* and are regarded as major obstacle for building precise classifiers [15]. The reason for that is that ML methods are designed to generalize the data in a suitable way [18]. Hence, they pay less attention to minority classes [18]. In addition to class imbalance, ML research also identifies the size of a dataset as further influencing factor [8,18]. However, a study in the process domain reveals, that the size of event logs nearly does not affect the quality of the classifier [12]. In general it is broad consensus in ML research that an approximately balanced dataset leads to better results. However, the imbalance must also be considered when selecting metrics for evaluating the classifiers, since some metrics are sensitive to the class distribution. From the application point of view the equal weighting of all errors is somehow unrealistic, since each error has its individual interpretation and consequences. Combining the problems of imbalance and equal weighting of misclassification errors, we identify three strongly related core issues:

Issue 1: Minority classes are not learned in an imbalanced dataset.
Issue 2: All misclassifications are treated and assessed equaly in an (approximately) balanced dataset.
Issue 3: Issues 1 and 2 jointly arise.

For an appropriate evaluation it is important to quantify class imbalance [15]. A frequently used metric is the so called *imbalanced ratio* (IR), which is defined as the number of examples of the largest class divided by the number of examples of the smallest class [8]. Hence, the IR is less suitable for multiclass problems, since only the largest and smallest classes are considered. In the following we use the *imbalance degree* (ID) that was proposed in [15] and considers all classes. The idea is to quantify the distance between the imbalanced class distribution ζ and a balanced class distribution e using statistical distance functions d_Δ (e.g., Kullback-Leibler divergence). The calculated distance is set in relation to the highest possible distance to e that a class distribution with the same number of classes and minority classes can have.

Definition 7. *The imbalanced degree is given by*

$$ID(\zeta) = \frac{d_\Delta(\zeta, e)}{d_\Delta(\iota_m, e)} + (m - 1),$$

where m is the number of minory classes, d_Δ a statistical distance function, and ι_m is the class distribution with exactly m minority classes and the highest distance to e.

From Definition 7 follows that the ID takes a value within $[0, k)$. A high ID indicates a stronger imbalance than small values. The ID also includes explicitly the number of minority classes, since a high number of minority classes aggravates the problem. For calculating the ID, it is necessary to determine ι_m. According to [15], this class distribution consists of exactly m minority classes with probability zero, $k - m - 1$ majority classes with probability $1/k$, and a majority class with the remaining probability of $1 - (k - m - 1)/k$.

3   Concept and Solution

3.1   General Idea

We address the issues described in Sect. 2.2 using *cost-sensitive learning* (CSL). While the conventional aim of training a ML model is to minimize the error (i.e., the number of incorrect predictions) of the model on a training dataset, CSL aims to minimize the cost of a model on a training dataset [8]. Therefore, each class is associated with a misclassification cost and the ML algorithms try to minimize the total misclassification cost [9]. The costs must be chosen in such a way, that the misclassification cost is always higher than correct classification [6]. The higher the costs of misclassification for a class, the more the training procedure must focus on examples of this class. We apply this concept to the three issues of the previous section as follows:

Issue 1: Misclassification errors of minority classes get assigned higher costs than those of majority classes.

Issue 2: Misclassification errors that produce high negative impact on the application get assigned high costs.

Issue 3: Costs stemming from Issue 1 and 2 must be combined.

Since, our concept is independent of a particular domain and imbalance is the prevalent problem, we focus on Issue 1 in the rest of the paper. However, the proposed concept is also applicable to the other issues and only requires an adaptation of the costs.

3.2   Calculating Class Weights

In the following, we assume that all classes are sequentially numbered from 1 to k. Let c_{ij} be the cost of predicting an example belonging to class j when in fact it belongs to class i. If $i = j$ the prediction is correct, otherwise the prediction is incorrect. This definition goes in hand with the assumption that the costs are only dependent on the classes and not on a concrete example [22]. The cost matrix $C = (c_{ij})_{1 \leq i,j \leq k}$ is defined as a $k \times k$ matrix with real values usually greater or equal to zero. In general, we can assume $c_{ii} = 0$, since a correct classification usually causes zero cost. The cost of the i-th class, i.e., the cost of misclassifying an example of class i regardless of the class predicted, is denoted by $C(i)$. Hence, a new example x should be classified to a class that minimizes the *conditional risk* (also called *expected cost*)

$$R(x|i) = \sum_{j=1}^{k} P(j|x)c_{ij}, \tag{1}$$

where $P(j|x)$ denotes the probability that the classifier classifies x as belonging to class j. In other words, R is the sum over all alternative possibilities for the true class of x [6]. Obviously, the cost matrix is crucial for the effectiveness of a CSL approach [8]. Wrong chosen costs hamper the learning procedure in several ways: *(i)* too low cost have no effect and the class boundaries are not adjusted, *(ii)* too high costs lead to overfitting, since they prevent a generalization of the model, and *(iii)* in the imbalanced case too high costs lead to an overcompensation and introduce a too strong bias towards the minority classes.

Hence, it would be optimal that domain experts create a cost matrix and derive the weight vector w. However, most of the time domain experts are not available, and the cost matrix must be created through heuristics. One option (called *Balanced Cost*) avoiding a cost matrix in the imbalanced case is to calculate class weights from the class distribution by $w_i = l/(k \cdot n(c_i))$, where $n(c_i)$ denotes the number of examples of the i-th class and l the number of examples of the dataset. Alternatively, we can use the IR for estimating a cost matrix. Therefore, we consider all pairs of classes seperately. That means, for two classes $i \neq j$, where i denotes the minority class and j the majority class, we calculate the IR between class i and class j and set c_{ij} to the calculated IR and c_{ji} to 1.

Given a cost matrix, we have two options for deriving the weight vector w:

Cost Sum: Here, all misclassfication costs related with a class are summed up [3,24], i.e., $w_i = C(i) = \sum_{j=1}^{k} c_{ij}$. However, the class weights are calculated isolated from other classes, why they are mostly not optimal. Hence, we try to derive the class weights simultanously for all classes from the cost matrix.

Optimized Costs: This procedure is based on the idea, that an optimal solution for w should also be an optimal solution in the binary case. Hence, we first study the binary case in more detail as it is done in [6,8,22,25]. An example x should be classified belonging to class 1, if and only if the expected costs (cf. Eq. 1) are lower than for classifying x as class 2, i.e.:

$$P(j = 1|x)c_{11} + P(j = 2|x)c_{12} \leq P(j = 1|x)c_{21} + P(j = 2|x)c_{22}.$$

Given $p = P(j = 1|x)$ this is equivalent to:

$$pc_{11} + (1 - p)c_{12} \leq pc_{21} + (1 - p)c_{22}.$$

If this inequality is in fact an equality, then the prediction is optimal. Assume that the optimal threshold is p^*. Then the following must hold:

$$p^*c_{11} + (1 - p^*)c_{12} = p^*c_{21} + (1 - p^*)c_{22}.$$

Since we can assume that c_{ii} is zero and $c_{ij} \neq 0$ for all $i \neq j$ the equation can be rearranged in the following way:

$$\frac{p^*}{1 - p^*} = \frac{c_{21}}{c_{12}} =: \lambda.$$

This means that the impact of the first class should be λ times of that of the second class. Hence, in the multiclass case this should hold for all $i, j \in \{1, ..., k\}$ with $i \neq j$:

$$\frac{w_i}{w_j} = \frac{c_{ji}}{c_{ij}} \Leftrightarrow w_i \cdot c_{ij} - w_j \cdot c_{ji} = 0.$$

Hence, we get a system of $\binom{k}{2}$ equations that is linear in w_i. This system has a non-trivial solution if and only if the rank of the corresponding coefficient matrix is smaller than k. In this case, we call the cost matrix *consistent* and calculate a solution with the QR decomposition. If the system does not satisfy this condition, it is not possible to derive an optimal weight vector.

3.3   Cost-Sensitive Transformation of Algorithms

Most ML algorithms are not cost-sensitive, i.e., they do not take the cost matrix into account [6,23]. Hence, it is necessary to use specialized CSL algorithms or transform existing algorithms into cost-sensitive ones. According to [8] we distinguish between *direct approaches* (DA) and *meta-learning approaches* (ME-A).

While direct approaches leverage the cost matrix directly in the training phase, the meta-learning approaches either modify the training data or the outputs of a trained classifier based on the cost matrix [8].

ME-A: Output Shift. In classifiers like neural networks that output a probability estimate $o = (o_1, ..., o_k)$, we can shift the output towards inexpensive classes [24]. For the probability estimate this equation holds: $\sum_{i=1}^{k} o_i = 1$ with $0 \leq o_i \leq 1$. Normally, the predicted class is determined by $\arg\max_i o_i$. For taking the cost matrix into account, we modify the probability estimate to

$$o_i^* = \delta \sum_{j=1}^{k} o_i c_{ij},$$

where δ is a normalization term such that $\sum_{i=1}^{k} o_i^* = 1$ and $0 \leq o_i^* \leq 1$. Hence, the cost matrix is only used in the test phase and does not modify the underlying data [24].

DA: Direct Approaches. If we want to use the cost matrix in the training phase directly, we must transform the classifiers individually into cost-sensitive ones. In case of neural networks as used for the process prediction, that means modifying the loss function used for training the neural network.

4   Related Work

Comparative studies reveal that deep learning approaches for next event prediction outperform classical ML techniques that use an explicit representation of the process model. Table 2 gives an overview of existing deep learning approaches for next activity prediction. Most of them apply Long-Short-Term Memory Neural Networks (LSTM) or Convolutional Neural Networks (CNN). Some other approaches are based on classical Deep Feedforward Networks (DFNN) or a specialized variant of a LSTM called Gated Recurrent Unit (GRU). These approaches take different event attributes into account (mostly the activity [ACT] and timestamp [Time]). Some of them also include derived event attributes, like roles or rules (i.e., process models). However, none of these approaches use CSL techniques or techniques for dealing with imbalance. Outside the process domain CSL gains attention since many years. In [6,25] the foundations of cost matrices are investigated and criteria for consistent cost matrices are derived. This concept is applied in [25] by an experimental study on 14 datasets of the *UCI ML repository* and 6 synthetic datasets (half of them imbalanced). In [8,9] the authors give an overview of the entire research of imbalanced learning. In [22] the concept of constant cost matrices is generalized to deal with cost intervals, that allows weighting particular examples differently. Also the usefulness of CSL in combination with different algorithms is thoroughly investigated: In [11] techniques of over- and undersampling are compared to CSL methods with regard to the class imbalance. Also they investigate, which classification algorithms are sensitive to class imbalance. In [24] methods (over- and undersampling, threshold moving, hard and soft ensembles) for imbalanced classification with neural networks are compared on UCI datasets. Although over- and undersampling are an alternative to CSL, they bear some problems. In case of oversampling, pure duplicates have barely value, often the generated data is noisy, and training time increases. On the other hand, information gets lost by undersampling.

Table 2. Overview of existing deep learning approaches for next event prediction.

Approach	Network	Input	Prediction
Evermann [7]	LSTM	ACT	ACT
Mehdiyev [14]	DFNN	ACT, DP	ACT
Tax [19]	LSTM	ACT, Time	ACT, Role, Suffix, Time
Al-Jebrni [2]	CNN	ACT	ACT
Schönig [17]	LSTM	ACT, ATTR	ACT, Resource
Camargo [4]	LSTM	ACT, Role, Time	ACT, Role, Time, Suffix
Lin [13]	LSTM	ACT, DP	ACT, DP, Suffix
Hinkka [10]	GRU	ACT, DP	ACT
Theis [21]	DFNN	ACT, Rules, Time, DP	ACT
Pasquadibisceglie [16]	CNN	ACT, Time	ACT
Mauro [5]	CNN	ACT, Time	ACT

5   Implementation and Evaluation

5.1   Dataset Description and Experimental Setup

We evaluate our concept by combining two state-of-the-art approaches for next activity prediction with the procedures described in Sect. 3. We select approaches [4] and [16], since they represent the most frequently used network architectures (LSTM and CNN respectively), and achieve good results in comparative studies. We build our implementation[1] upon those of the considered approaches using Python 3.7 and run the neural networks with Tensorflow, which we use at version 2.1.0 with GPU support via CUDA for expression evaluation. We perform our experiments using 4 real-life event logs from different domains with diverse characteristics (cf. Table 3) extracted from the *4TU Center for Research Data*[2]. However, the event logs Sepsis and Traffic do not record the performing resource,

---

[1] The source code can be accessed at https://github.com/mkaep/.
[2] https://data.4tu.nl/.

so they cannot be evaluated with approach [4]. The splitting of the event logs into training and testdata is done along the time dimension, by ordering the traces ascending by its first event timestamp and then selecting the first 70% of the traces for training. For evaluation, we use the following common metrics for imbalanced datasets that can be derived using the confusion matrix[3]: (i) Recall (also called sensitivity) defined as $R = TP/(TP + FN)$, (ii) Precision $P = TP/(TP + FP)$, and (iii) F-Measure defined as harmonic mean $F = 2RP/(R + P)$ of precision and recall. As recommended, we first calculate the metrics per class and afterwards, we calculate the arithmetic mean of the per class scores [8]. Hence, we weight the classes equally, independently of their probability. Additional, we use the geometric mean defined as $G = \sqrt{R_{avg}S_{avg}}$ the squared root of the averaged recall and specificity. The specificity can be calculated by $S = TN/(TN+FP)$. Furthermore, we quantify for each approach the imbalance of the dataset, by determining the number of minority classes (m), IR, and ID as recommended in [15]. The cost matrices are derived with the heuristic method described in Sect. 3.2. The experiments were run on a system equipped with a Windows 10 operating system, an Intel Core i9-9900K CPU3.60 GHz, 64 GB RAM, and a NVIDIA Quadro RTX 4000 having 6 GB of memory.

Table 3. Statistic of the used event logs.

Event log	BPIC12	Helpdesk	Sepsis	Traffic
Number of cases	9559	4580	1050	150370
Number of activities	36	14	16	11
Maximal case length	163	15	185	20
Minimal case length	3	2	3	2
Average case length	14.74	4.66	14.49	3.73

5.2 Overall Results and Further Analysis

We discuss now our results in detail. All measures are shown in Table 4. Note, that measuring the metrics does not primarily evaluate the concept of CSL itself rather than the quality of the cost matrices. Considering m and the IR, reveals that all event logs are multi-minority problems (cf. Definition 6) with strong imbalance. Normalizing the ID shows differences between the event logs and some minor ones between the applied approaches (caused by the different minimum length of event prefixes). Unfortunately, in all cases the cost matrix is inconsistent, so it is not possible to derive an optimal weight vector. Thus results for this method cannot be proclaimed. In case of the Sepsis log the *Balanced Cost* method lowers the performance. On all other logs the performance of

---

[3] i.e., it can be calculated from true positives (TP), true negatives (TN), false positives (FP), and true positive (TP).

Table 4. Overall results (best values are bold faced).

		BPIC12		Helpdesk		Sepsis		Traffic	
		[16]	[4]	[16]	[4]	[16]	[4]	[16]	[4]
Recall									
	Conventional	0.67	0.67	0.17	0.29	0.47	–	0.70	–
OS	*Balanced Cost*	0.65	0.66	0.21	0.28	0.36	–	0.63	–
	Cost Sum	0.75	0.73	**0.29**	**0.34**	**0.55**	–	**0.76**	–
DA	*Balanced Cost*	0.70	0.67	0.22	0.24	0.38	–	0.71	–
	Cost Sum	**0.76**	**0.75**	0.28	0.32	0.53	–	0.74	–
Precision									
	Conventional	0.72	0.76	0.13	0.26	0.48	–	0.74	–
OS	*Balanced Cost*	0.71	0.75	0.17	0.27	0.24	–	0.65	–
	Cost Sum	**0.78**	0.78	**0.21**	**0.31**	0.53	–	**0.79**	–
DA	*Balanced Cost*	0.68	0.75	0.15	0.25	0.29	–	0.70	–
	Cost Sum	**0.78**	**0.79**	**0.21**	0.30	**0.54**	–	0.78	–
F-Measure									
	Conventional	0.67	0.68	0.14	0.27	0.47	–	0.70	–
OS	*Balanced Cost*	0.65	0.61	0.09	0.28	0.12	–	0.51	–
	Cost Sum	0.74	**0.73**	**0.21**	**0.32**	0.52	–	**0.76**	–
DA	*Balanced Cost*	0.68	0.64	0.15	0.26	0.21	–	0.69	–
	Cost Sum	**0.75**	**0.73**	0.20	**0.32**	**0.55**	–	0.74	–
Geometric Mean									
	Conventional	0.81	0.82	0.40	0.53	0.67	–	0.83	–
OS	*Balanced Cost*	0.80	0.81	0.47	0.55	0.46	–	0.69	–
	Cost Sum	0.85	0.85	0.48	**0.59**	0.73	–	0.87	–
DA	*Balanced Cost*	0.83	0.81	0.48	0.51	0.57	–	0.85	–
	Cost Sum	**0.87**	**0.86**	**0.50**	0.58	**0.74**	–	**0.89**	
#Classes/m		34/18	37/21	13/9	14/9	16/10	–	10/6	–
Imb. Ratio (IR)		4011	2326	4702	3717	556	–	145	–
Imb. Degree (ID)		17.48	20.46	8.85	8.80	9.65	–	5.75	–
Normalized ID		0.51	0.55	0.68	0.63	0.60	–	0.57	–

the Balanced Cost method is for the output shift (OS) and the direct approach (DA) nearly identical to the performance of conventionally trained classifiers. Only some minor deviations can be observed. The *Cost Sum* method, however, achieves a respectable performance gain over all considered metrics between 3% and 10% and can be beneficially applied for both the output shift approach and the direct approach. Hence, this approach slightly outperforms the conventional ones. Furthermore, we observe usefulness of CSL for both predictive business process monitoring approaches. In general, we observe, that a less extreme imbalance results in better classification results. Hence, we can confirm the results of studies (cf. Sect. 4) in other domains also for the process domain. Since, the cost matrix is not consistent CSL has only a moderate but notable impact. Hence, we draw the following conclusions: The achieved performance seems promising for further research which should focus on the mining of consistent cost matrices. Furthermore, it should be investigated whether methods of numeric mathematic for calculating approximate solutions of overdetermined systems can be used to calculate a satisfactory weight vector.

6   Conclusion and Future Work

In this paper, we propose a concept for leveraging CSL for the prediction of categorical event attributes to cope the problem of an imbalanced class distribution. Our experiments reveal that CSL can be beneficially applied for next activity prediction. In addition to the above mentioned issues for future work, we plan to use CSL for binary problems like outcome oriented predictive process monitoring, since in the two class case the equation system for calculating an optimal weight vector is more often solvable. Also, it should be investigated how the performance differs from those of alternative options like over- and undersampling and, whether a combination of both, can achieve better results.

References

1. van der Aalst, W.: Process Mining - Discovery, Conformance and Enhancement of Business Processes. Springer, Heidelberg (2011)
2. Al-Jebrni, A.H., Cai, H., Jiang, L.: Predicting the next process event using convolutional neural networks. In: IEEE International Conference on PIC, pp. 332–338 (2018)
3. Breiman, L., Friedman, J., Stone, C., Olshen, R.: Classification and Regression Trees. Taylor & Francis (1984)
4. Camargo, M., Dumas, M., González-Rojas, O.: Learning accurate LSTM models of business processes. In: Hildebrandt, T., van Dongen, B.F., Röglinger, M., Mendling, J. (eds.) BPM 2019. LNCS, vol. 11675, pp. 286–302. Springer, Cham (2019). https://doi.org/10.1007/978-3-030-26619-6_19
5. Di Mauro, N., Appice, A., Basile, T.M.A.: Activity prediction of business process instances with inception CNN models. In: Alviano, M., Greco, G., Scarcello, F. (eds.) AI*IA 2019. LNCS (LNAI), vol. 11946, pp. 348–361. Springer, Cham (2019). https://doi.org/10.1007/978-3-030-35166-3_25
6. Elkan, C.: The foundations of cost-sensitive learning. In: Proceedings of IJCAI 2001, pp. 973–978. Morgan Kaufmann Publishers Inc., San Francisco (2001)
7. Evermann, J., Rehse, J.R., Fettke, P.: Predicting process behaviour using deep learning. Decis. Support Syst. **100**, 129–140 (2017)
8. Fernández, A., García, S., Galar, M., Prati, R.C., Krawczyk, B., Herrera, F.: Learning from Imbalanced Data Sets. Springer, Heidelberg (2018)
9. He, H., Ma, Y.: Imbalanced Learning: Foundations, Algorithms, and Applications, 1st edn. Wiley-IEEE Press (2013)

10. Hinkka, M., Lehto, T., Heljanko, K.: Exploiting event log event attributes in RNN based prediction. In: Welzer, T., et al. (eds.) ADBIS 2019. CCIS, vol. 1064, pp. 405–416. Springer, Cham (2019). https://doi.org/10.1007/978-3-030-30278-8_40

11. Japkowicz, N., Stephen, S.: The class imbalance problem: a systematic study. Intell. Data Anal. **6**(5), 429–449 (2002)

12. Käppel, M.: Evaluating predictive business process monitoring approaches on small event logs. arXiv (2021)

13. Lin, L., Wen, L., Wang, J.: A deep predictive model for multi-attribute event sequence. In: Proceedings of the International Conference on Data Mining 2019, pp. 118–126 (2019)

14. Mehdiyev, N., Evermann, J., Fettke, P.: A multi-stage deep learning approach for business process event prediction. In: IEEE 19th CBI, pp. 119–128 (2017)

15. Ortigosa-Hernández, J., Inza, I., Lozano, J.A.: Measuring the class-imbalance extent of multi-class problems. Pattern Recogn. Lett. **98**, 32–38 (2017)

16. Pasquadibisceglie, V., Appice, A., Castellano, G., Malerba, D.: Using convolutional neural networks for predictive process analytics. In: Proceedings of ICPM 2019 (2019)

17. Schönig, S., Jasinski, R., Ackermann, L., Jablonski, S.: Deep learning process prediction with discrete and continuous data features. In: Proceedings of ENASE 2018 (2018)

18. Sun, Y., Kamel, M.S., Wong, A.K., Wang, Y.: Cost-sensitive boosting for classification of imbalanced data. Pattern Recogn. **40**(12), 3358–3378 (2007)

19. Tax, N., Verenich, I., La Rosa, M., Dumas, M.: Predictive business process monitoring with LSTM neural networks. In: Dubois, E., Pohl, K. (eds.) CAiSE 2017. LNCS, vol. 10253, pp. 477–492. Springer, Cham (2017). https://doi.org/10.1007/978-3-319-59536-8_30

20. Teinemaa, I., Dumas, M., Rosa, M.L., Maggi, F.M.: Outcome-oriented predictive process monitoring: review and benchmark. TKDD **13**(2) (2019)

21. Theis, J., Darabi, H.: Decay replay mining to predict next process events. IEEE Access **7**, 119787–119803 (2019). https://doi.org/10.1109/ACCESS.2019.2937085

22. Wu, J., Wan, L., Xu, Z.: Algorithms to discover complete frequent episodes in sequences. In: Cao, L., Huang, J.Z., Bailey, J., Koh, Y.S., Luo, J. (eds.) PAKDD 2011. LNCS (LNAI), vol. 7104, pp. 267–278. Springer, Heidelberg (2012). https://doi.org/10.1007/978-3-642-28320-8_23

23. Zhang, X., Hu, B.: A new strategy of cost-free learning in the class imbalance problem. IEEE Trans. Knowl. Data Eng. **26**(12), 2872–2885 (2014)

24. Zhou, Z.-H., Liu, X.-Y.: Training cost-sensitive NN with methods addressing the class imbalance problem. IEEE Trans. Knowl. Data Eng. 63–77 (2006)

25. Zhou, Z.H., Liu, X.Y.: On multi-class cost-sensitive learning. In: Proceedings of the 21st National Conference on AI, pp. 567–572. AAAI Press (2006)

A Hybrid Data Model and Flexible Indexing for Interactive Exploration of Large-Scale Bio-science Data

Gajendra Doniparthi[1]([✉]), Timo Mühlhaus[2], and Stefan Deßloch[1]

[1] Heterogeneous Information Systems Group, University of Kaiserslautern, Kaiserslautern, Germany
{doniparthi,dessloch}@informatik.uni-kl.de
[2] Computational Systems Biology, University of Kaiserslautern, Kaiserslautern, Germany
muehlhaus@bio.uni-kl.de

Abstract. The advancement of high-throughput technologies has considerably increased the amount of research data generated from bioscience experiments. The integrated analysis of these large datasets provides opportunities to understand complex biological systems better. We present a novel research data management framework that uses a hybrid relational and NoSQL data model for interactively querying and exploring large-scale bio-science research data. Our framework uses a fast, scalable, space-efficient, and flexible indexing scheme leveraging bitmaps purpose-built for exploratory data analysis and supports containment, point, and range query types.

Keywords: Interactive data exploration · Relational JSON · Interval trees · Cross-omics · Research data management

1 Introduction

The technological advancements in the past decade have immensely increased the ease of collecting bio-science research data at a significantly reduced cost. As a result, more data scientists are attempting to integrate cross-domain data for explorative analysis to create new and meaningful scientific knowledge. Any given research data from a certain bio-science domain can be logically divided into two parts. Firstly, the structured *meta-data* which is the minimum required information to allow researchers to understand the biological and technical origins of the data for reproducibility. Secondly, the semi-structured *raw data* generated from the instruments used for conducting the experiment, which also includes contextual information, instrument parameters, etc. To perform exploratory analysis, not only both *meta-data* and *raw data* must follow an interoperable standard format, but the system also has to use a multi-model architecture to reduce complexity and maintain data consistency. Traditional RDBMS

© Springer Nature Switzerland AG 2021
L. Bellatreche et al. (Eds.): ADBIS 2021 Short Papers, Workshops and Doctoral Consortium,
CCIS 1450, pp. 27–37, 2021.
https://doi.org/10.1007/978-3-030-85082-1_3

with JSON datatype for schema-less development provides multi-modeling and can be an easy choice to store *meta-data* in relational format and *raw data* as JSON blobs. However, the indexing support on JSON data type is still not extensive [5] and the usual one-shot database querying approach is time-consuming and is impractical for interactive exploration. Similarly, NoSQL data stores have emerged as effective tools in recent times for managing high volumes of bioscience data [4,7]. They are ideal to maintain and index high volumes of semistructured *raw data*, however, the hierarchically structured *meta-data* will need to be de-normalized and duplicated to compensate for the lack of relational joins, which means, using more storage space in comparison.

In this paper, we introduce a scalable research data management and interactive exploration framework that offers a novel multi-step exploratory data analysis process to address the above challenges, in particular, the data model heterogeneity, indexing, and storage space requirements. The framework uses an integrated relational and NoSQL model for structured *meta-data* and semistructured *raw data* respectively, and a fast, scalable, space-efficient, and flexible indexing scheme leveraging bitmaps, used for exploratory data analysis. We discard the typical one-shot database querying approach and instead allow data analysts to explore research data progressively while interactively evaluating containment, point, and range queries on huge volumes of attribute-values, all in-memory. We designed our framework keeping in mind the typical research data management applications where *writes* and *reads* are frequent but *updates* and *deletes* are negligible. We developed the problem statement for this paper in collaboration with our Computational Biology group as part of DataPLANT[1], an NFDI project aimed at integration and exploration of large-scale cross-omics[2] research data.

In the following section, we introduce a typical bio-science experiment and analysis workflow to motivate our work and subsequently present the set of interactive exploration operations offered by the framework. We then discuss related work in Sect. 3. In Sect. 4, we introduce our system architecture and describe in detail the flexible indexing scheme for massive sets of attribute-value pairs using bitmaps, and how our segmentation approach helps maintain and tune the indexes. In our experiments presented in Sect. 5, we measure the interactive latencies with point and range queries executed using different types of flexible indexes. We also compare the storage space requirements and conclude by summarizing our experiment results.

2   Motivating Workflow

To begin with, we assume that the bio-science experiments conducted using high throughput technologies such as next-generation sequencing or mass spectrometry follow the standardized meta-data categories *Investigation* (the project

---

[1] https://nfdi4plants.de/.
[2] *omics* - an informal reference to a field of study in biology ending in -omics, such as Gen*omics*, Prote*omics*, Metabol*omics* etc.

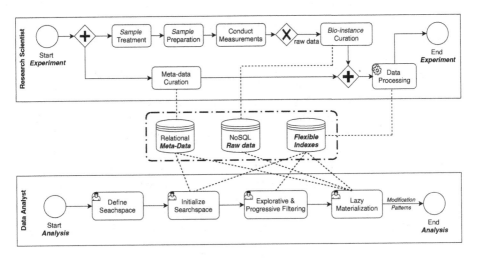

Fig. 1. Workflow of a bio-science experiment

context), *Study* (a unit of research), and *Assay* (analytical measurement), called the ISA specification, which uses a hierarchical tab/JSON file format [8]. Figure 1 shows the workflow which can be broadly seen as two logical parts. Note that we use BPMN only to describe the process flows and there is no workflow engine involved. The first part covers the combined process flow of curating meta-data, sample preparation, and performing the actual experiment. From the instrument-specific measurement data, the researcher then uses quantification and identification methods to generate output files in a semi-structured format along with the contextual information. In particular, the raw data is expected to be heterogeneous and diverse depending on the domain, type of the experiment, what is being measured, the kind of equipment used in the experiment, and the procedures followed. Controlled vocabularies or Ontologies are used to keep track of predefined and standardized contextual and instrument-specific parameter names. Depending on the type of scientific experiment, the attributes of any type i.e. numerical, categorical, etc., are taken from a relevant vocabulary.

The second part covers the analysis process, where a data analyst performs exploratory analysis on both the experimental meta-data and contextual information to extract resulting data sets satisfying given modification patterns (or refinement patterns). These are boolean combinations of attribute-value pairs that capture either the snapshot of the environment a particular experiment conducted in or the identification & quantification measurements recorded from the instrument. In our previous work [2], we have already outlined these novel interactive exploration steps in detail. Here, we present the main aspects of the exploration model to keep our paper self-contained. The analysis task includes four steps

1. Defining the search space, which is an imaginary n-dimensional space in main memory. Any relation from the *meta-data* hierarchy can be defined as one of its dimensions.
2. Initializing the search space by loading *signatures* of data tuples into main memory as per the search space definition, say, a set of samples and associated bio-instances with a known chemical property across the experiments from different domains. Here, the framework performs relational joins to load the list of *ids* of respective data tuples from *meta-data* into the search space.
3. Filtering the data tuple signatures progressively such as finding a common or comparable parameter of treatment. Here, the search space gets updated after evaluation of each predicate condition applied interactively on the corresponding attribute values from *raw data* using the integrated flexible indexes.
4. Materialization of snapshot of the search space at any stage of interactive exploration where the explored data in the search space is downloaded offline.

To that end, we introduce the data management and storage layer of our workflow comprising of both relational and document data stores for *meta-data* and *raw data* respectively. The flexible indexes integrate the structured meta-data from the relational side and the semi-structured contextual data from the NoSQL side. Depending on the stage of exploration, respective data stores and indexes are accessed to speed up the overall analysis. This approach gives us the flexibility to create different index structures for the same attribute across segments, thus, making it possible to tune and optimize the indexes at a granular level. We elaborate on the methods of segmentation, indexing, and optimizing the indexes in Sect. 4.

3 Related Work

In the relational space, while SQL can easily handle the structured portion and work seamlessly in standalone relational database systems, the schema-less semi-structured data does not comply with the n-ary storage model. A Decomposed Storage Model (DSM) [1] is one of the early solutions proposed to handle sparse attribute-value format data. The introduction of JSON data types in RDBMS platforms enabled multi-modeling by storing JSON data natively without the need for flattening it into relational format [5]. However, the schema-less JSON data can only be considered as text documents for indexing. In the NoSQL space, multi-model databases integrating different data models into a single database could help maintain structured *meta-data* and semi-structured *raw data* within the same database. However, these are single model NoSQL stores extended for multi-model data with no relational support, which also means the data need to be de-normalized and duplicated to compensate for the lack of relational joins and integrated indexes.

To handle terabytes of scientific data, array data models and polystore architectures were proposed in recent times. SciDB [9] is one such system where the logical object accessible to users is an N-dimensional array instead of a relational table. SciDB's array query language has facilities that allow users to extend the system with new scalar data types and array operators. BigDAWG [3] polystore architecture supports various database systems with different data models and uses a middleware layer that provides a uniform multi-model interface. However, these systems are primarily built for heavy analytic workloads rather than for interactive exploration. There is also related work on interactive query refinement, in which the input query is modified to increase or decrease the result cardinalities [6]. However, our approach of interactive exploration relies on loading the superset of data into the main memory and let the user drill down the data by progressively evaluating the predicate expressions.

4   Architecture

Figure 2 shows the high-level three-tier architecture of our framework. We developed a web-based user interface for both meta-data curation and analysis modules. The data layer comprises integrated relational and document databases to maintain structured meta-data and augmented measurement & contextual data respectively. The analyst manually curates *meta-data* through web-ui. However, after each analytical measurement of an experiment, the respective *raw data* files generated gets loaded into the staging area in a standard file format. At

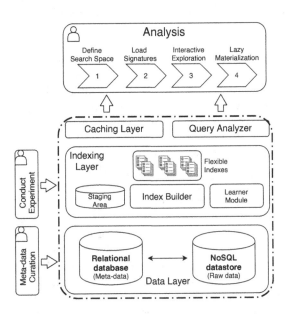

Fig. 2. System architecture

the end of the experiment, the analyst initiates the bulk ingestion of raw data into the document data store and also triggers the indexing process. To maintain data integrity, each relation in the hierarchical meta-data contains a cross-reference column to capture the respective document identifier of the corresponding document in the document database. The data in the NoSQL datastore is required only at the time of materializing the explored result sets offline, thus, avoiding the need for accessing high volume semi-structured data at the time of interactive exploration. The indexing layer is responsible for the creation and maintenance of indexes to support various exploratory query types such as containment, point, and range queries on the respective attribute-values. Each exploratory query operation issued by the analyst gets parsed by the query analyzer, which identifies the attributes and the query type to use the relevant index and its access method.

4.1   Flexible Indexes

At the time of ingestion, we divide the data tuples in the relations into non-overlapping subsets using a monotonically increasing primary key/tuple identifier. This approach is scalable and as such does not limit the number of attributes to be indexed or the number of data tuples in the relation. The segment size varies for each relation in the hierarchy and it can be individually optimized. It also makes sure that an index on a particular attribute is created in parts (per each segment), thus, making it faster to access when maintained in a distributed environment. The framework enforces each segment to contain *meta-data* from a particular bio-science domain. This offers flexibility in creating suitable indexes for an attribute of respective domain and cater to any query type. Figure 3, shows how the indexing layer creates and maintains integrated indexes per segment, per each measurement/contextual parameter. These indexes are used in the exploration & progressive filtering stage of analysis and are built for index-only scans. In our previous work [2], we restricted our scope only to Boolean containment queries and used Bloom filters for the same. We now extend the solution and complement Bloom filters with bitmaps and tree structures augmented with bitmaps, thus, making it possible to evaluate not only containment queries but also point and range queries. The idea is to encode the set of satisfying tuple ids of each segment in the relational side into a compressed data structure suitable for the given query type. For example, a simple containment query requires just a bitmap index of the size of the segment with encoded 1 and 0 bits, where a 1 bit at a certain position in the bitmap indicates that the respective tuple id in that segment contains the requested attribute. Such a set of bitmap indexes for all attributes and segments are good enough to evaluate any boolean combination of containment queries i.e. refinement patterns on any selected attributes during analysis without the need for referencing the semi-structured measurement/contextual data from the NoSQL data store. Similarly, a tree structure augmented with the tuple ids created on the values of the attribute in each segment is ideal for point queries. However, for range queries, we use interval trees which are tree structures similar to the ones used for point queries, except that the nodes hold the value ranges instead of a single value. Here, the nodes in the tree structure are augmented with the encoded bitmaps with 1s at certain positions where the respective tuple ids of the segment contain the value within the range. As expected, the interval ranges naturally introduce false positives in the query result and the number of false positives depends on the difference in the interval spans between the query interval and the interval range. We eliminate the false positives at the time of materializing the query result. Also, from historical user query accuracies we optimize the interval sizes of range index structures at each segment level as shown for segment 2, attribute k in Fig. 3.

Usually, the domain experts decide before-hand the index type for an individual known measurement/contextual attribute depending on the expected querying type. Although the point indexes handle range queries as well, approximate results are allowed during the analysis, particularly when querying for value

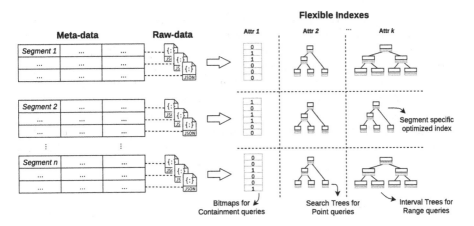

Fig. 3. Segmentation

ranges for certain attributes. Between the index size, search time, and accuracy, there exists a trade-off where the range index structures can be optimized and the interval trees can help reduce the interactive latencies in comparison to using point indexes for answering range queries.

4.2   Interval Trees

We use the simple case of interval trees where the intervals do not overlap. An interval $[x_1, x_2]$ is an ordered pair of real numbers which represents the set $\{x \in \mathbb{R} : x_1 \leq x \leq x_2\}$. For simplicity, we consider only closed intervals where both the endpoints are included in the set. Two intervals i and i' are considered as overlapping if $i \cap i' \neq \emptyset$ i.e. $i_1 \leq i'_2$ and $i'_1 \leq i_2$. These interval tree structures can support insert, delete and search operations and are useful to answer range queries, in particular, close-ended range queries such as $-5 \leq k \leq 5$. When a binary search tree is extended for non-overlapping interval ranges, each node n in the tree holds the element x which is assumed to be part of the interval $[x_1, x_2]$ with x_1 being the low endpoint and x_2 becoming the high endpoint. Given a query interval of $q.int$, the search operation to find the intersecting interval node is expected to take $O(\log n)$. For a given attribute k, consider a segment S_i with tuple id-value pairs $(p_i, v_i), (p_{i+1}, v_{i+1}), \ldots, (p_{i+x}, v_{i+x})$ where x is the segment size. After sorting by value, we use binning and the data is split into disjoint intervals of predefined value ranges. Also, we make sure that the repeated values are enclosed within the same interval, thus, avoiding any interval overlaps.

For each interval, individual bitmaps of size x with the set of corresponding tuple ids are generated before they are augmented with the nodes of the tree. The entire tree data structure is serialized to disk for the given attribute and segment S_i. To find the resulting interval $r.int$ intersecting with the query interval $q.int$, two conditions have to be checked. One, the start and/or endpoint of $r.int$ is in $q.int$; or $r.int$ completely encloses $q.int$. The search algorithm finds all the intervals intersecting the given query interval $q.int$ and returns one bitmap per entire segment i.e. it performs logical OR of the bitmaps from each intersecting interval. In scenarios where the analyst issues a complex query with multiple attributes, say, $(0.0045 \leq protein\,abundance \leq 0.0046)$ AND $(98 \leq mass/charge\,ratio \leq 100)$, the query gets parsed into individual queries per attribute. The respective indexes are then used to evaluate either the point or the range query to get the bitmaps per segment for each attribute. Lastly, the bit-wise OR or AND operation is performed on the individual bitmaps for each attribute to get the final bitmap per segment as the result of the complex query.

5   Experiments

We use uniformly distributed synthetic research data for our evaluation and the test database is of size ≈ 40 GB. While we are in the process of manually curating the standardized plant sample data from our high-throughput bio-science experiments, we believe that the dense synthetic data is the best-case scenario to evaluate our framework since our focus is on measuring latencies and disk space requirements. The test database is created with a simplified hierarchical data model where we only consider two levels of the standardized bio-science research data hierarchy (bio-samples and measured bio-instances). We simulate the research data by taking 8 different types of bio-instances i.e. *lipid, peptide, gene, etc.*. We generate a total of 500 bio-instance data tuples for each bio-sample of any type. Similarly, each bio-instance is enriched with 300 randomly chosen attributes from the test vocabulary containing a total of 2550 unique attributes. The database contains a total of 7000 bio-samples with 3.5 million associated bio-instances each containing an array of 300 contextual attribute-value pairs in JSON format, taking the total count of attributes in the database to a billion.

We use PostgreSQL version 11.7 for *meta-data* and MongoDB version 4.4.0 for *raw data* in all of our experiments performed on a server running Ubuntu 18.04.5 LTS 64bit. The server hardware configuration is given as Intel Core i7-8700 CPU @ 3.20 GHz ×12, 16 GB RAM, and 500 GB SSD. The application layer along with the front-end is developed in Spring Boot Framework 2.1.6 and also deployed in the same server.

We simulate refinement patterns by evaluating our analysis tasks with 5 interactive *update* operations per task. Each operation is evaluated on a set of 5 contextual attributes. The attributes for the experiments are selected randomly out of the test vocabulary and all the analysis tasks are repeated for each bio-instance type (*peptide, gene* etc.) and the results are averaged. The average number of signatures per each bio-instance type loaded into the memory is 0.4 million and it took ≈ 3.1 s to initialize the search space. As the data is uniformly distributed, each *update* operation proportionately reduces the size of the search space. We use the segmentation approach with a fixed segment size of $2^{15} = 32\,768$ for our experiments.

5.1   Query Evaluation

We primarily measure the performance of our bitmap index-based interval trees for the point and range queries. We investigate the following

1. How fast are interactive response times when bitmap index-based interval trees are used to evaluate point and range queries?
2. What are the storage space requirements across different index types?
3. What is the precision when using interval trees for point and range queries?

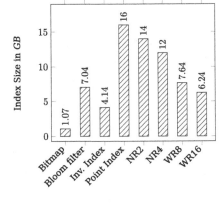

Fig. 4. Index size comparison

To address each of the above points, we have built five types of indexes from the given test dataset i.e. one *Point Index* and four range indexes *NR2*, *NR4*, *WR8*, *WR16* with varied interval sizes 2, 4, 8, and 16 respectively. For example, the narrow range index *NR4* for a certain attribute is built with interval value ranges where the resulting tuples per segment in that range are approximately 4, excluding duplicates. We use randomized binary search trees for all of the indexes and also the bitmaps are compressed, except that the nodes in the point index are augmented with the tuple ids. Figure 4 shows the index size comparisons. As expected, the wide range indexes only used ≈40% of the storage space in comparison to the point index for the same sized data set. Although it helps with faster query response, this comes at the expense of accuracy, where the wide range index used for a point query is expected to return a high number of false positives. For containment queries, bitmap indexes are the most space-efficient in comparison to Bloom filter and inverted indexes.

To measure both the interactive latencies and the accuracies, we run three different types of queries on these indexes. The point queries where we query with an equality predicate should return all tuples in a segment for an attribute with the exact matching value. We consider narrow range queries as the query intervals where the resulting tuples per segment are equal to or less than 8. For wide range queries, we consider the query intervals where the resulting tuples per segment are more than 8. We vary this input query interval ranges from 1 to 128, where 1 is the point query. Figure 5 shows the precision scores and latencies measured at the application layer at the end of each interactive exploration operation for each input query interval range. As expected, the point index achieves a perfect precision score of 100% and is the benchmark for other indexes. Although the range indexes fare poorly for point queries in comparison, however, it is evident that the higher the input query interval range, say 128, even the wide range index achieves higher precision. Concerning interactive latencies, the wide range index is the fastest overall. The reason being the smaller index structure, faster traversals, and hence faster query response. In comparison, the average latency measured with bitmap indexes for containment queries is still 3× faster than the range queries. By design, the materialization phase of the analysis removes any false positives before downloading the explored search space to the disk. This gives us leverage to tune and operate the multi-step analysis at slightly lower precision levels so that we achieve gains in the form of faster interactive query response and less storage space. Nevertheless, in environments where a higher level of accuracy is expected, the point indexes are still the best fit for any type of query at the expense of interactive latency and storage space.

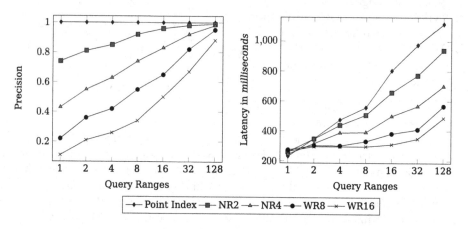

Fig. 5. Precision and latencies measured for point and range queries

6   Conclusion

We presented a novel research data management and interactive exploration framework in this work, offering a multi-step exploratory data analysis process. We resolved the underlying data model heterogeneity by following an integrated relational and NoSQL approach and proposed a segmentation method to use a fast, scalable, space-efficient, and flexible indexing scheme leveraging bitmaps. We measured the interactive latencies, precision, and storage space requirements for different indexes and query types in our experiments. Our measurements confirm that our multi-step exploratory analysis solution helps achieve faster interactive response times and maintains optimized indexes. As an avenue of future work, we plan on implementing a *learning* module which learns from the underlying data to find the optimal index for any given attribute.

References

1. Copeland, G.P., Khoshafian, S.: A decomposition storage model. In: Navathe, S.B. (ed.) Proceedings of the 1985 ACM SIGMOD International Conference on Management of Data, Austin, Texas, USA, 28–31 May 1985, pp. 268–279. ACM Press (1985). https://doi.org/10.1145/318898.318923
2. Doniparthi, G., Mühlhaus, T., Deßloch, S.: A bloom filter-based framework for interactive exploration of large scale research data. In: Darmont, J., Novikov, B., Wrembel, R. (eds.) ADBIS 2020. CCIS, vol. 1259, pp. 166–176. Springer, Cham (2020). https://doi.org/10.1007/978-3-030-54623-6_15
3. Gadepally, V., et al.: BigDAWG version 0.1. In: 2017 IEEE High Performance Extreme Computing Conference (HPEC), pp. 1–7 (2017). https://doi.org/10.1109/HPEC.2017.8091077
4. Kaur, K., Rani, R.: Managing data in healthcare information systems: many models, one solution. Computer **48**(3), 52–59 (2015). https://doi.org/10.1109/MC.2015.77
5. Liu, Z.H., Gawlick, D.: Management of flexible schema data in RDBMSs - opportunities and limitations for NoSQL. In: Seventh Biennial Conference on Innovative Data Systems Research, CIDR 2015, Asilomar, CA, USA, 4–7 January 2015 (2015). http://cidrdb.org/cidr2015/Papers/CIDR15_Paper5.pdf. Online Proceedings. www.cidrdb.org
6. Mishra, C., Koudas, N.: Interactive query refinement. In: Kersten, M.L., Novikov, B., Teubner, J., Polutin, V., Manegold, S. (eds.) 12th International Conference on Extending Database Technology, Saint Petersburg, Russia, EDBT 2009, 24–26 March 2009, Proceedings. ACM International Conference Proceeding Series, vol. 360, pp. 862–873. ACM (2009). https://doi.org/10.1145/1516360.1516459
7. Nti-Addae, Y., et al.: Benchmarking database systems for genomic selection implementation. Database J. Biol. Databases Curation **2019**, baz096 (2019). https://doi.org/10.1093/database/baz096
8. Sansone, S.A., Rocca-Serra, P., Field, D., Maguire, E., Taylor, C., et al.: Toward interoperable bioscience data. Nat. Genet. **44**(2), 121–126 (2012). https://www.nature.com/articles/ng.1054
9. Stonebraker, M., Brown, P., Zhang, D., Becla, J.: SciDB: a database management system for applications with complex analytics. Comput. Sci. Eng. **15**(3), 54–62 (2013). https://doi.org/10.1109/MCSE.2013.19

Relational Conditional Set Operations

Alexis I. Aspauza Lescano$^{(\boxtimes)}$ (ID) and Robson L. F. Cordeiro (ID)

University of São Paulo – USP, Av. Trab. São Carlense 400, São Carlos, SP, Brazil

Abstract. A set is a collection of different objects. Some basic operations from the Theory of Sets are the set membership ($\in$), subset ($\subset$), intersection ($\cap$), and difference ($-$). However, these operations have limitations because of the implicit use of the identity predicate. That is, a tuple is a member of a set if it is identical to any tuple in the set. Many applications need other comparison predicates that are not limited to identity. This paper presents the new Relational Conditional Set Operations, or **RelCond Set Operations** ($\in_c, \subseteq_c, \cap_c, -_c$) for short. Our operators are naturally suited to answer queries of conditional membership, subset, intersection, and difference with customized predicates. For example, they are potentially useful in applications of product sales with units and prices, job promotion, and internship. We validate our proposals by studying the first of these applications.

Keywords: Set operations · Relational algebra · Theory of sets

1 Introduction

The set membership ($\in$), subset ($\subseteq$), intersection ($\cap$), and difference ($-$) are basic operations from the Theory of Sets [12]. A set is a collection of **different** objects and the Relational Algebra [3–5] employs the set operations to work with relations. Their usability is very intuitive. For example, in Fig. 1.b we have a relation with Desired Products (DP) of a user, and another one with a Store's Products (SP) available. With the aforementioned operators, we can easily answer queries like **Q1:** *Can I buy a certain product X in the store?* with the set membership ($X \in SP$), **Q2:** *Can I buy all the desired products in the store?* with the subset operator ($DP \subseteq SP$), **Q3:** *Which desired products can be bought in the store?* with the intersection ($DP \cap SP$), and **Q4:** *Which desired products won't be found in the store?* with the difference ($DP - SP$).

The traditional set operators are very useful when elements are compared by identity. This is, using the implicit identity predicate illustrated in Fig. 1.a. Note that it could be expanded to support any custom predicate with **and** and **or** connectors, $=, <, \leq, >$, and $\geq$ logical operators, $+, -, *, /$ arithmetic operators, and negations $\neg$. As we discuss in this paper, the use of custom predicates have prompt applications in real life. However, operators that support them have not

This work has been partially supported by CAPES, FAPESP (2016/17078-0) and CNPq.

L. Bellatreche et al. (Eds.): ADBIS 2021 Short Papers, Workshops and Doctoral Consortium, CCIS 1450, pp. 38–49, 2021.
https://doi.org/10.1007/978-3-030-85082-1_4

been defined in the literature. This paper tackles the problem by presenting the new **RelCond Set Operators** $(\in_c, \subseteq_c, \cap_c, -_c)$. Our main contributions are:

C1 Operators Design and Usability – we identified severe limitations on the usability of the traditional set operations, which are caused by the use of implicit identity predicates. Thus, we tackled the problem by extending these operations into our RelCond Set Operators $(\in_c, \subseteq_c, \cap_c, -_c)$ that are naturally well suited to answer queries in sets with custom predicates.

C2 Formal Definition and Algorithms – we formally defined the new operators as the RelCond Set Membership $(\in_c)$, RelCond Subset $(\subseteq_c)$, RelCond Intersection $(\cap_c)$, and RelCond Difference $(-_c)$, thus enabling their use in queries along with the existing algebraic operators. Also, we designed novel algorithms to execute these new operations in a fast and scalable manner.

C3 Semantic Validation – we performed a case study by analyzing real data from thousands of toy products available for sale at Amazon. The results corroborate the practical usability of our operators in real-life applications.

C4 Generality and Usability – we present other applications where our operators are well suitable, thus corroborating their general usability.

The rest of this paper presents our case study (Sect. 2), related work (Sect. 3), proposed techniques (Sect. 4), experiments (Sect. 5), and conclusions (Sect. 6).

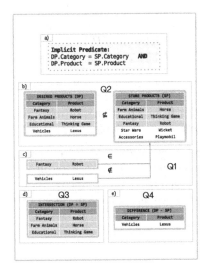

Fig. 1. Relational set operations.

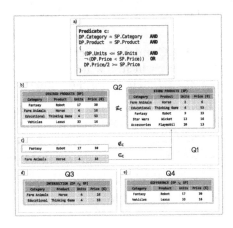

Fig. 2. Relational conditional set operations.

2 Case Study

To demonstrate the utility of the new RelCond Set Operators, let us expand our example of sales of products. Here, we add two more attributes to both sets DP and SP: units and price. In DP, these attributes refer to the user's desired units and maximum affordable price. In SP they refer to the store's available units and tag price. This is shown in Fig. 2.b. Let us assume also that the client can accept fewer units than the desired ones if the price is at most half of his/her budget. Now, to compare tuples by identity would not be useful. We need to employ a custom predicate to select products with the client's preferences, i.e., having enough units and acceptable price, or simply low price. The predicate c is shown in Fig. 2.a. This paper focuses on supporting custom predicates. Let us analyze our four previous queries aimed at demonstrating their utility.

Q1: *Can I buy a certain product X in the store?* – it could be answered with the **RelCond Set Membership** operator ($\in_c$). One tuple x **is a relcond$_c$ member** of one set T if, given a predicate c, the evaluation $c(x, y)$ is true for one tuple $y \in \mathsf{T}$. For example, in Fig. 2.c, tuple $x = (FarmAnimals, Horse, 4, 16)$ is a relcond$_c$ member of SP, since SP contains a tuple $y = (FarmAnimals, Horse, 2, 6)$ such that $c(x, y)$ is true. By contrast, tuple $w = (Fantasy, Robot, 17, 30)$ is not a relcond$_c$ member of SP because there is not a tuple z in SP where $c(w, z)$ is true. We can also follow this idea to define a **RelCond Set** $_c\mathsf{T}$ as a set in which $c(x, y)$ is false for any pair of tuples $x, y \in \, _c\mathsf{T}$.

Q2: *Can I buy all the desired products in the store?* – it could be answered by the **RelCond Subset** operator ($\subseteq_c$). A set is a relcond$_c$ subset of another set if all of its elements are relcond$_c$ members of the second set. For example, in Fig. 2.b, DP is not a subset of SP as there are tuples in DP that are not members of SP, i.e., (Fantasy, Robot, 17, 30) and (Vehicles, Lexus, 33, 16).

Q3: *Which desired products can be bought in the store?* – it could be answered with the **RelCond Intersection** operator ($\cap_c$). Figure 2.d illustrates the result of $DP \cap_c SP$, which is a new set with all tuples from DP that are relcond$_c$ members of SP. These are the desired products that are available in the store.

Q4: *Which desired products won't be found in the store?* – it could be answered with the **RelCond Difference** operator ($-_c$). Figure 2.e shows the result of $DP -_c SP$, which is a new set with all tuples from DP that are not relcond$_c$ members of SP. These are the desired products that are unavailable in the store.

3 Related Work

There are many works focused on implementing or expanding the set-based operations of the Relational Algebra. In the line of traditional set operations, a recent work [10] compares different strategies to implement the intersection, union, and difference. The authors start by describing naive algorithms, like those based on nested loops; then, they discuss improvements to reduce the number of computations. Consequently, the best algorithms for each operator are indicated. These ones still use nested loops, but also break and continue instructions, as well as removals from the relations, to reduce computations. Moreover, many modern applications require not only to store numeric and short character strings but also videos, photos, large texts, and other types of "complex" data elements. However, to compare complex elements by identity is usually senseless because an exact match almost never happens. Instead, it is more significant to evaluate their similarity. This is why some works [1,7–9] extend the relational set operations to handle complex data. Finally, there is an extensive body of work [2,11,13,14] on extensions of the Relational Algebra semantics to a fuzzy domain [6], which implies having fuzzy relations with vague values using linguistic labels, weighted tuples, and grades with different meanings for the attributes.

To summarize the related literature, there exist many works focused on the set-based operations, including those that are efforts to extend relational databases to support complex and fuzzy data. However, to the best of our knowledge, **no one** allows answering "conditional" queries with custom predicates.

4   Proposed Relational Conditional Operators

4.1   Formal Definitions

This section presents the formal definitions for our **RelCond Set Operators**. At first, preliminary definitions are given; then, the new operators are defined.

Definition 1. *An **arithmetic operator between values** $(\otimes)$ is represented by $a_1 \otimes a_2$, in which $a_1 \in \mathbb{R}$ and $a_2 \in \mathbb{R}$, and the result $a_1 \otimes a_2 \in \mathbb{R}$. By enumeration:*

$$\otimes \in \{+, -, *, /\} \tag{1}$$

Definition 2. *A **logical operator between attribute values** $(\odot)$ is represented by $a_1 \odot a_2$, in which $a_1 \in A_1$ and $a_2 \in A_2$ are attribute values and the result is either true or false. A_1 and A_2 are attributes that must have the same domain, $Dom(A_1) = Dom(A_2)$. By enumeration, the operator is expressed as:*

$$\odot \in \{<, \leq, >, \geq, =, \neq\} \tag{2}$$

Definition 3. *The **logical negation** $(\neg)$ is represented as $\neg b$. It reverses the state of truth of one boolean value b. That is:*

$$\neg b = \begin{cases} true, & if\ b\ is\ false \\ false, & if\ b\ is\ true \end{cases} \tag{3}$$

Definition 4. *A **logical connector between boolean values** $(\diamond)$ is represented by $b_1 \diamond b_2$, in which b_1 and b_2 are logical states of true or false, and the result of the connection can only be true or false as well. By enumeration:*

$$\diamond \in \{\wedge, \vee\} \tag{4}$$

Definition 5. *A **predicate between tuples** (c) is represented by $c(t_1, t_2)$, in which t_1 and t_2 are tuples from relations T_1 and T_2, respectively. Relations T_1 and T_2 must be union compatible. The predicate is a logical expression that can group different operands with logical connectors, negations, and parentheses; its result is always true or false. Formally, we have:*

$$c(t_1, t_2) = \begin{cases} t_1.a_1 \, [\,\otimes p\,] \; \odot \; t_2.a_2 \, [\,\otimes q\,] \\ (c(t_1, t_2) \diamond c(t_1, t_2)) \\ \neg c(t_1, t_2) \end{cases} \tag{5}$$

Here, $[\,]$ indicate optional operands, p and q are arithmetic expressions involving only constant values, and t.a is an attribute value of tuple t. $c(t_1, t_2)$ may not be commutative. Thus, $c(t_1, t_2)$ and $c(t_2, t_1)$ are not necessarily equivalent.

Definition 6. *A **relational conditional set** $(_c T)$ is a relation T in which all of its tuples are conditionally different. That is, given a predicate c, there are no pair of tuples in T that satisfies the predicate. Formally, we have:*

$$T \text{ is } _c T \Leftrightarrow \nexists \, t_i \in T, \, t_j \in T : \; c(t_i, t_j) \text{ is true } \wedge \; t_i \neq t_j \tag{6}$$

Definition 7. *The **relational conditional set membership** $(\in_c)$ is represented as $t \in_c \, _c T$, in which $_c T$ is a relational conditional set and t is a tuple from any relation T. Relations $_c T$ and T must be union compatible. Tuple t is a conditional element of $_c T$ if and only if there exists one tuple $t_j \in \, _c T$ that satisfies the predicate $c(t, t_j)$. Formally, we have:*

$$t \in_c \, _c T \Leftrightarrow \exists \, t_j \in \, _c T : c(t, t_j) \tag{7}$$

Following the same idea, t is not a conditional element of $_c T$ if and only if there is no tuple $t_j \in \, _c T$ that satisfies predicate $c(t, t_j)$. Formally, it is given by:

$$t \notin_c \, _c T \Leftrightarrow \nexists \, t_j \in \, _c T : c(t, t_j) \tag{8}$$

Definition 8. *The **relational conditional subset** $(\subseteq_c)$ is given by $_c T_1 \subseteq_c \, _c T_2$, where $_c T_1$ and $_c T_2$ are relational conditional sets, and the result is either true or false. Relations $_c T_1$ and $_c T_2$ must be union compatible. $_c T_1$ is a conditional subset of $_c T_2$ if and only if every tuple $t_i \in \, _c T_1$ is also a conditional element of $_c T_2$. Formally, its is defined by:*

$$_c T_1 \subseteq_c \, _c T_2 \Leftrightarrow \forall \, t_i \in \, _c T_1 : t_i \in_c \, _c T_2 \tag{9}$$

The conditional subset operation may not be commutative, so if $_c T_1 \subseteq_c \, _c T_2$ is valid we cannot affirm that $_c T_2 \subseteq_c \, _c T_1$ is also valid. However, one interesting property is that $_c T_1 \subseteq_c \, _c T_2$ can be valid even when the cardinality of $_c T_1$ is larger than that of $_c T_2$, since a single tuple of $_c T_2$ can satisfy many tuples of $_c T_1$.

Definition 9. *The **relational conditional intersection operation** ($\cap_c$) is a binary operation represented as $_cT_1 \cap_c {}_cT_2 = {}_cT_R$, in which $_cT_R$ has the result of the conditional intersection between $_cT_1$ and $_cT_2$. Relations $_cT_1$ and $_cT_2$ must be union compatible. The resulting relation $_cT_R$ has all tuples of $_cT_1$ that are also conditional members of $_cT_2$. Formally, we have:*

$$_cT_R = \{t_i : t_i \in {}_cT_1 \wedge t_i \in_c {}_cT_2\} \tag{10}$$

The relational conditional intersection may not be commutative. Thus, queries $_cT_1 \cap_c {}_cT_2$ and $_cT_2 \cap_c {}_cT_1$ do not necessarily produce the same results.

Definition 10. *The **relational conditional difference operation** ($-_c$) is a binary operation represented as $_cT_1 -_c {}_cT_2 = {}_cT_R$, in which $_cT_R$ has the result of the conditional difference between $_cT_1$ and $_cT_2$. Relations $_cT_1$ and $_cT_2$ must be union compatible. The resulting relation $_cT_R$ has all tuples of $_cT_1$ that are not conditional members of $_cT_2$. Formally, we have:*

$$_cT_R = \{t_i : t_i \in {}_cT_1 \wedge t_i \notin_c {}_cT_2\} \tag{11}$$

The relational conditional difference may not be commutative. Thus, queries $_cT_1 -_c {}_cT_2$ and $_cT_2 -_c {}_cT_1$ do not necessarily produce the same results.

4.2   Proposed Algorithms

In order to allow the execution of our conditional set operations, we developed algorithms to support them. Here, the following considerations must be made.

Relations: The algorithms read both relations from disk. One relation is iterated in a full table scan; the other is read through indexes for each column. In our motivational examples, the right relation is the largest one, i.e., SP; thus, this section assumes that it has the indexes. Note, however, that we documented in a complementary material[1] the case when the left relation is the indexed one.

Arithmetic Expressions: We assume that all expressions in the predicate refer to the table without indexes. Note that it is always possible, e.g., instead of writing $DP.Price \geq SP.Price * 2$ one may write $DP.Price/2 \geq SP.Price$.

Postfix Predicate: Our algorithms assume that the predicate was previously converted from the user-friendly format, i.e., the infix notation, to the postfix notation by using a stack. The stack stores the operations in the order that must be used for evaluation; the one to be evaluated first is at the top. For example, if we had $DP.Price/2 \geq SP.Price$, we would work with the following stack of tokens: ($DP.Price$, 2, /, $SP.Price$, $\geq$). The infix-to-postfix conversion of expressions is a known procedure in compilers, so we do not explain it. However, we included the algorithm and its description in the complementary material (See footnote 1).

---

[1] https://github.com/alivasples/RCSetOp/blob/main/Documentation/Complementary.pdf.

Algorithm 1. IndexTupleQuery($_cT$, t, c)

Input: $_cT$: relational conditional set, t: tuple of interest, c: predicate
Output: *result*: array of bits indicating each tuple of $_cT$ that satisfies c for t

1: create stack *operands*;
2: create stack *subresults*;
3: **for** each *token* in c **do**
4:     **switch** *token* **do**
5:         **case** relation's column $_cT.A$ **do**
6:             push *token* into *operands*;
7:         **case** tuple's column $t.A$ **do**
8:             push value of $t.A$ into *operands*;
9:         **case** constant value *val* **do**
10:            push *token* into *operands*;
11:        **case** arithmetic operator $\otimes$ **do**
12:            $valR$ = pop from *operands*;
13:            $valL$ = pop from *operands*;
14:            push value of $valL \otimes valR$ into *operands*;
15:        **case** logical operator $\odot$ **do**
16:            val = pop from *operands*;
17:            A = pop from *operands*;
18:            $subresult$ = IndexQuery($_cT$, A, val, $\odot$);
19:            push *subresult* into *subresults*;
20:        **case** negation $\neg$ **do**
21:            R = pop from *subresults*;
22:            push $\neg R$ into *subresults*;
23:        **case** logical connector $\diamond$ **do**
24:            R = pop from *subresults*;
25:            L = pop from *subresults*;
26:            push result of $L \diamond R$ into *subresults*;
27:    **end switch**;
28: **end for**;
29: **return** top from *subresults*;

Let us now focus on the algorithms. Algorithm 1 identifies which tuples of a relcond set $_cT$ satisfy a predicate c for a tuple of interest t. Remember that c is a predicate in postfix notation, and there exist index structures for each column of $_cT$. First, we need two stacks: one to store the next operations and another stack to store the subresults. Then, for each token read from predicate c, we perform: a) if the token refers to a relation's attribute $_cT.A$, push the token into *operands*; b) if the token refers to a tuple's attribute $t.A$, push its value into *operands*; c) if the token is a constant value, push it into *operands*; d) if the token is an arithmetic operator $\otimes$, pop from *operands* the right and left operands and push the result of the operation into *operands*; e) if the token is a logical operator $\odot$, pop the last two operands from the operands stack. One is a constant value val, and the other refers to an attribute $_cT.A$. Then, perform an indexed query with both operands. For example, if the logical operator is $<$, the query result is a bits vector indicating if each tuple has its value in column A lower than val. Then, push the bits vector into subresults; f) if the token is a negation $\neg$, negate the top of *subresults*, and; g) if the token is a logical connector ◇, pop the last two bits vectors from *subresults*, and push the result of the bitwise operation into *subresults*. Finally, return the remaining result in *results*.

Algorithm 2. IsCondMember($_cT$, t, c)

Input: $_cT$: relational conditional set, t: tuple of interest, c: predicate
Output: *true* if $t \in_c {_cT}$; *false*, otherwise

1: $result = $ IndexTupleQuery($_cT$, t, c);   // from Algorithm 1
2: **if** $result$ has any bit 1 **then**
3:     **return** *true*;
4: **end if**;
5: **return** *false*;

Algorithm 3. RelCondSetOp($_cT_1$, $_cT_2$, c, *SetOp*)

Input: $_cT_1$, $_cT_2$: relational conditional sets, c: predicate, *SetOp*: $\cap_c$ or $-_c$
Output: the result $_cT_R$ from $_cT_1 \cap_c {_cT_2}$ or from $_cT_1 -_c {_cT_2}$, according to *SetOp*

1: create an array of bits R of size $|_cT_1|$;
2: **if** *SetOp* is $\cap_c$ **then**
3:     initialize R with 0s;
4: **else** initialize R with 1s;
5: **end if**;
6: **for** each tuple $t_i \in {_cT_1}$ **do**
7:     **if** IsCondMember($_cT_2$, t_i, c) **then**   // from Algorithm 2
8:         **if** *SetOp* is $\cap_c$ **then**
9:             set $R[i]$ as 1;
10:         **else** set $R[i]$ as 0;
11:         **end if**;
12:     **end if**;
13: **end for**;
14: $_cT_R = $ get tuples from $_cT_1$ that refer to each bit 1 in R;
15: **return** $_cT_R$;

Algorithm 2 implements the relational conditional set membership $\in_c$ from Definition 7. Thus, it identifies whether or not a tuple t is a conditional member of a set $_cT$ according to a predicate c. The algorithm is twofold: a) use Algorithm 1 to get an array of bits representing the tuples of $_cT$ that satisfy the condition, and; b) return *true* if there is any bit 1 in the array; return *false*, otherwise.

Algorithm 3 implements both the conditional intersection $\cap_c$ and the conditional difference $-_c$ from Definitions 9 and 10, respectively. It receives as parameters the left relation $_cT_1$, the right relation $_cT_2$, the predicate c, and an indicator $SetOp \in \{\cap_c, -_c\}$ of the operation of interest. The result $_cT_R$ is either $_cT_1 \cap_c {_cT_2}$ or $_cT_1 -_c {_cT_2}$, according to *SetOp*. As it is shown in Lines 1-5, the algorithm begins by creating an array of bits R to be used latter to indicate the tuples of $_cT_1$ that should be in $_cT_R$. If $SetOp = \cap_c$, R is initialized with 0s; otherwise, it receives 1s. The main loop in Lines 6-13 iterates for each tuple $t_i \in {_cT_1}$. The loop updates array R only when the current tuple t_i is a conditional member of $_cT_2$. In this case, $R[i]$ receives 1 if $SetOp = \cap_c$; otherwise, $R[i]$ is set to 0. Finally, $_cT_R$ is obtained as being the tuples of $_cT_1$ that refer to each bit 1 in R.

Algorithm 4. IsCondSubset($_cT_1$, $_cT_2$, c)

    Input: $_cT_1$, $_cT_2$: relational conditional sets, c: predicate
    Output: *true* if $_cT_1 \subseteq_c {}_cT_2$; *false*, otherwise

1: **if** RelCondSetOp($_cT_1$, $_cT_2$, c, $\cap_c$) = $_cT_1$ **then**   // from Algorithm 3
2:     **return** *true*;
3: **end if**;
4: **return** *false*;

Algorithm 4 implements the relational conditional subset operator $\subseteq_c$ from Definition 8. Thus, it receives as parameters the left relation $_cT_1$, the right relation $_cT_2$, and the predicate c. This short algorithm simply executes the conditional intersection $_cT_1 \cap_c {}_cT_2$, and verifies it the result of this operation is equal to relation $_cT_1$. If so, the algorithm returns *true*; otherwise, it returns *false*.

Complexity Analysis: For the analysis, let us consider p to be the number of tokens that a predicate c has, while m and n are the cardinalities of relations $_cT_1$ and $_cT_2$, respectively. Algorithm 1 iterates each predicate token. Each iteration runs an indexed query over a given relation; let us assume it to be $_cT_2$. The algorithm takes advantage of existing index structures to perform the queries, so the use of state-of-the-art indexes allows each execution of function IndexQuery in Line 18 to cost $\mathcal{O}(\log n + s)$ time, where $s \leq n$ is the number of tuples selected by the query. Thus, the overall time complexity of Algorithm 1 is $\mathcal{O}(p(\log n + s))$. Algorithm 2 executes Algorithm 1 using as parameters a tuple t, a predicate c and a given relation; again, let us assume it to be $_cT_2$. Then, the algorithm looks at negligible cost for a bit 1 in the resulting array of bits. Thus, the overall time complexity of Algorithm 2 is $\mathcal{O}(p(\log n + s))$. Algorithm 3 initializes at negligible cost an array of bits. Then, the loop in Lines 6-13 executes Algorithm 2 for each of the m tuples in $_cT_1$, using $_cT_2$ as a parameter; so, the loop costs $\mathcal{O}(mp(\log n + s))$ in total. At last, in Line 14, relation $_cT_1$ is scanned at cost $\mathcal{O}(m)$. Thus, the overall cost of Algorithm 3 is $\mathcal{O}(mp(\log n + s) + m)$. Algorithm 4 runs Algorithm 3 and validates at cost $\mathcal{O}(m)$ if the result is equal to $_cT_1$. Thus, the overall complexity of Algorithm 4 is $\mathcal{O}(mp(\log n + s) + m)$. Finally, let us emphasize that p tends to be very small in practice because large predicates are rare, so our algorithms are fast and scalable as long as the query selectivity s is also small, just like it happens with any index-based access method.

5 Experiments

We evaluated the semantics and the usability of our operators, as well as the scalability of our algorithms by following the motivational example of product sales. The experiments were performed to answer two main Research Questions:

- **RQ1**: *How accurate are the new relational conditional set operations in the sense of returning what the users expect to receive?*
- **RQ2**: *How effective and scalable are the algorithms that we propose?*

Algorithms 1–4 were implemented in C++ with page buffer management. Library Arboretum[2] was used to run indexed queries. All experiments were performed with an Intel Core i7 processor working at 3.4 GHz and 8 GB of RAM.

Following our motivational example, we studied a dataset[3] of Amazon toys. The dataset was preprocessed to suit the purpose of our case study and it is freely available for download online[4]. It has $25,457$ tuples referring to $4,277$ unique toy products, and 5 columns: supplier, category, product, units, and price.

Observation: for the purpose of reproducibility, all codes, results, and preprocessed datasets studied in this paper are freely available for download online (See footnote 4).

5.1 Semantic Validation

This section investigates Research Question **RQ1** by validating the semantics of our operators in the case study of sales of products. To make it possible, we generated 200 conditional sets by sampling at random the Amazon Toys dataset; they represented 100 pairs of relations DP and SP to be given as input for the queries. As this experiment intended to validate semantics only, the numbers of tuples, i.e., products, in the relations are small, varying from 3 to 6, so that we could manually verify if the results are meaningful. For each pair DP and SP, we ran the 4 motivational queries Q1, Q2, Q3, and Q4 that were described previously in Sect. 2. In summary, correct results were obtained for all 100 pairs of relations; thus, we argue that they validate the semantics of our proposals. From the pairs of relations studied, one specific case is illustrated in Fig. 2, showing the results for the relcond membership in Fig. 2.c, relcond subset in Fig. 2.b, relcond intersection in Fig. 2.d, and relcond difference in Fig. 2.e.

5.2 Scalability

This section investigates Research Question **RQ2** by evaluating the scalability of our algorithms. To make it possible, we generated random samples of varying sizes from dataset Amazon Toys to represent pairs of conditional sets $_cT_1$ and $_cT_2$. Specifically, we created 100 pairs by varying the cardinality of $_cT_1$ from 40 to $4,000$ and using a fixed cardinality of $1,000$ for $_cT_2$. Other 300 pairs were created in a similar way: 100 pairs with $|_cT_2| = 2,000$, 100 pairs with $|_cT_2| = 3,000$ and 100 pairs with $|_cT_2| = 4,000$. Then, operators $\cap_c$, $-_c$ and $\subseteq_c$ were executed 10 times for each of these pairs of relations to obtain average runtime results. A distinct procedure was necessary for operator $\in_c$, since it receives a tuple t and a single conditional set $_cT$ as input; not, two sets. Thus, we took each of the $4,000$ tuples from our largest relation $_cT_1$ to be tuple t and executed $t \in_c \ _cT_2$ to obtain the average runtime. The distinct versions of relation $_cT_2$ were considered, whose cardinalities go from $1,000$ to $4,000$. Figure 3 reports

---

[2] https://bitbucket.org/gbdi/arboretum/.

[3] https://www.kaggle.com/PromptCloudHQ/toy-products-on-amazon.

[4] https://github.com/alivasples/RCSetOp.

the results obtained from the aforementioned procedure. Each individual plot reports average runtime versus the cardinality of $_cT_2$ for the four cardinality variations of $_cT_1$: with $1,000$, $2,000$, $3,000$, and $4,000$ tuples. As it can be seen, the results corroborate our theoretical analysis of complexity from Sect. 4.2, by indicating that all of our algorithms are fast and scalable.

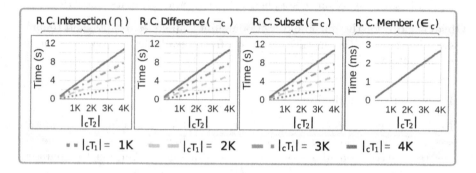

Fig. 3. Scalability of our conditional set operators in dataset Amazon Toys.

6   Conclusion

In this work, we demonstrated that the set-based operations of the Relational Algebra have severe limitations that prevent their use in several real-world applications. Thus, we tackled the problem by presenting the new **RelCond Set Operators** that are naturally well suited to answer queries using custom predicates. We also formally defined the new operators as the RelCond Set Membership $\in_c$, RelCond Subset $\subseteq_c$, RelCond Intersection $\cap_c$, and RelCond Difference $-_c$, thus enabling their usage in queries along with the existing algebraic operators. Additionally, we designed novel algorithms that execute these operators in a fast and scalable manner. Finally, we performed a case study by analyzing real data from thousands of toy products available for sale at Amazon. The results corroborate the practical usability of our operators in real-life applications. We also include in our complementary material (See footnote 1) other cases of use where our operators are well suitable, thus corroborating their generality and usability. Among the operators presented, let us emphasize that we did not include a conditional union; this is because we could not identify practical utility for it. However, we include a discussion with definition and algorithm for it in our complementary material (See footnote 1). Additionally, in future work we intend to implement our algorithms in a DBMS, propose SQL implementations for them, and compare both approaches.

References

1. Al Marri, W.J., et al.: The similarity-aware relational database set operators. Inf. Syst. **59**, 79–93 (2016)
2. Bosc, P., Pivert, O.: On a fuzzy bipolar relational algebra. Inf. Sci. **219**, 1–16 (2013)
3. Codd, E.F.: Relational completeness of data base sublanguages. IBM Research Report RJ987 (1972)
4. Codd, E.F.: The Relational Model for Database Management: Version 2, p. 538. Addison-Wesley Longman Publishing Co., Inc., Boston (1990)
5. Elmasri, R., Navathe, S.B.: Fundamentals of Database Systems. Pearson, 7th edn. (2015)
6. Galindo, J., Urrutia, A., Piattini, M.: Fuzzy Databases: Modeling, Design and Implementation: Modeling, Design and Implementation. Idea Group (2005)
7. Marri, W.J.A., Malluhi, Q., Ouzzani, M., Tang, M., Aref, W.G.: The similarity-aware relational intersect database operator. In: Traina, A.J.M., Traina, C., Cordeiro, R.L.F. (eds.) SISAP 2014. LNCS, vol. 8821, pp. 164–175. Springer, Cham (2014). https://doi.org/10.1007/978-3-319-11988-5_15
8. Pola, I.R., et al.: Similarity sets: a new concept of sets to seamlessly handle similarity in database management systems. Inf. Syst. **52**, 130–148 (2015)
9. Pola, I.R.V., Cordeiro, R.L.F., Traina, C., Traina, A.J.M.: A new concept of sets to handle similarity in databases: the SimSets. In: Brisaboa, N., Pedreira, O., Zezula, P. (eds.) SISAP 2013. LNCS, vol. 8199, pp. 30–42. Springer, Heidelberg (2013). https://doi.org/10.1007/978-3-642-41062-8_4
10. Red'ko, V.N., Buy, D.B., Kanarskaya, I.S., Senchenko, A.S.: Precise estimates for the time complexity of implementing algorithms of set-theoretic operations in table algebras. Cybern. Syst. Anal. **53**(1), 1–11 (2017)
11. Sharma, A.K., et al.: An extended relational algebra for fuzzy multidatabases. In: 7th I-SPAN, pp. 445–450 (2004)
12. Stoll, R.R.: Set Theory and Logic. W.H. Freeman, New York (1963)
13. Tang, X., Chen, G.: A complete set of fuzzy relational algebraic operators in fuzzy relational databases. FUZZ-IEEE **1**, 565–569 (2004)
14. Zhao, F., et al.: A vague relational model and algebra. In: 4th FSKD 2007, vol. 1, pp. 81–85 (2007)

GASP: Graph-Based Approximate Sequential Pattern Mining for Electronic Health Records

Wenqin Dong[1], Eric W. Lee[2]($\boxtimes$) (iD), Vicki Stover Hertzberg[2] (iD),
Roy L. Simpson[2], and Joyce C. Ho[2]($\boxtimes$) (iD)

[1] Carnegie Mellon University, Pittsburgh, USA
wenqind@andrew.cmu.edu
[2] Emory University, Atlanta, USA
{ewlee4,vhertzb,roy.l.simpson,joyce.c.ho}@emory.edu

Abstract. Sequential pattern mining can be used to extract meaningful sequences from electronic health records. However, conventional sequential pattern mining algorithms that discover all frequent sequential patterns can incur a high computational and be susceptible to noise in the observations . Approximate sequential pattern mining techniques have been introduced to address these shortcomings yet, existing approximate methods fail to reflect the true frequent sequential patterns or only target single-item event sequences. Multi-item event sequences are prominent in healthcare as a patient can have multiple interventions for a single visit. To alleviate these issues, we propose GASP, a graph-based approximate sequential pattern mining, that discovers frequent patterns for multi-item event sequences. Our approach compresses the sequential information into a concise graph structure which has computational benefits. The empirical results on two healthcare datasets suggest that GASP outperforms existing approximate models by improving recoverability and extracts better predictive patterns.

Keywords: Sequential pattern mining · Healthcare data

1 Introduction

An increasing amount of electronic healthcare records (EHRs) are collected. Sequential pattern mining (SPM) can help discover important or useful patterns in such data [5] such as the sequence of health interventions that resulted in an unfavorable outcome. Various SPM algorithms have been proposed to discover all the frequent sequential patterns that satisfy a user-defined threshold, or support count (we refer the reader to a survey on the topic [5]). Unfortunately, there are several notable limitations that prevent the widespread usage of these algorithms: computational complexity (in terms of time and memory), generation of noisy frequent patterns, and development for single-item event sequences (or one item

L. Bellatreche et al. (Eds.): ADBIS 2021 Short Papers, Workshops and Doctoral Consortium,
CCIS 1450, pp. 50–60, 2021.
https://doi.org/10.1007/978-3-030-85082-1_5

per event sequences). However, EHRs are characterized by noisy, multi-item event sequences (i.e., 1 or more items per event sequences). For example, a patient can have multiple treatments for the same visit and the documentation process can be prone to human errors. Thus exact SPMs are not always desirable.

Approximate SPM was proposed to alleviate limitations, by clustering similar sequences together to obtain representative patterns [7,12] or utilizing different data structures such as trees or graphs to approximate the subsequent patterns [1,8] to minimize the number of passes through the data. Unfortunately, there are several limitations of existing approximate SPM algorithms. The majority of the algorithms are developed only for single-item event sequences and are not easily generalizable to multi-item event sequences. Moreover, the empirical results can fail to improve computational efficiency or suffer from poor recall.

We propose a **G**raph-based **A**pproximate **S**equential **P**attern mining algorithm, GASP, for multi-item sequential databases (SDBs) constructed from EHRs. Our approach approximates the database as a new weighted graph structure. A sampling-based approach is then utilized to efficiently identify frequent subsequences in the database while providing reasonable recall with the true patterns. Our evaluation showcases that GASP requires comparable computational time and memory footprints to state-of-the-art exact SPM methods. Moreover, the approximate patterns contain better predictive power than the exact patterns.

2 Preliminaries

2.1 Notation

Let $I = \{i_1, i_2, \ldots, i_m\}$ be the set of unique items (i.e., symbols or alphabets) in the sequential database, SDB. An event or itemset, X, is an unordered collection of items, and denoted as $X = \{i_1, i_2, \ldots, i_k\}$, where i_j is an item from I. A sequence s is an ordered list of itemsets such that $s = \langle X_1, X_2, \ldots, X_n \rangle$. A sequence database contains a list of sequences and is denoted as $SDB = \langle s_1, s_2, \ldots, s_p \rangle$, with p unique sequence identifiers. Table 1 provides an example of SDB which contains four sequences ($p = 4$). A sequence $s_a = \langle a_1, a_2, \ldots, a_m \rangle$ is a subsequence of another sequence $s_b = \langle b_1, b_2, \ldots, b_n \rangle$ if and only if there exist integers $i_1, i_2, \ldots, i_m$ such that $1 \leq i_1 \leq i_2 \leq \ldots \leq i_m \leq n$ and $a_1 \subseteq b_{i_1}, a_2 \subseteq b_{i_2}, \ldots, a_m \subseteq b_{i_m}$. In other words, s_a is contained in s_b. From the first sequence s_1 in Table 1, one potential subsequence is $\langle \{53\}, \{98\} \rangle$.

2.2 Exact Sequential Pattern Mining

Given a sequential database, SDB, the goal of exact SPM is to find all the frequent subsequences (i.e., sequential patterns) that occur in at least some user-specified number of sequences in the SDB. Given the computational challenges of SPM, CM-SPAM and CM-SPADE [11] have been proposed to achieve better time and space scalability by pruning the candidate patterns. Although these algorithms are relatively efficient, they can be susceptible to noise in the data and can fail to deal with long, multi-event sequences.

Table 1. An example of a sequential database (SDB).

SID	Sequences
1	$\langle\{53,98\},\{58,98\}\rangle$
2	$\langle\{257,53\},\{257,58\}\rangle$
3	$\langle\{10,53\},\{257,259,58\},\{98\}\rangle$
4	$\langle\{10\},\{259,53,58\}\rangle$

2.3 Approximate Sequential Pattern Mining

Approximate SPM was proposed to identify "similar" patterns while reducing noise in the patterns and improving computational efficiency. Since patterns may not have a direct one-to-one correspondence to the exact SPMs, minimizing the percentage of dissimilarity between the patterns have been proposed as the objective [12]. Unfortunately, this is limited to single-event patterns and requires specification of the error tolerance. We propose the following approximate SPM framework based on average Levenshtein distance for multi-item SDB.

Definition 1. (Levenshtein distance). *Given two sequences a, b, the Levenshtein distance is the minimum number of single-item edits, including insertions, deletions, and substitutions, required to change a to the b or vice versa.*

$$lev(a,b) = \begin{cases} |a| & \textit{if } |b| = 0 \\ |b| & \textit{if } |a| = 0 \\ lev(tail(a), tail(b)) & \textit{if } a[0] = b[0] \\ 1 + min \begin{cases} lev(tail(a), b) \\ lev(a, tail(b)) \\ lev(tail(a), tail(b)) \end{cases} & \textit{otherwise.} \end{cases} \tag{1}$$

Given a string x, $tail(x)$ refers to a string excluding the first character of x, and starting with the index of 0, $x[n]$ refers to the nth character of x.

Problem Statement. Let s be a frequent subsequence as defined above, and s_1, s_2 be two arbitrary subsequences. s_1 is a better approximation of s than s_2 if $lev(s, s_1) < lev(s, s_2)$. Thus, the goal of approximate SPM is to discover a list of subsequences, $\mathcal{S}_A$, that minimizes the average Levenshtein distance to a pattern in the exact pattern list $\mathcal{S}_E$:

$$\min \frac{1}{|\mathcal{S}_A|} \sum_{s \in \mathcal{S}_A} \min_{s_i \in \mathcal{S}_E} lev(s, s_i). \tag{2}$$

Existing approximate SPM methods can be grouped into two approaches. The clustering approaches, such as ApproxMap [7] and a Hamming Distance-based model [12], mine consensus patterns by grouping the frequent patterns based on similarity. Yet these algorithms produce poor recall and require additional parameters (i.e., number of clusters). Another area of work tackle online data streams to identify patterns using a single pass of the data such as GraSeq [8], a graph-based approximate SPM algorithm. GraSeq transformed sequences into a directed weighted graph structure with only one scan of data and introduced a non-recursive depth-first search algorithm to acquire approximate sequential patterns. Unfortunately, these works are developed only for single-item sequences and a naïve extension of single-item sequences to multi-item sequences does not yield desirable results (as demonstrated by our empirical results).

3 GASP

We introduce GASP, a graph-based approximate SPM model, to address the limitations of existing approximate SPM algorithms. GASP transforms the SDB into a Markov chain graph, $\mathcal{G}$ and uses a probabilistic generative model to extract the sequential patterns. $\mathcal{G}$ can be viewed as a random sample of the original SDB and thereby retains the same bounds on accuracy of the discovered patterns [10].

3.1 Subsequence Generation

Our graph $\mathcal{G}$ captures the order and relation between all the items I in the SDB. Since the SDB can have multiple items per event, GASP distinguishes between the two scenarios where the two items occur in the same event (type 1), and two items occur in chronological order (type 2).

Definition 2. *(1-subsequence).* *For a sequence s, a 1-subsequence is $y = \langle i_k \rangle$ for all $i_k \in X_1 \cup X_2 \cup ... \cup X_n$.*

Definition 3. *(2-subsequence-type-1).* *For a sequence s, a 2-subsequence-type-1 is $z^{(1)} = \langle i_k, i_j \rangle$ for all $i_k, i_j \in X_p$ such that $1 \le p \le n$.*

Definition 4. *(2-subsequence-type-2).* *For a sequence s, a 2-subsequence-type-2 is $z^{(2)} = \{\langle i_k \rangle, \langle i_j \rangle\}$ for all $i_k \in X_p$, $i_j \in X_q$ and $p < q$.*

GASP scans all the sequences in SDB exactly once to determine all the frequent item sets $\mathcal{Y} = \{y_1, y_2, ...\}$, $\mathcal{Z}^{(1)} = \{z_1^{(1)}, z_2^{(1)}, ...\}$, and $\mathcal{Z}^{(2)} = \{z_1^{(2)}, z_2^{(2)}, ...\}$ and its frequency. As all supersets of infrequent patterns are infrequent, subsequences in $\mathcal{Y}$, $\mathcal{Z}^{(1)}$, $\mathcal{Z}^{(2)}$ that fall below the support count are pruned.

3.2 Graph Construction

GASP constructs a mixed-type graph, $\mathcal{G} = (V, E)$, where V is the set of vertices that represent an item, and E is the set of edges (directed and undirected) to represent the ordering or relation between two items. Each vertex, V_i, corresponds to the ith 1-subsequence in $\mathcal{Y}$. Since items that occur in the SDB are

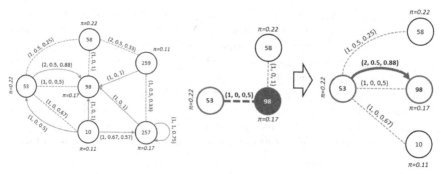

(a) A example of the partially   (b) A example of random walk to extract
   constructed graph. $\{\langle 53, 98 \rangle, \langle 98 \rangle\}$.

Fig. 1. A simplified example of the graph constructed and one iteration of the random walk. Only partial edges are shown in (a). Each node refers to the item in SDB and has a starting item probability (π). A blue dotted undirected edge denotes a type-1 edge and a green directed edge denotes a type-2 edge. Each edge contains the edge weight, event transition weight, and ending probability (w, α, β). For (b), the selection is node 98, type-1 edge with node 53, type-2 edge with node 98 and then terminated to obtain $\{\langle 53, 98 \rangle, \langle 98 \rangle\}$. (Color figure online)

more likely to be part of a frequent pattern, the start probability is set to reflect the likelihood of the item occurring in the SDB: $\pi_i = \frac{freq(y_i)}{\sum_j freq(y_j)}$.

An undirected edge, $(v_i \leftrightarrow v_j)$, represents two items occurring in the same event (or an item in $\mathcal{Z}^{(1)}$). A directed edge, $(v_i \rightarrow v_j)$, denotes the sequential relationship between v_i and v_j, such that v_i occurs in an event prior to v_j (or an item in $\mathcal{Z}^{(1)}$). A weight function, w, is associated with each edge based on the frequency of the particular item set, i.e., $w(v_i \leftrightarrow v_j) = \frac{freq(<v_i, v_j>)}{\sum_\ell freq(<v_i, v_\ell>)}$. The likelihood of staying in the same event is also a function of how many items typically occur in the same event with a specific item.

Proposition 1. *The number of items in an event, X_k with item i_j, is bounded by the maximum number of items in any X_j that contains the item i_j across all the sequences in the SDB, $\langle s_1, s_2, \ldots, s_p \rangle$.*

Given Proposition 1, we introduce a new event transition weight function, α, to capture the likelihood that the next item will be from the same event conditioned on sampling item i_j. Let $|X_k|$ denote the number of items in the event k. For a sequence s, if v_j occurs in the kth event, α_s is defined as:

$$\alpha_s(v_i \leftrightarrow v_j) = \frac{|X_k| - 2}{|X_k| - 2 + \sum_{z=k+1}^{n} |X_z|}$$

$$\alpha_s(v_i \rightarrow v_j) = \frac{|X_k| - 1}{|X_k| - 1 + \sum_{z=k+1}^{n} |X_z|} \tag{3}$$

Then, α is the average weight across all sequences with the item i_j.

Another limitation of existing approximate SPM algorithms is the need to provide a user-defined length for the extracted patterns. We propose to model the length of a candidate pattern as a random variable L. We first introduce two propositions to bound the length of the pattern.

Proposition 2. *The length of a frequent subsequence, ℓ is bounded by the maximum length of all subsequences in the SDB, $\ell \leq \max\limits_{i=1,\ldots,p} |s_i|$.*

Thus the empirical cumulative distribution, $P(L \leq \ell)$, can serve as an upper bound for the maximum number of events in a candidate pattern. Yet, this is independent of the items in the pattern.

Proposition 3. *Given the presence of an item, i_j, in the frequent subsequence, the maximum length of the subsequence is bound by the length of the sequences in the SDB that contain i_j: $\ell \leq \max\limits_{i_j \in s_i, \forall i=1,\ldots,p} |s_i|$.*

Conceptually, if some items occur towards the end of a sequence in the SDB, their presence can be used to terminate the candidate pattern. We introduce a new end weight function, β, to calculate the likelihood that it will terminate the pattern. If v_i, v_j occurs in the sequence s, β_s is defined as:

$$\beta_s(v_i \leftrightarrow v_j) = 1 - \frac{\sum_{z=k+1}^{n} |X_z| + |X_k| - 2}{\sum_{z=1}^{n} |X_z|}$$

$$\beta_s(v_i \rightarrow v_j) = 1 - \frac{\sum_{z=k+1}^{n} |X_z| + |X_k| - 1}{\sum_{z=1}^{n} |X_z|}. \tag{4}$$

The final weight, β is then calculated as the average of the β_s weights across all sequences in the SDB. Figure 1(a) shows the graph for Table 1.

3.3   Random Walk

Random walk was introduced to simulate the likely paths through the graph [9]. Using the same premise, edges, and vertices that have higher weights (or likelihoods) in $\mathcal{G}$, should be traversed more often as they occurred more frequently in the original SDB. To account for the new weight functions and ending probability, GASP uses a modified random walk algorithm. The random walk edge weight, d, is determined by the edge weight w and the event transition weight α:

$$d(v_i, v_j) = w(v_i, v_j) \times \alpha(v_i, v_j) \tag{5}$$

The stopping criteria for random walk is also adapted to reflect the number of items currently in the pattern, $\tilde{\ell}$ and the sampled edge, (v_i, v_j). The iteration is stopped based on a Bernoulli random variable

$$L(\tilde{\ell}, (v_i, v_j)) \sim Bernoulli(\frac{1}{2}P(L \leq \tilde{\ell}) + \frac{1}{2}\beta(v_i, v_j)) \tag{6}$$

Algorithm 1 RandomWalk

1: $s = $ Draw v_i randomly using π_i and set $v = v_i$
2: **while** True **do**
3:      Calculate the edge weight for all outgoing edges $d(v_i, v_j) = \alpha(v_i, v_j)w(v_i, v_j)$.
4:      Choose the new vertex v_j based on edge weight, $d(v_i, v_j)$.
5:      Append v_j to the sequence s and set $v = v_j$.
6:      Sample pattern end using Eq. (6)
7: **end while**
8: Return candidate pattern s

The detailed steps of our customized random walk are summarized in Algorithm 1 with an example provided in Fig. 1(b). Upon the completion of a pattern, the weights of all the edges traversed are summed up to yield the final weight of this particular sequence. These cumulative weights are used for the final candidate pattern ranking.

3.4   Algorithm Complexity

Let P represent the number of sequences in the SDB, N the maximum number of items for a subsequence, I the number of unique items, and L the number of random walk iterations. Since GASP only requires a single scan through the SDB, the graph generation has a computational complexity of $O(PN^2)$. For the random walk, the complexity is $O(ILN)$. Hence, the computational complexity of GASP is $O(PN^2 + ILN)$. The memory complexity of GASP is dominated by the graph and the generated patterns. Only 2-subsequences along with their weight and various probabilities are stored in memory ($O(I^2)$). In the random walk stage, the worst memory scenario is a distinct pattern for each iteration ($O(LN)$). Thus, the memory complexity is $O(I^2 + LN)$.

4   Experiment Setup

4.1   Dataset

We employed two healthcare datasets to assess the performance of GASP. CMS is a synthesized and publicly available dataset provided by the Centers for Medicare and Medicaid Services[1]. This dataset contains information about the patients' diagnosis on their visits between the period 2008 to 2009. To construct the SDB, the patient visits are sorted in chronological order and the International Classification of Diseases (ICD-9) billing diagnosis codes are extracted. Clinical Classifications Software (CCS) codes [6] are used to group ICD-9 into broader categories. The Nursing Electronic Learning Lab (NELL) dataset includes electronic health records (EHRs) from Emory Healthcare for type 2 diabetes patients with new

---

[1] https://www.cms.gov/research-statistics-data-and-systems/Downloadable-Public-Use-Files/SynPUFs/DE_Syn_PUF.

Table 2. Characteristics of each SDB.

Dataset	$|P|$	$|I|$	Avg $|s_i|$	Avg $|X_k|$
CMS	68,185	283	40.96	2.22
NELL	12,576	260	12.65	8.55

onset of cardiovascular disease (CVD) and matched controls. It includes 2,112 cases and 10,464 controls. We extracted diagnosis codes for all patients prior to the CVD index date and group them using CCS codes. The characteristics of the SDBs are summarized in Table 2.

4.2 Experimental Design

All the experiments were run on a single machine, an Amazon EC2 r5.4xlarge instance, with 16 CPU cores and 128GB memory.

4.3 Baseline Methods

We compared GASP with the following SPM algorithms: (1) CM-SPAM [11], (2) CM-SPADE [11], (3) GraSeq (fixed), a modified approximate algorithm based on GraSeq [8] to support multi-item event sequences where items in the same event are considered as a single vertex in the graph, (4) GraSeq (variable), an extension of the GraSeq (fixed) to use our proposed random walk algorithm combined with the sequential pattern ending probability to generate variable-length patterns. GASP, GraSeq (fixed), and GraSeq (variable) are implemented in Python 3.6. The random walk utilizes multiple threads to further reduce running time on machines with multiple CPUs. The code will be open-sourced in Github upon acceptance[2]. For CM-SPAM and CM-SPADE, we used the SPMF library [4] implementations in Java[3]. FAST [3] and ApproxMap were considered, but the results are omitted due to the poor performance. Other approximate SPM algorithms were not publicly released and thus not compared.

4.4 Evaluation Metrics

We compared the SPM algorithms from multiple perspectives:

- *Computation Time*: The total running time of the algorithm.
- *Memory Usage*: The maximum memory consumed by the algorithm.
- *Levenshtein distance*: The measure defined in Eq. (2).
- *Precision & Recall*: Two measures to capture the relevance of the patterns extracted from the approximate SPM.

$$Prec = \frac{|\mathcal{S}_A \cap \mathcal{S}_E|}{|\mathcal{S}_A|}, \quad Rec = \frac{|\mathcal{S}_A \cap \mathcal{S}_E|}{|\mathcal{S}_E|}$$

---

[2] https://github.com/cynthiadwq/GASP.
[3] https://www.philippe-fournier-viger.com/spmf/.

Table 3. Comparison of SPM algorithms on the two datasets. The memory is reported in megabytes and time is in seconds.

Model	CMS					NELL				
	Time.	Mem.	Prec.	Rec.	Lev.	Time.	Mem.	Prec.	Rec.	Lev.
CM-SPAM	3798	1937	1.0	0.975	0.034	110	1280	1.0	0.749	0.202
CM-SPADE	815	11008	–	–	–	113	2276	–	–	–
GraSeq (fixed)-5M	304	591	0.106	0.085	1.656	110	481	0.101	0.076	1.697
GraSeq (variable)-5M	311	653	0.147	0.130	1.347	101	352	0.122	0.107	1.458
GASP-5M	322	1036	0.195	0.381	0.527	109	533	0.125	0.255	0.914
GASP-10M	426	1491	0.230	0.507	0.409	219	957	0.172	0.396	0.716

5   Experimental Results

5.1   Pattern Recoverability

CMS The exact SPM algorithms were run using a support threshold of 20% and yielded 127,941 and 124,776 frequent patterns for CM-SPADE and CM-SPAM, respectively. Since CM-SPADE resulted in more patterns, it was used as the ground truth. Table 3 summarizes the performance of the SPM methods on the CMS dataset. GASP can achieve a reasonable approximation of the *exact* frequent patterns generated by CM-SPADE in terms of Levenshtein distance without requiring a trade-off in terms of time or memory. To extract the frequent patterns using CM-SPADE, it uses almost 10× more memory than GASP with 10 M iterations. Moreover, for CM-SPAM, it requires almost 10× the computational time than GASP to produce similar patterns to CM-SPADE. The results also illustrate the importance of variable-length pattern as GraSeq (variable) outperforms GraSeq (fixed) in terms of pattern recoverability. Moreover, GASP (5M) outperforms GraSeq (variable) across all three measures, highlighting the benefit of modeling the different types of two-item subsequences.

NELL The exact SPM algorithms were run using a support threshold of 1% and yielded an average of 1,459,820 and 1,085,201 frequent patterns for CM-SPADE and CM-SPAM, respectively. The results from Table 3 show that GASP-10M is able to identify almost 40% of the original patterns of CM-SPADE, whereas CM-SPAM extracts almost 75% of the original patterns. Moreover, the Levenshtein distance between the original patterns and patterns generated by GASP is less than 1 whereas the two variants of GraSeq have Levenshtein distance greater than 1 and identifies only 10% of the original patterns. While the computation time is similar across CM-SPAM, CM-SPADE, and GASP-5M, GASP-5M requires almost half the memory of CM-SPAM and quarter of the memory of CM-SPADE.

5.2   Pattern Usefulness

We evaluate the usefulness of the extracted patterns as a feature for risk prediction of CVD on NELL. We performed 5 random, stratified 70-30 train-test splits where frequent patterns are extracted using the train set, and then the top 500 patterns are used to construct binary features (i.e., the occurrence of the pattern). An XGBoost model [2] is trained and the performance is evaluated using the receiver operating characteristic (ROC) curve and the area under the ROC curve (AUC) shown in Fig. 2. GASP-5M and GASP-10M outperform exact SPM models in terms of AUC. This indicates that exact SPM models identify many noisy patterns while GASP generates patterns that are more useful for risk prediction. The results also demonstrate the insensitivity to the specification of the random walk iterations (5 M versus 10 M) as there is a limited difference in predictive power. Finally, the results illustrate the impact of approximation to combat the noise inherent in EHRs.

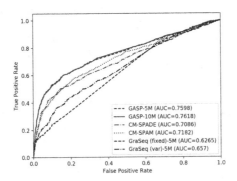

Fig. 2. The ROC curve and AUC score for risk prediction of CVD.

6 Conclusions

In this paper, we propose GASP, a new approach for approximate SPM of EHRs. We present a new weighted graph structure using both directed and undirected edges which compresses the sequential information. We also introduce a variant of a random walk model to extract variable-length sequential patterns. Empirical evaluations on two EHR databases suggest that GASP reduces the noise in patterns and can enhance pattern usefulness without sacrificing computational and memory efficiency. As approximate SPM is applicable to many other applications, future work can focus on evaluation across multiple domains.

Acknowledgements. This work was supported by the National Science Foundation award IIS-#1838200 and the National Institutes of Health (NIH) awards 1R01LM013323 and 5K01LM012924.

References

1. Chang, J.H., Lee, W.S.: Efficient mining method for retrieving sequential patterns over online data streams. J. Inf. Sci. **31**(5), 420–432 (2005)

2. Chen, T., Guestrin, C.: Xgboost: a scalable tree boosting system. In: Proceedings of the KDD, pp. 785–794 (2016)
3. Fournier-Viger, P., Gomariz, A., Campos, M., Thomas, R.: Fast vertical mining of sequential patterns using co-occurrence information. In: Proceedings of the PAKDD, pp. 40–52 (2014)
4. Fournier-Viger, P., et al.: The spmf open-source data mining library version 2. In: Proceedings of the ECML/PKDD, pp. 36–40 (2016)
5. Fournier-Viger, P., Lin, J.C.W., Kiran, R.U., Koh, Y.S., Thomas, R.: A survey of sequential pattern mining. Data Sci. Pattern Recognit. **1**(1), 54–77 (2017)
6. Geraci, J.M., Ashton, C.M., Kuykendall, D.H., Johnson, M.L., Wu, L.: International classification of diseases, 9th revision, clinical modification codes in discharge abstracts are poor measures of complication occurrence in medical inpatients. Medical care, pp. 589–602 (1997)
7. Kum, H.C., Pei, J., Wang, W., Duncan, D.: Approxmap: approximate mining of consensus sequential patterns. In: Proceedings of the SDM, pp. 311–315 (2003)
8. Li, H., Chen, H.: Graseq: a novel approximate mining approach of sequential patterns over data stream. In: Proceedings of the ADMA, pp. 401–411 (2007)
9. Pearson, K.: The problem of the random walk. Nature **72**(1867), 342–342 (1905)
10. Raïssi, C., Poncelet, P.: Sampling for sequential pattern mining: from static databases to data streams. In: Proceedings of the ICDM, pp. 631–636 (2007)
11. Salvemini, E., Fumarola, F., Malerba, D., Han, J.: Fast sequence mining based on sparse id-lists. In: Proceedings of the ISMIS, pp. 316–325 (2011)
12. Zhu, F., Yan, X., Han, J., Philip, S.Y.: Efficient discovery of frequent approximate sequential patterns. In: Proceedings of the ICDM, pp. 751–756 (2007)

Semi-synthetic Data and Testbed for Long-Distance E-Vehicle Routing

Andrius Barauskas⬛, Agnė Brilingaitė⬛, Linas Bukauskas⬛, Vaida Čeikutė, Alminas Čivilis, and Simonas Šaltenis$^{(\boxtimes)}$⬛

Institute of Computer Science, Vilnius University, Vilnius, Lithuania
{andrius.barauskas,agne.brilingaite,linas.bukauskas,vaida.ceikute, alminas.civilis,simonas.saltenis}@mif.vu.lt

Abstract. Electric and autonomous mobility will increasingly rely on advanced route planning algorithms. Robust testing of these algorithms is dependent on the availability of large realistic data sets. Such data sets should capture realistic time-varying traffic patterns and corresponding travel-time and energy-use predictions. Ideally, time-varying availability of charging infrastructure and vehicle-specific charging-power curves should be included in the data to support advanced planning.

We contribute with a modular testbed architecture including a semi-synthetic data generator that uses a state-of-the-art traffic simulator, real traffic distribution patterns, EV-specific data, and elevation data to generate time-dependent travel-time and energy-use weights in a road-network graph. The experimental study demonstrates that the testbed can reproduce travel-time and energy-use patterns for long-distance trips similar to commercially available services.

Keywords: Semi-synthetic data · Data generation · Testbed · Electric vehicle · Long-distance EV routing · Time-dependent road network

1 Introduction

Transportation is currently undergoing a profound transformation. This is driven by the emergence of new automotive technologies, such as electric (EV) and autonomous vehicles, new business models such as ridesharing, and the continued digitalization of all aspects of transportation. For example, the efficiency of a fleet of autonomous electric vehicles will be highly dependent on effective routing and scheduling algorithms and these will, in turn, depend on data-driven predictions of travel time and energy use. Furthermore, as real-world routing problems are often formulated as multi-objective optimization involving multiple constraints, the optimal algorithms are intractable; thus, only heuristic algorithms are possible [2]. The efficiency and the efficacy of such algorithms can only be tested through extensive experimental studies on large datasets and workloads.

© The Author(s) 2021
L. Bellatreche et al. (Eds.): ADBIS 2021 Short Papers, Workshops and Doctoral Consortium,
CCIS 1450, pp. 61–71, 2021.
https://doi.org/10.1007/978-3-030-85082-1_6

To understand the complexity of the data required by real-world routing algorithms, consider a long-distance EV routing query. It has to take into account the predicted traffic to estimate both the expected travel time and the expected energy use. To plan charging stops, this information is combined with the information about the availability and the power of chargers. Both the traffic and the availability of chargers are *time-dependent* (TD). Furthermore, we argue that any realistic long-distance routing system has to work with the inherent uncertainties of predictions. Thus, the travel time, the used energy, and the time waiting for charging are all modeled as *intervals* of expected values.

Research studies that explore advanced routing problems expend much effort to prepare their experiments. For example, to implement Eur-PTV and Ger-PTV benchmarks [2], road network data, elevation information, energy consumption data, traffic data, and charging station data are preprocessed and integrated. Åkerblom et al. [1] extend the simulation framework of Russo et al. [15] taking the traffic patterns from the LuST Scenario data [5]. Several studies apply statistical and machine learning methods to forecast travel time and future congestion along the route using data collected from Google Maps Platform API. Traffic conditions can be identified by capturing traffic layer image and identifying *color* data on monitored road segments [14,18], or using Estimated Time of Arrival [17]. Suggested methodologies are time consuming—to apply machine learning algorithms, a substantial amount of data has to be collected during an extended period of time. Brinkhoff [3] pioneered a framework to generate moving objects on a road network. The framework did not consider traffic models and resulting vehicle movements were not very realistic. In contrast, the open-source microscopic traffic simulation tools, such as SUMO [12], used in this work, and GeoSparkSim [16], were designed to handle realistic traffic simulation on large-scale road networks.

This paper aims to do the necessary legwork for the road-network algorithms community. While there are a few traffic simulators and general-purpose spatial and graph data generators, we provide, to the best of our knowledge, the first testbed for experimentation with advanced routing algorithms, in particular, algorithms for EVs.

The paper contributes with a modular architecture and a data preparation workflow to generate realistic semi-synthetic EV-specific TD traffic data that captures uncertainty. We provide a layer of services on top of the generated data to be used as building blocks of future advanced routing algorithms. The experiments indicate that the proposed environment provides data patterns similar to commercial ones, and it can be used to test the TD routing of EVs.

The work is structured as follows. Section 2 introduces the testbed architecture. Sections 3 and 4 present implementation details and the experimental evaluation of the testbed, respectively. Section 5 concludes the paper.

2   Semi-synthetic Data Generation and Testbed API

2.1   Testbed Architecture and Functionality

Generating and managing the test data introduced above calls for a multi-component architecture (see Fig. 1). First, driving speed depends on the traffic at a particular time. Therefore, the TD Traffic Information component requires Traffic Simulation data and TD Traffic Statistics to define parameters of road edges. Second, the Energy Consumption component is dependent on elevation data and the consumption function that uses the EV properties as its parameters. Finally, long-distance EV routing requires charging stops along the road. Hence, the component of Charging Stations is supported by TD availability data of charging stations and charging function that uses the parameters of EV type.

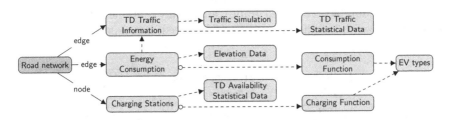

Fig. 1. Components of the testbed architecture

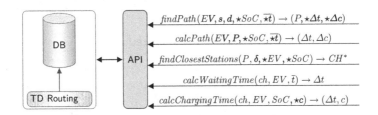

Fig. 2. Testbed API

While the main contribution and focus of this work is the generation of semi-synthetic data, a thin layer of services is proposed as well. Such services query and aggregate the data and can be used as the building elements of advanced routing algorithms. Figure 2 presents five API functions with optional parameters marked by $\star$. Function *findPath* uses a TD router to construct a path P and to estimate the expected trip duration interval Δt and the expected energy consumption interval Δc when traveling from start s to destination d and starting the trip some time during $\bar{t}$ time interval. The starting time is given as an interval, which is useful if the function computes a leg of a longer route. If the initial state of charge of the EV battery SoC is given, the returned Δc is the final expected interval of the state of charge of the battery, rather than the consumed energy.

Function *calcPath* is used to calculate the same travel estimates on an already known path P. Function *findClosestStations* returns a set of charging stations CH^* containing the stations within a Euclidean buffer δ around path P and reachable by EV when starting on the path with SoC. Finally, functions *calcWaitingTime* and *calcChargingTime* return waiting-time and charging-time intervals Δt at charging station ch for EV. A waiting time interval depends on the daytime interval when the EV reaches ch. Also, a charging time interval depends on the SoC before starting the charging process. The required SoC c can be provided and the reached SoC c is returned.

2.2 Traffic Data Simulation and Calibration

Traffic data preparation process is shown in Fig. 3. To prepare the road network (RN), first, map data is filtered leaving only car roads. Next, the road network graph is made routable (a directed graph), and finally routable network segments are augmented with length data and free-flow speed data (speed-limit data).

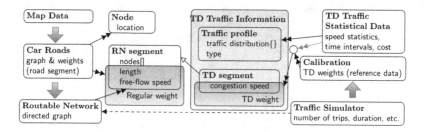

Fig. 3. Traffic data preparation

Two main sources are used to generate the semi-synthetic TD weights of road-network segments. TD traffic statistical data for a given region describes how traffic at large changes relatively to the time of day. This is used to derive a traffic profile. Then, network segments are augmented with congestion speed data—either real statistical data, if available, or synthesized data generated by traffic simulators. Finally, the results of simulations are calibrated using commercial traffic data providers.

Semi-synthetic TD segment weights are composed of edge-specific minimum traffic speed, edge-specific maximum traffic speed, and region-wide TD traffic distribution. Given a time of day, they are used to calculate edge-specific traffic speed as a weighted average of the minimum and the maximum traffic speeds of a segment. We use maps from the OpenStreetMap project (OSM, [7]). Thus, the maximum speed is the free-flow speed from OSM, the minimum speed is derived from congestion modeling using open-source traffic simulator SUMO [8], and the TD traffic distribution is sourced from TomTom's (TT) Traffic Index.

SUMO takes a routable network as data input for traffic simulation and augments it with simulated traffic data. The testbed's routable network is fed to

SUMO using the `netconvert` tool. The output of the simulation is a congestion-hour travel time for each segment on the routable network. To perform a simulation, the whole map is divided into regions and each region is simulated separately. Random traffic generation method of the SUMO tool *randomTrips* is used. This method allows choosing different weights affecting the probability of selecting a segment for routing. Segment length is used as a weight; thus, dense regions like city centers get more traffic. Finally, the number of trips is calculated proportionally to the population size of the region and distributed evenly in an interval from 0 to 3600 s.

Assuring realistic generated data requires *calibration* of both the free-flow and congestion travel times. The calibration is implemented via two coefficients for the congestion speed and the free-flow speed. The coefficients are calculated by comparing simulated travel times with Google Maps travel times. First, two sets of routes are generated—inside cities and out of cities—for congestion and free-flow travel time calibration, respectively. Then, travel times are calculated at peak hours for inside-cities set and off-peak hours for out-of-cities set.

2.3   Data for Energy Consumption Estimation

Energy consumption (EC) along a given route is estimated by adapting the Vehicle Energy Model (VEM) as introduced in the SUMO simulator [11]. In addition, the EC model considers traffic information to estimate TD energy use along the route.

Energy consumption calculation uses two types of parameters—vehicle specific and road-network dependent. The following EV characteristics are employed: battery, vehicle mass, front surface area, air drag coefficient, internal moment of inertia, radial drag coefficient, roll drag coefficient, propulsion efficiency, recuperation efficiency, and constant power intake. While we are currently using a predefined set of these values, the constant power intake parameter could be extended and vary based on weather conditions for more precise modeling. Such EV data can be collected from various sources, including car manufacturers and EV enthusiasts that try to measure various parameters of their vehicles under specific conditions.

The core road-dependent parameters, the slope and the radius, are precomputed for each EC segment and stored in the database. In addition, a segment inherits free-flow speed, length, congestion speed, and contains node coordinates, as it extends the TD segment. Segment geometry is used to compute the length and the radius of each segment. We deem the slope and the radius as the essential terrain approximation parameters.

2.4   Charging and Waiting Times at Charging Stations

Each charging station contains a set of chargers (see Fig. 4), a TD availability profile, and geographic location to be mapped to the road network. The model could be extended with other features, e.g. connection fee or charging price. Each

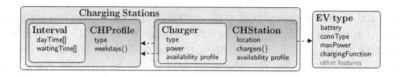

Fig. 4. Domain model of charging stations

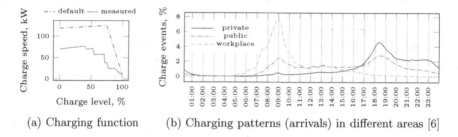

(a) Charging function   (b) Charging patterns (arrivals) in different areas [6]

Fig. 5. Charging functions of two different 65 kWh batteries [4] and charging demands

charger of the station is described by a connector type, power, and its own availability profile. Figure 5a presents two charging functions for 65 kWh battery—the
default one and one measured by observation. The charging function depends
on EV features and charger properties. First, some EV types are limited by
their own maximum charging power. Second, charging process is slower when
the battery's SoC is below 20% and above 80%, especially in the case of rapid
charging. Charging functions are retrieved from open data available on the internet, e.g. [4], to get EV maximum power and power at different points of charging
(piece-wise linear function) for different chargers. Availability of a charging station or an individual charger can be represented by a piece-wise linear function
of time. Features like type, e.g. rural areas, and weekdays, e.g. Sunday and
Saturday, define a particular availability profile. Figure 5b shows percentage of
charging cases throughout a 24-h period for private, public, and workplace charging points on work days [6]. For example, at 9:00 the need for power is very high
at workplaces. At night a number of charging cases is low in all cases. Therefore,
the availability profile can be constructed based on observation data with a high
probability of a waiting time that depends on the charger power.

3   Testbed Implementation

Figure 6 summarizes the process of semi-synthetic data preparation using open
tools and data sources. Germany map was retrieved from OSM and filtered for
vehicle roads using `osmfilter`. The `osm2po` tool generated a routable network
and it was stored in PostgreSQL with PostGIS extension. SUMO tool was used
to simulate traffic flow. Congestion index was obtained from TomTom. Network-
segment slopes were calculated based on CGIAR-CSI SRTM 90 m Digital Ele-

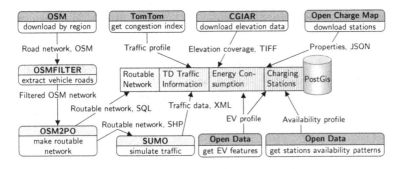

Fig. 6. Data sources and processing

vation Data [10]. EV features and charging stations [13] with their availability patterns were set up using publicly available data.

At the core of *findPath* and *calcPath* functions (see Fig. 2) is the computation of the total energy and the total driving time on a route. For each segment, the EC segment properties at a given time are used to calculate the travel time and energy required to traverse the segment. For some segments, it might result in a negative value e.g. going downhill. If so, the calculation has to make sure that the battery is not charged more than its capacity. Finally, the estimations for each route segment are added up to get the total energy and travel time.

Travel speed $v(t, seg)$ along the segment seg is a time function as it varies based on traffic conditions. Each TD segment has an estimated free-flow speed, $v_{freeflow}(seg)$, and congestion speed, $v_{congestion}(seg)$. The speed along the segment varies between the two extremes. This is modeled via the $cost(t)$ function defined by the traffic profile:

$$v(seg, t) = v_{freeflow}(seg) - (v_{freeflow}(seg) - v_{congestion}(seg)) \cdot cost(t).$$

The testbed simulates the uncertainty of prediction by assuming that the timing of peaks in the traffic profile might slightly shift from day to day. The testbed calculates the minimal cost value and the maximal cost value for each segment using the time window defined by t and uncertainty ϵ:

$$_{min}cost(t) = \min(cost([t - \epsilon, t + \epsilon])), \; _{max}cost(t) = \max(cost([t - \epsilon, t + \epsilon])).$$

The default ϵ value is set to 30 minutes, but can be adjusted. Note that t is a time when an EV reaches a given segment seg_i along the route. Thus, it depends on the travel speed and departure time of previous $i - 1$ segments. Let us assume the trip start time is t_{start} and Δt_i is the time required to pass segment i, then seg_i entrance time t_i is $t_i = t_{start} + \sum_{n=1}^{i-1} \Delta t_n$. For the whole route, the bounds of the estimated energy and time intervals are calculated in two iterations. The first iteration uses $_{min} cost(t)$ as the cost function for the lower bound and the second iteration—$_{max} cost(t)$ for the upper bound.

The testbed contains synthetic data to estimate waiting times at charging stations at different times of the day. Various profiles, e.g. business premises, were integrated to follow statistical data on charging patterns [6]. Then, waiting time intervals were constructed as $wt = (wt_{min}, wt_{max}) = (0, \Delta ch)$ where Δch is the time needed to charge from 10% to 100% SoC and 80% at AC and DC chargers, respectively. For example, $\Delta ch = 15$min in the case of rapid charging. Afterwards, the intervals were shifted based on the number of chargers of the same type, location of the charging station in relation to the highway, and charging patterns (similar to TD travel speed). The calculation and tuning details are left out of the scope of this paper.

4   Experiments and Results

KaTCH [9] implementation of time-dependent contraction hierarchies was integrated into the testbed as a routing engine, and several tests were run to illustrate realistic results and appropriate scalability within the testbed.

As a case study, sources and destinations were chosen for 8 representative trips in Germany. Then, the travel-time and energy-consumption intervals were calculated for all of them when traveling from a source to a destination and back—16 individual trips in total. Also, the departure times were set to 00:00 and 16:00 as non-congestion and congestion-time representatives. To estimate energy consumption in both testing environments, the energy use curve was constructed as a sequence of pairs (kWh per 100 km, km/h)—$(24.94, 10)$, $(15.91, 20)$, $(12.73, 30)$, $(11.65, 40)$, $(12.06, 60)$, $(15.25, 80)$, $(19.36, 100)$, $(22.50, 120)$. The prototype vehicle had the following characteristics: mass 1785 kg, battery 62 kWh, 1.5 kW constant power consumption, 0.28 air drag coefficient and 2.44 m^2 front surface area, 0.8 propulsion efficiency and 0.8 recuperation efficiency.

Figure 7 plots estimated travel time and energy consumption for different departure times on the testbed with the results from TomTom shown for reference. The results show that the testbed is more conservative regarding travel time and energy consumption when leaving at non-congestion time. For the congestion hour, the testbed is more optimistic regarding travel time and energy consumption, and the generated uncertainty intervals are longer—for long trips interval length is approximately half an hour. The testbed provides different results for forward and backward trips as the model considers elevation details and recuperation.

Figure 8 plots results of scalability tests, which were run on a Linux workstation with Intel(R) 16 Core(TM), i9-9880H CPU @ 2.30 GHz, 32 GB RAM with an equivalent remote database server. For each different trip length (air distance), 1000 source-destination pairs were generated, their routing was executed, and routes were saved in the database. The region was loaded into the main-memory KaTCH data structure, with approx. 21GB RAM used. The difference in the air distance and route length is app. 30%. The results show that the query cost without energy consumption calculation is almost constant, whereas energy computation grows linearly to the length of the path.

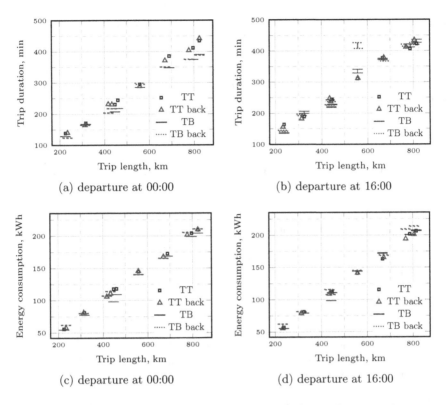

Fig. 7. Travel time and energy consumption for different departure times

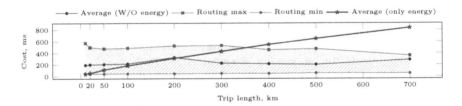

Fig. 8. Scalability tests

5   Conclusions and Future Work

Motivated by the inherent complexity of testing advanced routing algorithms, the paper proposes a testbed that integrates state-of-the-art tools and provides a systematic approach to available open-source data. We believe the provided insights and the testbed itself will shorten the preparation phase of future experimental studies. The scalability and reference-based tests demonstrate the merits of the testbed. The work can be extended in several directions, e.g. to enable functions to switch among EV energy consumption and life-cycle profiles or to enrich the environment with a flexible setup for experiments.

Acknowledgement. This project has received funding from European Regional Development Fund (project No 01.2.2-LMT-K-718-02-0018) under a grant agreement with the Research Council of Lithuania (LMTLT).

References

1. Åkerblom, N., Chen, Y., Chehreghani, M.H.: An online learning framework for energy-efficient navigation of electric vehicles. In: IJCAI, pp. 2051–2057. ijcai.org (2020). https://doi.org/10.24963/ijcai.2020/284
2. Baum, M., Dibbelt, J., Wagner, D., Zündorf, T.: Modeling and engineering constrained shortest path algorithms for battery electric vehicles. Transp. Sci. **54**(6), 1571–1600 (2020). https://doi.org/10.1287/trsc.2020.0981
3. Brinkhoff, T.: A framework for generating network-based moving objects. GeoInformatica **6**(2), 153–180 (2002). https://doi.org/10.1023/A:1015231126594
4. Chargeprice: Open EV Data. https://github.com/chargeprice/open-ev-data. Accessed 7 Mar 2021
5. Codeca, L., Frank, R., Faye, S., Engel, T.: Luxembourg SUMO traffic (lust) scenario: traffic demand evaluation. IEEE Intell. Transp. Syst. Mag. **9**(2), 52–63 (2017). https://doi.org/10.1109/MITS.2017.2666585
6. ElaadNL: Open data sets. https://platform.elaad.io. Accessed 12 Mar 2021
7. Geofabrik GmbH: Openstreetmap data extracts (2020). http://download.geofabrik.de/. Accessed 5 Sept 2020
8. German Aerospace Center (DLR) and others.: Sumo – simulation of urban mobility (2020). https://sumo.dlr.de/docs/. Accessed 24 Feb 2020
9. Institut fuer Theroretische Informatik, Karlsruher Institut fuer Technology (KIT): KaTCH - Karlsruhe Time-Dependent Contraction Hierarchies (2016). https://github.com/GVeitBatz/KaTCH. Accessed 16 Mar 2021
10. Jarvis, A., Reuter, H.I., Nelson, A., Guevara, E.: Hole-filled seamless SRTM data v4 (2008). http://srtm.csi.cgiar.org. Accessed 24 Feb 2020
11. Kurczveil, T., López, P.Á., Schnieder, E.: Implementation of an energy model and a charging infrastructure in SUMO. In: Behrisch, M., Krajzewicz, D., Weber, M. (eds.) SUMO 2013. LNCS, vol. 8594, pp. 33–43. Springer, Heidelberg (2014). https://doi.org/10.1007/978-3-662-45079-6_3
12. López, P.Á., et al.: Microscopic traffic simulation using SUMO. In: ITSC, pp. 2575–2582. IEEE (2018). https://doi.org/10.1109/ITSC.2018.8569938
13. Open Charge Map: The Open Charge Map API. https://openchargemap.org/site/develop/api. Accessed 7 Mar 2021
14. Pramanik, A., Rahman, M., Anam, I., Ali, A.A., Amin, A., Rahman, M.: Modeling traffic congestion in developing countries using google maps data (2020). preprint at arXiv:2011.02359
15. Russo, D., Roy, B.V., Kazerouni, A., Osband, I., Wen, Z.: A tutorial on Thompson sampling. Found. Trends Mach. Learn. **11**(1), 1–96 (2018). https://doi.org/10.1561/2200000070
16. Yu, J., Fu, Z., Sarwat, M.: Dissecting GeoSparkSim: a scalable microscopic road network traffic simulator in apache spark. Distrib. Parallel Databases **38**(4), 963–994 (2020). https://doi.org/10.1007/s10619-020-07306-x

17. Zafar, N., Haq, I.U.: Traffic congestion prediction based on estimated time of arrival. PLoS ONE **15**(12) (2020). https://doi.org/10.1371/journal.pone.0238200
18. Zhao, X., Spall, J.C.: Modeling traffic networks using integrated route and link data (2018). preprint at arXiv:1811.01314

Looking for COVID-19 Misinformation in Multilingual Social Media Texts

Raj Ratn Pranesh[1], Mehrdad Farokhnejad[2], Ambesh Shekhar[1],
and Genoveva Vargas-Solar[3(✉)]

[1] Birla Institute of Technology, Mesra, Ranchi, India
[2] Univ. Grenoble Alpes, CNRS, LIG, Grenoble, France
`Mehrdad.Farokhnejad@univ-grenoble-alpes.fr`
[3] CNRS, LIRIS-LAFMIA, Lyon, France
`genoveva.vargas-solar@liris.cnrs.fr`

Abstract. This paper presents the Multilingual COVID-19 Analysis Method (CMTA) for detecting and observing the spread of misinformation about this disease within texts. CMTA proposes a data science (DS) pipeline that applies machine learning models for processing, classifying (Dense-CNN) and analyzing (MBERT) multilingual (micro)-texts. DS pipeline data preparation tasks extract features from multilingual textual data and categorize it into specific information classes (i.e., 'false', 'partly false', 'misleading'). The CMTA pipeline was experimented with multilingual micro-texts (tweets), showing misinformation spread across different languages. We performed a comparative analysis of CMTA with eight monolingual models used for detecting misinformation. The comparison shows that CMTA has surpassed various monolingual models and suggests that it can be used as a general method for detecting misinformation in multilingual micro-texts.

Keywords: Misinformation · Multilingual analysis · Micro-text analysis · COVID-19

1 Introduction

The COVID-19 pandemic has highlighted the extent to which the world's population is interconnected through the Internet and social media. Social media provides particularly fertile ground for the spread of information and misinformation [7]. Misinformation may have intense implications for public opinion and behaviour, positively or negatively influencing the viewpoint of those who access it [3,10]. The study introduced in this paper aims to answer two research questions: (Q_1) determine whether it is possible to develop a method that can provide a general multilingual classification pipeline? (Q_2) Is it possible to build conclusions about how misinformation on COVID-19 spreads in different language speaking communities by analysing micro-texts published in social media?

© Springer Nature Switzerland AG 2021
L. Bellatreche et al. (Eds.): ADBIS 2021 Short Papers, Workshops and Doctoral Consortium,
CCIS 1450, pp. 72–81, 2021.
https://doi.org/10.1007/978-3-030-85082-1_7

Therefore, this paper proposes CMTA, a multilingual tweet analysis and information (misinformation) detection method for observing social media misinformation spread about the COVID-19 pandemic within communities with different languages. CMTA proposes a data science pipeline with tasks that rely on popular artificial intelligence models (e.g., multilingual BERT and CNN) to process texts and classifying them according to different misinformation classes ('false', 'partly false' and 'misleading'). The paper describes the experimental setting implemented for validating CMTA that uses datasets of COVID-19 related tweets. The paper presents an illustrative statistical representation of the findings insisting on the insights discovered in our study to discuss results. To assess the performance of CMTA, we compared CMTA with eight monolingual BERT models. The comparison shows that CMTA has surpassed various monolingual models and suggests that it can be used as a general method for detecting misinformation in multilingual micro-texts.

The remainder of the paper is organised as follows. Section 2 introduces works that have addressed misinformation detection about COVID-19 on social media datasets. Section 3 describes the general approach behind the method CMTA that we propose. It describes the experiment setting that we used for validating CMTA. Section 4 compares the performance of CMTA against mono-lingual analytics performed on the same dataset. It also discusses the study results about misinformation spread through micro-texts (i.e., tweets) across different languages. Finally, Sect. 5 concludes the paper and discusses future work.

2   Related Work

The COVID-19 pandemic has resulted in studies investigating the various types of misinformation arising during the COVID-19 crisis [2,6,10,15]. Studies investigate a small subset of claims [15] or manually annotate Twitter data [10]. In [2] authors analyse different types of sources for looking for COVID-19 misinformation. Pennycook et al. [11] introduced an attention-based account of misinformation and observed that people tend to believe and share false claims about COVID-19. Kouzy et al. [10] annotated about 600 messages containing hashtags about COVID-19, they observed that about one-fourth of messages contain some form of misinformation, and about 17% contain some unverifiable information. The work in [9] examined the global spread of information related to crucial disinformation stories and "fake news" URLs during the early stages of the global pandemic on Twitter.

The work in [14] focused on topic modelling and designed a dashboard to track Twitter's misinformation regarding the COVID-19 pandemic. Cinelli et al. [15] track misinformation flow across 2.7M tweets and compare it with infection rates. They noticed a significant Spatio-temporal connection between information flow and new COVID-19 instances. The work in [8] introduced the first example of a causal inference method to find and measure causal relationships between pandemic features (e.g. the number of infections and deaths) and Twitter behaviour and public sentiment. The work in [13] used text mining on Twitter data to demonstrate the epidemiological effect of COVID-19 on press publications in Bogota, Colombia. They intuitively note a strong correlation between the number of tweets and the number of infected people in the area.

Wani et al. [16] performed an evaluation study on deep learning-based text classification algorithms for the task of fake news detection approaches on Cons-traint@-AAAI 2021 Covid-19 Fake news detection dataset. The techniques included Convolutional Neural Networks(CNN), Long Short Term Memory (LSTM), and Bidirectional Encoder Representations from Transformers (BERT). Their study shows that transformer-based models outperform other basic models, and by using language model pretraining on BERT, they could achieve maximum accuracy (%) of 98.41.

3   The CMTA Method

CMTA implements a data science pipeline consisting of four phases: (1) tokenizing, (2) text features extraction, (3) linear transformation, and (4) classification. The first phases (tokenizing, text feature extraction, linear transformation) correspond to a substantial data-preparation process intended to build a multi-lingual vectorized text representation. The objective is to achieve a numerical pivot representation of texts agnostic of the language. CMTA classification task uses a dense layer and leads to a trained network model that can be used to classify micro-texts (e.g. tweets) into three misinformation classes: 'false', 'partly false' and 'misleading'.

1. Text Tokenization. Given a multilingual textual dataset consisting of sentences, CMTA uses the BERT multilingual tokeniser to generate tokens that BERT's embedding layer will further process. CMTA uses MBERT[1] to extract contextual features, namely word and sentence embedding vectors, from text data[2]. In the subsequent CMTA phases that use NLP models, these vectors are used as feature inputs with several advantages. (M)BERT embeddings are word representations dynamically informed by the words around them, meaning that the same word's embeddings will change in (M)BERT depending on its related words within two different sentences.

For the non-expert reader, the tokenisation process is based on a WordPiece model. It greedily creates a fixed-size vocabulary of individual characters, subwords, and words that best fit a language data (e.g. English)[3]. Each token in a

---

[1] https://github.com/google-research/bert/blob/master/multilingual.md.

[2] Embeddings are helpful for keyword/search expansion, semantic search and information retrieval. They help accurately retrieve results matching a keyword query intent and contextual meaning, even in the absence of keyword or phrase overlap.

[3] This vocabulary contains whole words, subwords occurring at the front of a word or in isolation (e.g., "em" as in the word "embeddings" is assigned the same vector as the standalone sequence of characters "em" as in "go get em"), subwords not at the front of a word, which are preceded by '##' to denote this case, and individual characters [18].

tokenised text must be associated with the sentence's index: sentence 0 (a series of 0s) or sentence 1 (a series of 1s). After breaking the text into tokens, a sentence must be converted from a list of strings to a list of vocabulary indices. The tokenisation result is used as input to apply BERT that produces two outputs, one pooled output with contextual embeddings and hidden-states of each layer. The complete set of hidden states for this model are stored in a structure containing four elements: the layer number (13 layers)[4], the batch number (number of sentences submitted to the model), the word/token number in a sentence, the hidden unit/feature number (768 features)[5].

In CMTA, the tokenisation is more complex because it is done for sentences written in different languages. Therefore, it relies on the MBERT model that has been trained for this purpose.

2. Feature Extraction Phase. This phase is intended to exploit hidden layers' information due to applying BERT to the tokenisation phase result. The objective is to get individual vectors for each token and convert them into a single vector representation of the whole sentence. For each token of our input, we have 13 separate vectors, each of length 768. Thus, to get the individual vectors, it is necessary to combine some of the layer vectors. The challenge is to determine which layer or combination of layers provides the best representation.

3. Linear Convolution. The hidden states from the 12th layer are processed in this phase, applying linear convolution and pooling to get correlation among tokens. We apply a three-layer 1D convolution over the hidden states with consecutive pooling layers. The final convolutional layer's output is passed through a global average pooling layer to get a final sentence representation. This representation holds the relation between contextual embeddings of individual tokens in the sentence.

4. Classification. A linear layer is connected to the model in the end for the CMTA classification task. This classification layer outputs a Softmax value of vector, depending on the output, the index of the highest value in the vector represents the label for the given sequence: 'false', 'partly false' and 'misleading'.

3.1 Experiment

To validate CMTA, we designed experiments on Google Colab with 64 GB RAM and 12 GB GPU. For implementing the method, we calibrated pre-trained models provided by hugging face[6]. We extracted annotated misinformation data from multiple publicly available open databases. We also collected many multilingual tweets consisting of over 2 million tweets belonging to eight different languages. *Misinformation datasets.* We collected data from an online fact-checker website called Poynter [12]. Poynter has a specific COVID-19 related misinformation detection program named 'CoronaVirusFacts/DatosCoronaVirus Alliance

---

[4] It is 13 because the first element is the input embeddings, the rest is the outputs of each of BERT's 12 layers.

[5] That is 219,648 unique values to represent our one sentence!.

[6] https://huggingface.co/.

Database[7]'. This database contains thousands of labelled social media information such as news, posts, claims, articles about COVID-19, which were manually verified and annotated by human volunteers (fact-checkers) from all around the globe. The database gathers all the misinformation related to COVID-19 cure, detection, the effect on animals, foods, travel, government policies, crime, lockdown. The misinformation dataset is available in 2 languages- 'English' and 'Spanish'.

We crawled through the content of two websites using Beautifulsoup[8], a Python library for scraping information from web pages. We scraped 8471 English language false news/information belonging to nine classes, namely, 'False', 'Partially false', 'Misleading', 'No evidence', 'Four Pinocchios', 'Incorrect', 'Three Pinocchios', 'Two Pinocchios' and 'Mostly False'. Studies have shown that Twitter users usually face difficulties while differentiating false and true information [19]. Therefore, we focused on collecting fine-grain misinformation data with different degrees of false information, excluding accurate information. We gathered the article's title, its content, and the fact checker's misinformation-type label for each article.

For Spanish[9], we collected 531 misinformation articles. The collected data contains the misinformation published on social media platforms such as Facebook, Twitter, Whatsapp, YouTube. Posts were mostly related to political-biased news, scientifically dubious information and conspiracy theories, misleading news and rumours about COVID-19. We also used a human-annotated fact-checked tweet dataset [1] available at a public repository[10]. The dataset contained true and false labelled tweets in English and Arabic language. We used only false labelled tweets consisting of 500 English. We compiled 9,502 micro-articles (i.e., tweets) distributed across 9 misinformation classes in English and Spanish. In English the classes were: false - 2,869 tweets from [12] and 500 tweets from [1]; partially false - 2,765 tweets; and misleading - 2,837 tweets. In Spanish, the classes were: false - 191 tweets; partially false - 161 tweets; and misleading - 179 tweets.

Dataset Pre-processing. The datasets contained noise such as emojis, symbols, numeric values, hyperlinks to websites, and username mentions that needed to be removed. To preprocesses the training and inference datasets, we used simple regular expressions to remove URLs, special characters or symbols, blank rows, re-tweets, user mentions. We did not remove the hashtags from the data as hashtags might contain helpful information. We removed stop words using NLTK[11]. For preprocessing the Hindi dataset, we used CLTK (Classical Language Toolkit)[12]. For removing Thai stop words from Thai tweets, we used PyThaiNLP [17].

---

[7] https://www.poynter.org/covid-19-poynter-resources/.
[8] https://pypi.org/project/beautifulsoup4/.
[9] https://chequeado.com/latamcoronavirus/.
[10] https://github.com/firojalam/COVID-19-tweets-for-check-worthiness.
[11] NLTK https://www.nltk.org/ is a Python library for natural language processing.
[12] https://docs.cltk.org/en/latest/index.html.

Attribute Engineering of the Training Dataset. The collected training data is unevenly distributed across nine classes: 'No evidence', 'Four Pinocchios, 'Incorrect', 'Three Pinocchios', 'Two Pinocchios' and 'Mostly False' (the smallest group). Most collected articles were labelled either as 'False', 'Partially false' and 'Misleading'. We performed an attribute engineering phase for preparing the dataset. We produced a uniformly distributed dataset reorganised the initial dataset as follows. The classes 'Four Pinocchios' and 'Incorrect' were merged with the class 'False'. The classes 'Three Pinocchios' and 'Two Pinocchios' were merged into the class 'Partially false'. The classes 'No evidence' and 'Mostly False' were merged with the class 'Misleading'.

Inference Dataset. We collected around 2,137,106 multilingual tweets. We focused on the tweets in eight major most commonly used languages used on Twitter: 'English', 'Spanish', 'Indonesian', 'French', 'Japanese', 'Thai', 'Hindi', and 'German'. Therefore, we used a dataset of tweets IDs associated with the novel coronavirus COVID-19 [4]. Starting on January 28, 2020, the current dataset contains 212,978,935 tweets divided into groups based on their publishing month[13]. The dataset was collected using multilingual COVID-19 related keywords and contained tweets in more than 30 languages. We used tweepy[14] which is a Python module for accessing Twitter API. We decided to retrieve the tweets using the tweet IDs published in the past five months (February, March, April, May and June) for our analysis. The distribution of tweets across eight languages corresponds to most English items, almost 1 and 1/8 of the whole data set, then Spanish (1/4 of the total number of tweets) and the rest for French, Japanese, Indonesian, Thai, and Hindi.

3.2   Model Setup and Training

Training Setting. For training our model, we divided the data into training, validation and testing datasets in the ratio of 80%/10%10% respectively. The final count for the train, validation and test dataset was 7,602, 950, 950. We fine-tuned the Sequence Classifier from HuggingFace based on the parameters specified in [5]. Thus, we set a batch size of 32, learning rate 1e-4, with Adam Weight Decay as the optimizer. We ran the model for training for 10 epochs. Then, we saved the model weights of the transformer, helpful for further training.

Hyperparameters' Setting. For training and testing, our model included both the average pooling and max pooling set to 8, the dropout probability set to 0.36, the number of dense layers set to 4, the text length set to 128 characters, the batch size set to 32, epochs set to 10, optimizer set to Adam and learning rate set to 1×10^{-4}. The calculations and selection of hyperparameters were made based on tests and the model's best output.

---

[13] English - 1,472,448, Spanish - 353,294, Indonesian - 80,764, French - 71,722, Japanese - 71,418, Thai - 36,824, Hindi - 27,320 and German - 23,316.
[14] Python module is available at http://www.tweepy.org.

3.3   Experiment and Results

We experimented with the multilingual data with their respective linguistic-based BERT models. We set the model with the same training parameters as the CMTA model and preprocessed the data as stated previously. Each monolingual model was fine-tuned for 10 epochs with a batch size of 32, and it was applied to the classification dataset of their respective language. Our model achieved an accuracy(%) of **82.17** and F_1 score (%) of **82.54** on the test dataset. The precision and recall reported by the model were **82.07** and **82.30** respectively.

4   Results Assessment

We used two strategies for assessing CMTA results according to the research questions we wanted to prove. The first research question was to determine (Q_1) whether it was possible to develop a method that could provide a general multi-lingual classification pipeline? The second question was (Q_2) whether it is possible to build conclusions about how misinformation on COVID-19 spreads in different language speaking communities by analysing micro-texts published in social media. For Q_1, we conducted a comparative study of CMTA with different independent mono-lingual misinformation classifiers. Thereby, CMTA's classification performance for a given set of micro-texts written in a given language was compared against a classification model targeting only that language. For Q_2 we analysed and plotted CMTA's results, and we proposed intuitive arguments according to our observations.

4.1   CMTA vs. Monolingual Classification

We conducted a comparative performance study of various monolingual BERT models concerning our proposed multilingual CMTA model for comparing their performance for the misinformation detection task. We investigated eight monolingual BERT model[15], namely, 'English', 'Spanish', 'French', 'Germann', 'Japanese', 'Hindi', 'Thai[16]' and 'Indonesian'. We used the same 9,502 tweets distributed across three misinformation classes for training the monolingual models. Our dataset consisted of English and Spanish tweets. Therefore, we translated the tweets into eight languages to train each of the eight monolingual models. We used Google Translator API[17] for converting the tweets into a particular language. The results scores vary from an F1-score of 77.9% for English tweets and 71.69% for Indonesian ones, recall rate from 74.18% for English and 65.36% for Japanese, and precision rate of 80.9% for Spanish and 65.66% for Indonesian. Based on these experiment results, we strongly suggest that the multilingual CMTA model could generalize smoothly on the dataset because its performance was equivalent to the monolingual models.

---

[15] Pretrained model available at https://huggingface.co/models.

[16] ThaiBERT is available at https://github.com/ThAIKeras/bert.

[17] Please refer https://cloud.google.com/translate/docs.

4.2   Multilingual Misinformation Analysis

We provide a detailed analysis of misinformation distribution across multilingual tweets. This analysis responds to the initial question: *how is misinformation about COVID-19 spread in communities speaking different languages.* Our survey studied and analyzed the distribution of COVID-19 misinformation across eight significant languages (i.e. 'English', 'Spanish', 'Indonesian', 'French', 'Japanese', 'Thai', 'Hindi' and 'German') for five months (i.e. February, March, April, May and June).

We used our trained, multilingual model, CMTA, to predict and categorize the misinformation type in tweets. We conducted our sequential misinformation analysis on a collection of over 2 million multilingual tweets. We could observe that for February, March and June months, our model predicted a large number of tweets as 'False', followed by 'Misleading', which is the second largest and the number of 'Partially false' was the least (see Fig. 1). Our model discovered that the number of 'Partially false' tweets are more than 'Misleading' tweets and 'False' tweets were again in the majority for the tweets generated during April and May.

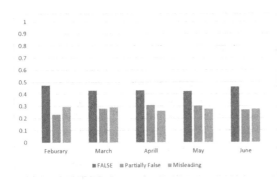

Fig. 1. Month-wise disinformation distribution.

The following specific observations were made concerning the languages. The misinformation distribution for English data indicates that there is a majority of **False** tweets during the five months, whereas the distribution of **Misleading** labelled data is slightly less than as compared to **False** labelled data. **Partially False** labelled tweets are moderately distributed, as in the month April, we can see that there is a more significant number concerning other months.

Spanish tweets have a greater frequency of **False** labelled tweets, whereas the **Misleading** tweets and **Partially False** tweets shows the almost identical number of tweet across the five months. There was a surge of **Misleading** labelled tweets during February, and the count remained the same throughout the five months. There was also an increase in **Partially False** tweets in March, but it decreased in successive months, leading to minor **False** labelled tweets.

On average, throughout the five months, approx 20% of Japanese tweets are labelled **False**. Similarly, approx 30% of the Japanese tweets are labelled **Partially False**, leading to the majority of 50% data are labelled as **Misleading**. We can also see a considerable increase in **Misleading** tweets in March, tweeted in the Japanese language.

Indonesian tweets, approximately 10% of tweets are labelled as **Misleading**, and on the contrary, there is a large distribution of **False** labelled tweets. Approximately 34% of the Indonesian dialect data is labelled as **Partially False** throughout the five months. The largest majority of the French tweets were classified as **False** misinformation. Among **Partially false** and **Misleading**, the least number of tweets were labelled as **Misleading**.

The frequency of Hindi tweets is low in the dataset used in our experiment. However, our model can predict or label Hindi tweets. Tweets in Hindi have low numbers of **Misleading** tweets, whereas the **Partially False** tweets class has a great frequency. **False** labelled tweets are slightly low compared to **Partially False** tweets in this dialect. The distribution of Thai tweets shows that our model prediction is majorly oriented towards the **Misleading** tweets. The distribution of **Misleading** labelled tweets is the greatest among the labelled classes, in contrast to **Partially False** tweets. **False** labelled tweets are comparatively moderate in this language.

5   Conclusion and Future Work

This paper introduced CMTA, a multilingual model for analyzing text applied to classify COVID-19 related multilingual tweets into misinformation categories. We demonstrated that our multilingual CMTA framework performed significantly well compared to the monolingual misinformation detection models used independently. Experimental validation of CMTA detected misinformation distribution across eight significant languages. The paper presented a quantified magnitude of misinformation distributed across different languages in the last five months. Our future work aims to collect more annotated training data and analyse a larger multilingual dataset to gain a deeper understanding of misinformation spread. We are currently improving our model's robustness and contextual understanding for better performance in the classification task. We hope that researchers could gain deeper insights about misinformation spread across major languages and use the information to build more reliable social media platforms through our work.

References

1. Alam, F., et al.: Fighting the COVID-19 infodemic: modeling the perspective of journalists, fact-checkers, social media platforms, policy makers, and the society (2020)
2. Brennen, J.S., Simon, F., Howard, P.N., Nielsen, R.K.: Types, sources, and claims of COVID-19 misinformation. Reuters Institute **7** (2020)

3. Brindha, M.D., Jayaseelan, R., Kadeswara, S.: Social media reigned by information or misinformation about COVID-19: a phenomenological study (2020)
4. Chen, E., Lerman, K., Ferrara, E.: Tracking social media discourse about the COVID-19 pandemic: development of a public coronavirus twitter data set. JMIR Public Health Surveill. **6**(2), e19273 (2020)
5. Devlin, J., Chang, M.W., Lee, K., Toutanova, K.: Bert: Pre-training of deep bidirectional transformers for language understanding. arXiv preprint arXiv:1810.04805 (2018)
6. Dharawat, A.R., Lourentzou, I., Morales, A., Zhai, C.: Drink bleach or do what now? covid-hera: a dataset for risk-informed health decision making in the presence of covid19 misinformation (2020)
7. Frenkel, S., Alba, D., Zhong, R.: Surge of virus misinformation stumps Facebook and Twitter. The New York Times (2020)
8. Gencoglu, O., Gruber, M.: Causal modeling of twitter activity during COVID-19. arXiv preprint arXiv:2005.07952 (2020)
9. Huang, B., Carley, K.M.: Disinformation and misinformation on twitter during the novel coronavirus outbreak. arXiv preprint arXiv:2006.04278 (2020)
10. Kouzy, R., et al.: Coronavirus goes viral: quantifying the COVID-19 misinformation epidemic on twitter. Cureus **12**(3) (2020)
11. Pennycook, G., McPhetres, J., Zhang, Y., Lu, J.G., Rand, D.G.: Fighting COVID-19 misinformation on social media: experimental evidence for a scalable accuracy-nudge intervention. Psychol. Sci. **31**(7), 770–780 (2020)
12. Poynter Institute: The international fact-checking network (2020). https://www.poynter.org/ifcn/
13. Saire, J.E.C., Navarro, R.C.: What is the people posting about symptoms related to coronavirus in Bogota, Colombia? arXiv preprint arXiv:2003.11159 (2020)
14. Sharma, K., Seo, S., Meng, C., Rambhatla, S., Liu, Y.: Covid-19 on social media: analyzing misinformation in twitter conversations. arXiv preprint arXiv:2003.12309 (2020)
15. Singh, L., et al.: A first look at COVID-19 information and misinformation sharing on twitter. arXiv preprint arXiv:2003.13907 (2020)
16. Wani, A., Joshi, I., Khandve, S., Wagh, V., Joshi, R.: Evaluating deep learning approaches for covid19 fake news detection. arXiv preprint arXiv:2101.04012 (2021)
17. Phatthiyaphaibun, W., Korakot Chaovavanich, C.P.A.S.L.L.P.C.: PyThaiNLP: Thai natural language processing in python (2016). https://doi.org/10.5281/zenodo.3519354
18. Wu, Y., et al.: Google's neural machine translation system: bridging the gap between human and machine translation. arXiv preprint arXiv:1609.08144 (2016)
19. Zubiaga, A., Liakata, M., Procter, R., Wong Sak Hoi, G., Tolmie, P.: Analysing how people orient to and spread rumours in social media by looking at conversational threads. PloS ONE **11**(3), e0150989 (2016)

Semantic Discovery from Sensors and Image Data for Real-Time Spatio-Temporal Emergency Monitoring

Ilia Triapitcin[1]([✉]) [ID], Ajantha Dahanayake[1] [ID], and Bernhard Thalheim[2] [ID]

[1] Lappeenranta-Lahti University of Technology, Yliopistonkatu 34, Lappeenranta, Finland
ilia.triapitcin@student.lut.fi, ajantha.dahanayake@lut.fi
[2] Christian Albrechts Universität Zu Kiel, Kiel, Germany
bernhard.thalheim@email.uni-kiel.de

Abstract. Modern district heating (DH) systems are essential contributors to large-scale city heating infrastructures. They consist of sensors, nodes, and methods for monitoring the status of the DH network. Sensing, processing, and analyzing data to locate an actual problematic emergency is a complicated task. This article presents the Spatio Temporal Emergency Monitoring (STEM) method that processes the semantic from multivariable sensors and infrared image data to locate real-time heating water leak emergencies in the DH network system. The sensory data includes multi-parameter DH network sensor data, such as water temperature, sweat rate, energy delivered, etc. The multimedia data is the infrared image data from a camera mounted on an unmanned aerial vehicle (UAV) for monitoring the environment at the underground DH network locations. The DH network monitoring system's primary purpose is to automate the central heating network's emergency monitoring. The presented real-time emergency monitoring approach processes the correlations between semantic and context of incoming data sets to determine an emergency's actual risk of a water leak. The STEM approach processes the network's sensors' data and the corresponding multimedia image data into single observations to monitor real-time Spatio-temporal emergency events.

Keywords: Semantic · Context · Real-time data · Sensors · Emergency event monitoring · Spatio temporal mapping · Heat images · District heating network

1 Introduction

District heating (DH) network provides heating to many buildings, such as heating multiple apartment houses, office buildings, industrial facilities, and large settlements. There are many methods for monitoring the stability (to have valid deviation for the parameters) of the DH network. This research presents a new approach for analyzing multivariate monitoring data, including multivariable sensor data and infrared images, for the DH network's emergency monitoring system.

Underground district heating (DH) pipelines are difficult to inspect. The available approaches to determining district heating pipelines' conditions are often expensive and

L. Bellatreche et al. (Eds.): ADBIS 2021 Short Papers, Workshops and Doctoral Consortium, CCIS 1450, pp. 82–92, 2021.
https://doi.org/10.1007/978-3-030-85082-1_8

do not provide adequate data for assessing pipelines' quality [1]. The district heating industry is continuously developing intelligent solutions to check the DH network's condition, reliably control the production process, and ensure the desired quality of heat supplied to consumers. Based on significantly improved visualization and image analysis methods, it becomes possible to analyze pipelines' conditions and offer more effective alternatives to the human experience. This study focuses on developing a thermal image analysis approach to determine the type of thermal leak. A possible system is combining existing data sources with satellite imagery, analyzing and processing the output data, and then using it in preventive maintenance processes. While this is a hazardous solution with the inherent risks of replacing pipes only when a leak or other failure is detected, unnecessarily replacing well-functioning district heating pipes has little economic benefit. Usually, old district heating pipes are typical for the low degree of insulation. Reducing heat losses from pipeline renovation is not a sufficiently significant financial incentive for first repair lines. Companies are interested in finding the most cost-effective service option and reducing the risk of network failures [1].

Since traditional quality control takes time and does not allow direct control of the process, and the requirements for decreasing carbon emissions are continually increasing, the industry is interested in transferring the planned measurements to the online process [1, 2]. In this context, real-time means that the time between event and action is limited depending on the process. The amount of heating medium supplied on an industrial scale is significant. A real-time solution can offer substantial benefits to the industry, such as improved response times to a network failure.

This research introduces a new approach for determining thermal leakages in underground DH network pipelines. The research motivation is to automate the engineer's routine to identify DH network water leakages. The scientific impact is developing a new data-centered direction for analyzing and abstracting the meaning from the network's sensor and thermal image data to increase the accuracy of determining the type of emergency in the underground DH network. Therefore, the significant contribution is that it presents a new approach for implementing a semantic space model's logical computation to determine the type of emergency in DH networks.

The research focus is on determining: how to monitor the heating water network's emergencies by correlating the semantic of sensor and thermal image data to real-time emergency events?

The research method followed consists of a literature study of related works, development of the real-time Spatio-temporal emergency monitoring (STEM) approach, and the analysis and validation of the STEM approach's effectiveness.

The rest of this document is organized as follows: Sect. 2 gives an overview of related works; Sect. 3 contains a summary of standard DH network; Sect. 4 describes the Spatio-temporal emergency method approach; Sect. 5 explains the application of the STEM in the DH network; Sect. 6 contains conclusion and discussion.

2 Related Work

2.1 DH Network Monitoring Systems

Research [3] explores the current leak detection technologies. A summary of the various types of accessible technologies for leak detection is in Fig. 1. The methods represented by yellow blocks have some limitations. Those methods require the closure of the system or branch of the system for maintenance to carry on. Methods in green blocks are often helpful. Passive systems require direct visual inspection or monitoring of networks. Active systems are the next generation of strategies and provide intelligent analysis of data collected from the sensors.

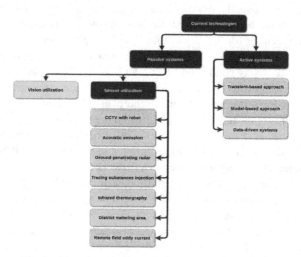

Fig. 1. Overview of current leak detection technologies

The oldest, and perhaps most unsystematic, passive leak detection system is to search for signs of water on the ground or irregular plant development that suggests a possible heat or water leak.

Manual tools or portable measurement instruments are in use in **The sensor utilization** methods. This group contains: **Closed-Circuit Television (CCTV)** combined with the robot method; **The acoustic emission technique** based on elastic waves emitting from active sources; **Ground-penetrating radar (GPR),** a geophysical method that uses radar pulses to image a pipe; **Tracing substances injection** based on sensor cable technique; **Infrared thermography** or **thermal imaging,** a scientific method of obtaining a thermogram - an image in infrared rays showing a picture of the distribution of temperature over the field; **The district metering area** method splits the heating water network into clusters; **Remote field eddy current** based on low-frequency alternating current for inspecting the inside and outside defects of steel pipes; **A transient-based approach and model-based approach** are not helpful for a real-time application. The most critical drawback of a model-based approach is the instability of model parameters, including pipe condition. Methods are often primarily based on dynamic models or large-scale sensor investments. Data-driven systems rely on sensors to capture and process information. For classification or predictive models, it necessitates a considerable volume of data. These devices may use data from a few different sensors to eliminate false signals in the DH network [4].

2.2 Environmental Event Monitoring Approaches

Many scholars have suggested semantics as a way to explain what meaning entails. Y. Kiyoki et al. have developed an approach and meta-database framework for extracting suitable images to user perception and image content using a mathematical model [5]. Y. Kiyoki and S. Ishihara suggest a metadatabase information structure with semantic associations on a mathematical model as context and a search for semantic associations by integrating semantic space [6]. Y. Kiyoki and M. Kawamoto have developed medical semantic spaces in the medical information field. Y. Kiyoki et al. have created a space or multi-space to evaluate the meaning of terms, phrases, numeric, and environmental change simulation. Semantic computation is extended to biological, chemical event monitoring [7], GIS system as 5D World Map System for environment disaster visualization, and medical emergencies [8–10].

3   DH Network System

District heating is a complicated and extensive engineering network. It is responsible for generating heating and transporting hot water from a source to various buildings and structures through a trunk pipeline.

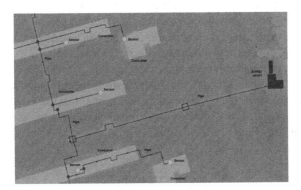

Fig. 2. DH network schema.

This system includes several structural elements (Fig. 2):

– **The heat source** or **energy center** is Waste-to-energy plants or combined heat and power plants. The first is to transfer heat to heated premises, heat water by burning gas, fuel oil, and coal. Initially, in heating plants, steam is produced, which is used for rotating the turbines, becomes a source of electricity, and after cooling down, it is used to heat water. Thus, the heated water is supplied to the heating systems of consumers;

- **The pipeline** and **pipes** transport water from the source to the consumer. This system is a complex and extended network of two large-diameter heating water pipes (supply and return), which are laid underground or overground;
- **Heat energy consumers** are equipment that uses a heating water carrier to transfer heat to a heated room;

The DH network uses sensors to control the status in real-time. The sensors are integrated into the DH network at the construction or during the modernization stage. A typical project contains information about the spatial location of all network elements and their parameters.

4 Spatio-Temporal Emergency Monitoring (STEM)

Several sensors in the DH network control points capture the heating water network's temperature, pressure, and flow velocity. The sensor readings are current values and stored in real-time. The DH network monitoring system does calculations on the sensor data by tallying with expected parameters already available for the sensors. When there are deviations higher than permissible values, an alarm is triggered. A warning may indicate a water leak from a hole in the pipeline or a heat leak from the insulation. When all three sensors (temperature, pressure, and flow velocity) give alarms, it indicates a water leak and a pipe hole. That is an emergency that requires immediate attention and maintenance. But the standard practice of detecting the exact location of the bust in the pipeline involves digging a considerable length of the pipe segment.

The Spatio-Temporal Emergency Monitoring (STEM) approach brings in a significant improvement to the present-day practice. The method combines sending a UAV over the heating network area when an alarm indicates an emergency/water leak. UAV takes thermal images of the network area above the ground.

STEM is a data-centered approach. It uses fact modeling on the sensor data collection process already available and the thermal image data to identify the water leak's exact location. The emergency of the DH network is an essential fact. An emergency means a water leak in the pipeline. The associated data from the sensor alarms and thermal image defines the emergency (Fig. 3).

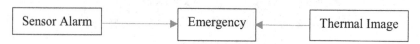

Fig. 3. The fact model of the emergency event

STEM method has the following main steps:

1. Collecting and processing the sensor data in the DH network. This step realizes the sensory data matrices for deviations of the temperature, pressure, and flow velocity input and output sensors;

2. The collection and processing of thermal images to implement the thermal image anomalies and spatial data matrix.
3. The development of a matrix to lay correspondence between fact and derived semantics of sensor and thermal image data.
4. The processing of the correlation between the semantics of sensor data and thermal image anomaly to the emergency context; This step implements the matrix in table 1.
5. Alert the identification of emergencies at exact locations on the DH network.

Figure 4 illustrates the architecture of the system and the relations between modules. STEM method contains spatial (1st, 2nd, 3rd dimensions), temporal (4th dimension), and semantic dimension (5th dimension, which represents multiple-dimensional semantic space).

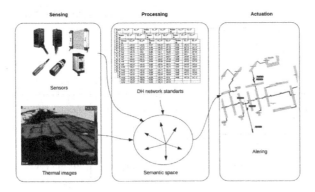

Fig. 4. The architecture of the STEM modules

Sensors data create the semantic context for the thermal images of a surface to determine the type of thermal anomaly.

4.1 Sensor's Data Population

The DH network monitoring system uses a unique identifier (UUID) to identify a sensor. Each sensor sends data in real-time to a database $DB_{sensors}$. $DB_{sensors}$ database stores A collection of tuples $S_\lambda := \{V, D, UUID, L, T\}$ as follows:

$$DB_{sensors} = \prod_{\alpha \in A} S_\lambda$$

The sensor's data are defined as follows:

V = {v1, v2,..., vn} – sensor data value in numeric.
D = {d1, d2,..., dn} – sensor type. This field can be one of the following values: input temperature sensor; output temperature sensor; input or output pressure sensor; input or output flow velocity sensor; It shows what type sensor value is stored in V.

UUID = {uuid1, uuid2,..., uuidp} – sensor's unique identifier.
L = {*latitude, longitude*} – GPS-coordinates, where the sensor is located.
T = {t1, t2,..., tn} – a timestamp of a measurement.

The following formulas show semantic computing a leak in a DH network:

$$\theta_T = \begin{cases} 1 & T - T_{\min} < 0 \\ 0 & T - T_{\min} \geq 0 \end{cases}, \theta_P = \begin{cases} 1 & P - P_{\min} < 0 \\ 0 & P - P_{\min} \geq 0 \end{cases}, \theta_{FV} = \begin{cases} 1 & FV - FV_{\min} < 0 \\ 0 & FV - FV_{\min} \geq 0 \end{cases}$$

Where $T, T_{min}, P, P_{min}, FV, FV_{min}$ represent temperature, pressure, flow velocity, them minimal deviation, θ_T, θ_P and θ_{FV} represent heat leak and water, respectively. Then if $\theta_T = 0$, $\theta_P = 0$ and $\theta_{FV} = 0$ then the DH network works correctly, or if θ_{FV} and θ_P equal 1, then the DH network has a water leak. Otherwise, the DH network has a heat leak.

Sensors installed at the test points measure the temperature in the supply and return pipes. For example, an alarm signal on the $T1 - T1_{min}$ means there is an emergency in the supply pipes (a pipe coming from the heating center). An alarm signal on the $T2 - T2_{min}$ means there is an emergency in the return pipes (a tube that goes into the heating center). If there is an outage in the network, all network branches' sensors begin to register deviations from the specified parameters. When there is a deviation in the temperature readings in the supply pipe and return pipe, respectively

$$\gamma_1 = T1 - T1_{\min}, \gamma_2 = T2 - T2_{\min}$$

Deviation of pressure readings in the supply and return pipes, respectively

$$\delta_1 = P1 - P1_{\min}, \delta_2 = P2 - P2_{\min}$$

Deviation of flow rate readings in the supply and return pipes, respectively

$$\varepsilon_1 = FV1 - FV1_{\min}, \varepsilon_2 = FV2 - FV2_{\min}$$

4.2 Thermal Images Data

Very often, finding hidden leaks underground is a difficult task. DH network condition monitoring sensors can be located at a great distance from each other. This makes it difficult to find the leak. This study uses the aerothermography method. As a result, a heat map of the surface is obtained. It is possible to find thermal anomalies associated with the DH network, probably water leakage from the DH network.

As a result of image processing, the spatial coordinates of the boundaries of thermal anomalies (spatial objects) are calculated. Further, each received object is checked for intersection with the problematic section of the DH network. As a result, thermal anomalies remain, which are spatially related to the DH network.

The next step is to determine the location of a leak in the pipe. The heated spot can reach large sizes, up to several tens of meters [11]. It is necessary to calculate the hole coordinates in the pipe as accurately as possible to minimize the useless work of digging a section of the pipe. This study uses finding the center of a blob (centroid) algorithm to determine the likely location of a pipe leak. A blob is a group of linked pixels in the image that shares some common color (e.g., grayscale value). The centroid of a shape is the arithmetic mean (i.e., the average) of all the points in a shape. Suppose a shape consists of n distinct points $x_1 \ldots x_n$, then the centroid is given by

$$c = \frac{1}{n} \sum_{i=1}^{n} x_i$$

In the context of image processing, each shape has been made by pixels, and the centroid is the weighted average of all the pixels of the shape. It is possible to find the center of the drop using the moments of the image. An image moment is a specific weighted average of the intensity of pixels in an image. To find the centroid of an image, you need to convert the image to binary and then find its center using the formula. The formula determines the centroid:

$$C_x = \frac{M_{10}}{M_{00}}, C_y = \frac{M_{01}}{M_{00}}$$

C_x is the x coordinate and C_y is the y coordinate of the centroid, and M denotes the moment [12]. This algorithm is a set of spatial coordinates of the proposed leakage location in the pipe. Figure 4 shows the result of the calculations. An orange cross indicates the suspected leak at the thermal anomaly spot (Fig. 5).

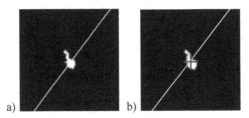

a)   b)

Fig. 5. a) Thermal anomaly spot with pipeline and b) detected place of the hole (red cross) in a pipe, thermal spot, and pipeline (Color figure online)

4.3 DH Network States

Any industrial district heating network is designed based on accepted state standards. One of the standard requirements is the availability of calculated parameters for the normal functioning of the network. The parameters calculated for the DH network sensors are expected values. The sensors are monitoring those values. The locations of the sensors are the control points of the district heating network (Fig. 2). This data is available in the DH network database. The database contains all information about the DH network. These data sets assign DH network states. The DH network has three states: working correctly; has heat leakages; has water leakage;

- The DH network has a correct working state when all sensor values are within Δ.
- The DH network has a heat leakages state when values of water temperature sensors register less temperature than T_{min}. This statement is true for both input and output pipelines.
- The DH network is in the water leakages state when values of pressure and flow velocity sensors have deviations than P_{min} and FV_{min}. This statement is true for both input and output pipelines.

5   Application of STEM to the DH Network

Each segment of the DH network is assessed according to an emergency, problematic state, and good condition. For example, if there is a hole in the pipeline in a network section, the central heating network goes into an emergency state. The DH network becomes a problematic state when the DH network doesn't have water leakages but has thermal insulation problems, heat loss. If a section or the entire DH network is within acceptable values, such a segment or the whole DH network is in a proper state. The number of states the DH network lands in depends on the state of the event's influence. In the STEM method and for this research, only the emergency state is considered, defined as water leakage, and other conditions are not emergencies.

5.1   Understanding the Semantic of Context Regarding an Event

Each sensor is installed in a specific area according to the design documentation of the central heating network. Accordingly, the spatial coordinates of the sensor are already known. It is also possible to select all thermal anomalies in the thermograms by checking their intersection with a given radius r using the sensors' coordinates within alarms getting activated. Thus, a spatial relationship is obtained between the emergency zone and the temperature anomaly on the thermogram. Table 1 presents the semantic matrix that maps to determine sensory data's meaning to an emergency event. Heat leak has 0 anomaly type because only a water leak is an emergency.

Table 1. Matrix results for semantic corresponding to an event context

Event	γ	δ	ε	Anomaly type
Heat leak	1	0	0	0
Water leak	1	1	1	1
No issues	0	0	0	0

6 Conclusions

The STEM approach is novel; it uses the fact model to associate an emergency to sensor and thermal image data semantics to locate the DH network's water leakages. The approach integrates thermal images to the DH networks sensor data to produce location coordinates of water leakages in pipelines. Both sensor data and thermal image data are necessary to identify water leakage emergencies precisely. It is not that one could compensate for the other. The STEM results are promising and significant. The location coordinates of leakages are within the margin of error of the DH network maintenances log location coordinates of water leakages. The main requirement for using the STEM method is data available on the DH network and its parameters. More sensors on the network help narrow down the initial location of the crash. If there are only two sensors in the network at the heating plant, it means that it is necessary to analyze the entire network using a thermal camera. The time estimate for finding a leak depends on the length of the survey area. The time spent searching in this study was approximately 40–50 min: flying around the area with a leak and photographing; assembly of a heat map and binding to the coordinates of the central heating network; search for leaks.

The precision of water leakage location coordinates' needs to improve before introducing daily use at DH networks even though the STEM approach is successful. Therefore, future research concentrates on fine-tuning the image processing technique and matching water leak coordinates with DH maintenance log data. Also, refinements are required of the UAV thermal imaging technique. Future research will also include the visualization of emergency events and pipelines' ware and tear status in a real-time Spatio-temporal map.

References

1. Sernhed, K., Jönsson, M.: Risk management for maintenance of district heating networks. Energy Procedia **116**, 381–393 (2017). https://doi.org/10.1016/j.egypro.2017.05.085
2. Malm, A., Hosrtmark, A., Jansson, E., Larsson, G., Meyer, A., Uusijärvi, J.: Handbok I förnyelseplanering av VA-ledningar (In English: Handbook for renovation planning of sewerage) Stockholm (2011). http://vav.griffel.net/filer/Rapport_2011-12.pdf
3. Chan, T.K., Chin, C.S., Zhong, X.: Review of current technologies and proposed intelligent methodologies for water distributed network leakage detection. IEEE Access **6**, 78846–78867 (2018). https://doi.org/10.1109/ACCESS.2018.2885444
4. Triapitcin, I., Dahanayake, A., Thalheim, B.: A sensing, processing, and analytical actuation functions semantic computing method for district heating network analysis (2020)
5. Kiyoki, Y., Kitagawa, T., Hayama, T.: A metadatabase system for semantic image search by a mathematical model of meaning. ACM SIGMOD Rec. **23**(4), 34–41 (1994). https://doi.org/10.1145/190627.190639
6. Kiyoki, Y., Ishihara, S.: A semantic search space integration method for meta-level knowledge acquisition from heterogeneous databases. Inf. Model. Knowl. Bases **14**, 86–103 (2003)
7. Kiyoki, Y., Chen, X.: Contextual and differential computing for the multi-dimensional world map with context-specific spatial-temporal and semantic axes. Inf. Model. Knowl. Bases XXV **260**, 82 (2014). https://doi.org/10.3233/978-1-61499-361-2-82

8. Rungsupa, S., Chawakitchareon, P., Hansuebsai, A., Sasaki, S., Kiyoki, Y.: Photographic assessment of coral stress: effect of low salinity to Acropora sp. Goniopora sp. and Pavona sp. at Sichang Island, Thailand. Inf. Model. Knowl. Bases XXIX **301**, 137 (2018). https://doi.org/10.3233/978-1-61499-834-1-137

9. Dubnov, S., Burns, K., Kiyoki, Y.: A 'Kansei' multimedia and semantic computing system for cross-cultural communication. In: 7th IEEE International Conference on Semantic Computing, Keynote Speech, pp. 1–20 (2016)

10. Sasaki, S., Kiyoki, Y.: Analytical visualization function of 5D world map system for multi-dimensional sensing data. Inf. Model. Knowl. Bases **29**, 71–89 (2018). https://doi.org/10.3233/978-1-61499-834-1-71

11. Lah, A.A.A., Dziyauddin, R.A., Yusoff, N.M.: Localization techniques for water pipeline leakages: a review. Appl. Therm. Eng. (2018)

12. OpenCV documentation (2021). https://docs.opencv.org/3.1.0/dd/d49/tutorial_py_contour_features.html

ADBIS 2021 Workshop: Intelligent Data – From Data to Knowledge – DOING

DOING: Intelligent Data – From Data to Knowledge

Mírian Halfeld Ferrari[1] and Carmem S. Hara[2]

[1] Université d'Orléans, INSA CVL, LIFO EA, France
[2] Universidade Federal do Paraná, Brazil

Description. Texts are important sources of information and communication in diverse domains. The intelligent, efficient, and secure use of this information requires, in most cases, the transformation of unstructured textual data into data sets with some structure, organized according to an appropriate schema that follows the semantics of an application domain. Indeed, solving the problems of modern society requires interdisciplinary research and information cross-referencing, thus surpassing the simple provision of unstructured data. There is a need for representations that are more flexible, subtle, and context-sensitive, which can also be easily accessible via consultation tools and evolve according to these principles. In this context, consultation requires robust and efficient processing of queries, which may involve information analysis, quality assessments, consistency checking, and privacy preservation. Knowledge bases can be built as new generation infrastructures to support data science queries with a user-friendly framework. They can provide the required machinery for advised decision-making.

The 2nd Workshop on Intelligent Data - From Data to Knowledge (DOING 2021) focused on transforming data into information and then into knowledge. The workshop gathers researchers from natural language processing (NLP), databases (DB), and artificial intelligence (AI). This edition featured works in two main areas: (1) information extraction from textual data and its representation on knowledge bases and (2) intelligent methods for handling and maintaining these databases. Overall, the purpose of the workshop was to focus on all aspects concerning modern infrastructures to support these areas, giving particular, but not sole, attention to data on health and environmental domains.

DOING 2021 received nine submissions, out of which three were accepted as full papers and three as short papers, resulting in an acceptance rate of 50%. Each paper received three reviews from members of the Program Committee.

This workshop is an event supported by the French network MADICS[1]. More specifically, it is an event of the action DOING[2] within MADICS and of the DOING working group in the regional network DIAMS[3]. The workshop is the result of the collective effort of a large community, which we gratefully acknowledge. We thank the ADBIS 2021 conference chairs, who worked hard to support the workshop

---

[1] https://www.madics.fr/.

[2] https://www.madics.fr/actions/doing/.

[3] https://www.univ-orleans.fr/lifo/evenements/RTR-DIAMS/.

organization. We are also grateful to the members of the Program Committee, who did an outstanding job, providing timely and thoughtful reviews. Finally, we are grateful to the authors who submitted their work to DOING 2021.

Selected Papers. Three full papers were presented at DOING 2021. The first paper, entitled "LACLICHEV: Exploring the History of Climate Change in Latin America within Newspapers", proposes a platform to expose and study the history of climate change in Latin America. LACLICHEV combines data collection exploration techniques with information retrieval, data analytics, and geographic querying and visualization. It aims at understanding newspaper contents and at identifying patterns in order to build a climatologic event history.

The second paper is entitled "COVID-19 Portal: Machine learning techniques applied to the analysis of judicial processes related to the pandemic". COVID-19 Portal presents quantitative and qualitative data about texts of decisions or lawsuits addressed to the Supreme Federal Court in Brazil. Three NLP-based components are employed in the portal: word frequencies, context graphs, and clustering of legal documents. It aims at helping lawyers develop arguments for judicial processes related to the pandemic.

The third paper, "Standard Matching-Choice Expressions for Defining Path Queries in Graph Databases", proposes standard matching-choice expressions and uses them as the basis for the specification of non-regular languages for querying graph databases.

Three short papers completed the program of DOING 2021. "The Formal-Language-Constrained Graph Minimization Problem" focuses on computing the minimal sub-graph that preserves utility queries. The goal is to formalize this minimization problem and study its complexity. In "Public Health Units - Exploratory analysis for decision support", we find a study whose goal is to understand the dynamics of healthcare needs and usages in the city of Curitiba, Brazil. The paper also presents a prototype for data visualization. Finally, "Interpreting decision-making process for multivariate time series classification" proposes the combination of classifiers for univariate time series, and explainability methods for these classifiers, as a way to address the question of modeling explainability in multivariate time series problems.

Organization

Chairs

Mírian Halfeld Ferrari   Université d'Orléans, INSA CVL, LIFO EA,
France

Carmem S. Hara   Universidade Federal do Paraná, Brazil

Program Committee

Cheikh Ba   Université Gaston Berger, Senegal
Karin Becker   Universidade Federal do Rio Grande do Sul,
Brazil
Javam de Castro Machado   Universidade Federal do Ceará, Brazil
Laurent d'Orazio   IRISA, Université de Rennes, France
Vasiliki Foufi   University of Geneva, Switzerland
Michel Gagnon   Polytechnique Montréal, Canada
Sven Groppe   University of Lubeck, Germany
Mingda Li   Pinterest, USA
Jixue Liu   University of South Australia, Australia
Anne-Lyse Minard-Forst   LLL, Université d'Orléans, France
Damien Novel   ERTIM, INALCO, France
Fathia Sais   LRI Université Paris-Saclay, France
Roberto Santana   University of the the Basque Country, Spain
Agata Savary   LIFAT, Université de Tours, France
Rebecca Schroeder Freitas   Universidade Estadual de Santa Catarina, Brazil
Aurora Trinidad Ramirez Pozo   Universidade Federal do Paraná, Brazil

Standard Matching-Choice Expressions for Defining Path Queries in Graph Databases

Ciro Medeiros[1,2(✉)], Umberto Costa[1], and Martin Musicante[1]

[1] Universidade Federal do Rio Grande do Norte, Natal, Brazil
cirommed@ppgsc.ufrn.br, {umberto,mam}@dimap.ufrn.br
[2] University of Orléans, Orléans, France
ciro.morais-medeiros@etu.univ-orleans.fr

Abstract. In the context of graph databases, regular expressions are usually the basis for the definition of property paths. However, regular expressions are not enough to specify some important properties, such as same generation queries. In this paper, we introduce *Standard Matching-Choice Expressions* and use them as the basis for the specification of non-regular languages for querying graph databases. SM Expressions allow us to specify path queries in terms of a meaningful subset of context-free languages. We define the syntax and semantics of SM Expressions, and we formalize the translation of SM Expressions into a set of rules of context-free grammars. The translation of SM Expressions into context-free grammars allow us to use a more natural specification language on the existing implementations for non-regular path queries. Finally, we illustrate the usage of SM Expressions in the context of another work that supports non-regular path queries for graph databases.

Keywords: Graphs databases · Non-regular path queries · Matching-choice sets

1 Introduction

In a graph database, the entities of data can be seen as vertices of the graph, whereas the relationships among those entities correspond to the edges of the graph. The data in a graph database is usually represented as sets of triples, written in RDF [13] (Resource Description Framework). Each subject-predicate-object triples (s, p, o) defines how the vertex/resource s relates to the vertex/resource o by means of an edge/property p. In this manner, sequences of edges connecting source and target vertices can be used to query graph databases. This kind of route is described by *Property Paths*. SPARQL [9] is the standard language used to specify path queries over graph databases.

SPARQL queries use *Regular Expressions* to define property paths [12]. However, Regular Expressions are not enough to specify some important properties

© Springer Nature Switzerland AG 2021
L. Bellatreche et al. (Eds.): ADBIS 2021 Short Papers, Workshops and Doctoral Consortium,
CCIS 1450, pp. 97–108, 2021.
https://doi.org/10.1007/978-3-030-85082-1_9

over graph databases, such as same generation queries [1], where one searches for nodes equidistant with respect to a central node. Queries of this kind are known as *Context-Free Path Queries* (CFPQs) [4]. In Sect. 4 we present several examples of queries in this class. The extension of expressiveness has been investigated in the context of graph databases to encompass CFPQs [5]. The majority of such initiatives concentrates on the implementation of CFPQs, treating the definition of query languages as a secondary concern.

The definition of specification languages for CFPQs is important since it impacts the practical usage of graph databases. A few works have been driven to the proposal of languages for the definition of non-regular query languages. In one extension[1] of [2] the authors introduce context-free grammars in the query language, allowing the usage of non-terminal symbols in the writing of property paths. In [7], authors extend SPARQL in a similar way. Another extension of SPARQL is found in [8], where authors extend Regular Expressions with *Nested Regular Expressions*. Although these initiatives have contributed to the pragmatic of the query languages, they require the knowledge of context-free grammars, which are not as simple and well-known as Regular Expressions, preventing a larger adoption of such proposals as specification languages.

In this paper, we propose *Standard Matching-Choice Expressions (SM Expressions)* and use them as the basis for the specification of non-regular languages for querying graph databases. Our proposal is inspired by the families of context-free languages presented in [14]. Specifically, our SM Expressions allows us to specify path queries in terms of languages of the family of *Standard Matching-Choice Sets* [14]. The family of Standard Matching-Choice Sets is a meaningful subset of context-free languages. We define the syntax and semantics of SM Expressions, and we formalize the translation of SM Expressions into a set of rules of a context-free grammar. The translation of SM Expressions to context-free grammars fills the gap between a more natural specification language for querying graph databases and the existing query engines for CFPQs.

This paper is organized as follows. Section 2 gives an overview of the Matching-Choice Sets presented in [14], as well as presents their basic operators. In Sect. 3 we present the syntax and semantics of our Standard Matching-Choice Expressions, together with the mechanism proposed to translate them into context-free grammars. In Sect. 4 we show how SM Expressions can be used as part of the query language adopted in the implementation proposed in [6]. Section 5 is devoted to the presentation of related work. Finally, we conclude the paper in Sect. 6 with important remarks and pointing directions for future work.

2   Regular Matching Choice Languages

In [14], the author defines an inclusion hierarchy of classes of context-free languages that can be defined using operations over sets of strings. Given an alphabet Σ, the proposal is based on the definition of two sets: (*i*) A set of pairs of strings $S \subseteq \Sigma^* \times \Sigma^*$ and (*ii*) A set of strings $C \subseteq \Sigma^*$.

---

[1] https://github.com/thobe/openCypher/blob/rpq/cip/1.accepted/CIP2017-02-06-Path-Patterns.adoc#153-compared-to-context-free-languages.

The set of pairs S can be seen as a generalization of open and close parentheses. The languages built from these sets contain strings $\alpha\,\beta\,\gamma$ such that $(\alpha,\gamma) \in S$ and $\beta \in C$. Given sets of pairs of strings S_1 and S_2, Yntema introduces the notion of *Matching-Choice Sets* defined over S_1 and S_2 as follows:

$$S_1 \oplus S_2 = \{(x,y) \mid (x,y) \in S_1 \vee (x,y) \in S_2\}; \tag{1}$$

$$S_1 S_2 = \{(xz, wy) \mid (x,y) \in S_1 \wedge (z,w) \in S_2\}; \tag{2}$$

$$S_1* = \{(\varepsilon, \varepsilon)\} \cup \{(x_1 \cdots x_n, y_n \cdots y_1) \mid (x_i, y_i) \in S_1, i \leq n, n \in \mathbb{N}^+\} \tag{3}$$

Notice that these operations define pairs of matching strings. They will be used as enclosing parentheses in the next definition. Given a matching choice set S (*i.e.*, a set of pairs built using the operations above) and a set of strings C, the *Matching-Choice Language* $S \circ C$ is defined as: $S \circ C = \{xzy \mid (x,y) \in S \wedge z \in C\}$.

Given a finite number of sets of strings A_i, B_i $(1 \leq i \leq n)$, we can build a set S of pairs of strings by *(i)* defining the sets of pairs $A_i \times B_i = S_i$, and *(ii)* recursively applying the union, sequence and closure operations over the sets S_i. The sets of strings A_i, B_i and C are called the *underlying sets* of the expression $S \circ C$. Notice that the elements in S can be seen as pairs of matching parentheses while the strings in C correspond to strings placed between parentheses. The union, sequencing and star operations ensure that the set $S \circ C$ is formed by strings containing well-formed parenthesized expressions.

Yntema [14] defines the class of *Standard Matching-Choice Languages*, a proper subset of context-free languages that can be described using the above-mentioned operations.

In the next section, we define a set of expressions to denote Standard Matching Choice Sets. We call these expressions Standard Matching Expressions.

3   Standard Matching Choice Expressions

In this section, we propose Standard Matching-Choice (SM) Expressions to define Standard Matching-Choice Languages [14]. We provide the semantics of these expressions by means of Standard Matching-Choice Languages and show that the class of languages defined by SM Expressions is exactly the class of Regular Matching Choice Sets. We also provide a way of defining context-free grammars that generate the language denoted by an SM Expression.

Regular expressions are the basis for the definition of SM Expressions. Given an alphabet $\Sigma = \{t_1, \ldots, t_n\}$, the set of regular expressions over Σ is the set of strings defined by $R \rightarrow () \mid t_1 \mid \ldots \mid t_n \mid (R) \mid RR \mid R \mid R \mid R*$.

Definition 1 (Syntax of SM Expressions). *The set of Standard Matching Choice Expressions over an alphabet Σ is inductively defined as follows:*

1. Any regular expression E over Σ is an SM Expression.

2. *If E, E', E_0 are SM Expressions over Σ, then $<E>E_0<E'>$ is an SM Expression:*

$$\frac{E, E_0, E' \text{ are SM Expressions}}{<E>E_0<E'> \text{ is an SM Expression}}$$

This case defines the base case of SM Expressions.

3. *If $<P_1>E_0<P_1'>$ and $<P_2>E_0<P_2'>$ are SM Expressions, then $<P_2.P_1>E_0<P_1'.P_2'>$ is an SM Expression.*

$$\frac{<P_1>E_0<P_1'> \text{ is an SM Expression} \qquad <P_2>E_0<P_2'> \text{ is an SM Expression}}{<P_2.P_1>E_0<P_1'.P_2'> \text{ is an SM Expression}}$$

This case defines SM Expressions that contains sequences of nested parentheses.

4. *If $<P_1>E_0<P_1'>$ and $<P_2>E_0<P_2'>$ are SM Expressions, then $<P_2+P_1>E_0<P_1'+P_2'>$ is as SM Expression.*

$$\frac{<P_1>E_0<P_1'> \text{ is an SM Expression} \qquad <P_2>E_0<P_2'> \text{ is an SM Expression}}{<P_2+P_1>E_0<P_1'+P_2'> \text{ is an SM Expression}}$$

The SM Expressions defined above contain choice of parentheses.

5. *If $<P>E_0<P'>$ is an SM Expression, then $<\,:\,P\,:\,>E_0<\,:\,P'\,:\,>$ is an SM Expression.*

$$\frac{<P>E_0<P'> \text{ is an SM Expression}}{<\,:\,P\,:\,>E_0<\,:\,P'\,:\,> \text{ is an SM Expression}}$$

The pair of operators $:\,\_\,:$ will denote a Kleene star operator over matching parentheses.

Notice that the strings P_i in the cases above are, in general, not *SM expressions, but they denote a language of opening parentheses. Analogously, P_i' represent sets of closing parentheses. These strings may contain characters like ".", "+" and ":", that are used to build the languages that will denote opening and closing parentheses.*

The next definition establishes the semantics of SM Expressions as a set of strings over an alphabet Σ. The language defined for each expression is a Standard Matching-Choice Set [14].

Definition 2 (Semantics of SM Expressions). *The language $\mathcal{L}(M)$, denoted by an SM Expression M is inductively defined by the following rules:*

1. *The Standard Matching Choice language defined by a regular expression is the (regular) language denoted by this expression:*

$$\overline{\mathcal{L}(R) = \mathcal{L}_{RE}(R)}$$

The function $\mathcal{L}_{RE}(R)$ defines the language denoted by a regular expression. Notice that this language is an Standard Matching-Choice set of rank 0 (according to the definition in [14]).

2. *The language defined by an SM Expression $<E>E_0<E'>$ over Σ is defined in terms of the languages defined by the SM Expressions E, E', and E_0, as follows:*

$$\frac{L = \mathcal{L}(E), \qquad L_0 = \mathcal{L}(E_0), \qquad L' = \mathcal{L}(E')}{\mathcal{L}(<E>E_0<E'>) = (L \times L') \circ L_0}$$

The languages L, L' and L_0 are the underlying sets of $\mathcal{L}(<E>E_0<E'>)$. If n is the maximum rank among the ranks of L, L' and L_0, then $\mathcal{L}(<E>E_0<E'>)$ is a Standard Matching-Choice set of rank $n + 1$.

3. *The language denoted by an SM Expression $<P_2.P_1>E_0<P_1'.P_2'>$ containing sequences of nested parentheses is defined as follows:*

$$\frac{(S_1 \times S_1') \circ C = \mathcal{L}(<P_1>E_0<P_1'>), \qquad (S_2 \times S_2') \circ C = \mathcal{L}(<P_2>E_0<P_2'>)}{\mathcal{L}(<P_2.P_1>E_0<P_1'.P_2'>) = (S_2\ S_1 \times S_1'\ S_2') \circ C}$$

where $S_i S_j$ represents the concatenation of languages S_i and S_j.

4. *The language denoted by an SM Expression containing choices of parentheses is defined as:*

$$\frac{(S_1 \times S_1') \circ C = \mathcal{L}(<P_1>E_0<P_1'>), \qquad (S_2 \times S_2') \circ C = \mathcal{L}(<P_2>E_0<P_2'>)}{\mathcal{L}(<P_2 + P_1>E_0<P_1' + P_2'>) = ((S_1 \times S_1') \oplus (S_2 \times S_2')) \circ C}$$

It is easy to verify that

$$((S_1 \times S_1') \oplus (S_2 \times S_2')) \circ C = ((S_1 \times S_1') \circ C) \cup ((S_2 \times S_2') \circ C).$$

5. *The language denoted by an SM Expression containing the synchronized repetition of matching parentheses is given by the rule:*

$$\frac{(S \times S') \circ C = \mathcal{L}(<P>E_0<P'>)}{\mathcal{L}(<: P : >E_0<: P' : >) = (S \times S') * \circ C}$$

It is easy to see that, for all SM Expressions M, the set $\mathcal{L}(M)$ is an SM language. The following property states that every language denoted by an SM Expression is a Standard Matching-Choice set [14].

Property 1 ($\mathcal{L}(SMExp) \subseteq SMCLang$). Given an SM Expression M, there exists an SM language Y such that $\mathcal{L}(M) = Y$.

Proof. This result is immediate, by structural induction on SM Expressions M and the languages $\mathcal{L}(M)$ defined for them. ∎

The next property shows that there exists at least one SM Expression that defines each Standard Matching-Choice set.

Property 2 ($SMCLang \subseteq \mathcal{L}(SMExp)$). Given an Standard Matching-Choice language L [14], there exists an SM Expression M such that $\mathcal{L}(M) = L$.

Proof. The proof of this property proceeds in two cases:

Case 1: L is a regular language. In this case, there exists a regular expression R such that $\mathcal{L}_{RE}(R) = L$.

In this case, the SM Expression that defines L is also R (see Definition 2).

Case 2: L is a Standard Matching Choice set of rank $n > 0$. In this case we have that $L = T \circ C$, where T and C are built from Standard Matching Choice sets of rank of at most n.

The proof of this case is by induction on the rank of L:

Base Case ($n = 1$), we have that the underlying sets from which $L = T \circ C$ is built are regular languages.

In this case, we have that $L = (L_1 \times L_2) \circ C$, being L_1, L_2, C regular languages. Let be R, R', R_0, respectively, regular expressions that define L_1, L_2, and C. By using cases (2.) and (1.) of Definition 2, we can see that $L = \mathcal{L}(\texttt{<}R\texttt{>}R_0\texttt{<}R'\texttt{>})$.

Inductive Hipothesis ($n \leq k$). Let us suppose that for any Standard Matching Choice set L of rank $n \leq k$ there exists an SM Expression E such that $L = \mathcal{L}(E)$.

Inductive case ($n = k + 1$).

Since L is a Standard Matching-Choice set of rank $n > 1$, then $L = T \circ C$, such that T and C are built using Standard Matching Choice sets of ranks less than n. In this way, we can proceed by cases on the construction of T:

Suppose the Standard Matching Choice sets $L_1 = T_1 \circ C$ and $L_2 = T_2 \circ C$. By the induction hypothesis, there exist SM Expressions $S_1 = \texttt{<}P_1\texttt{>}S_0\texttt{<}P_1'\texttt{>}$ and $S_2 = \texttt{<}P_2\texttt{>}S_0\texttt{<}P_2'\texttt{>}$ such that $\mathcal{L}(S_1) = T_1 \circ C$ and $\mathcal{L}(S_2) = T_2 \circ C$.

Case $T = (T_1\ T_2)$: We need to build an SM Expression for $(T_1\ T_2) \circ C$. By Definition 2.3, it is easy to see that the language associated to $M = \texttt{<}P_1.P_2\texttt{>}S_0\texttt{<}P_2'.P_1'\texttt{>}$ is indeed $(T_1\ T_2) \circ C$.

Case $T = (T_1 \oplus T_2)$: By Definition 2.4, it can be verified that the language associated to the SM Expression $M = \texttt{<}P_2 + P_1\texttt{>}S_0\texttt{<}P_1' + P_2'\texttt{>}$ is $(T_1 \oplus T_2) \circ C$.

Case $T = T_1*$: We need to build an SM Expression for $T* \circ C$. By Definition 2.5, it is easy to verify that $\mathcal{L}(\texttt{<} : P_1 : \texttt{>}R_0\texttt{<} : P_1' : \texttt{>}) = T* \circ C$. ∎

These results allow us to enunciate the following property:

Corollary 1. *The class of SM Sets is denoted by SM Expressions.*

Proof. Immediate from properties 1 and 2. ∎

Let us now define context-free grammars to generate the language denoted by an SM Expression. The function φ_{SM} takes an SM Expression and returns a set of production rules. For the sake of clarity, we denote as $\{S \to \alpha, \dots\}$ a set of production rules of a grammar such that the symbol S is the start non-terminal. We also use the function φ_{RE}, to obtain a grammar from a regular expression.

Definition 3 (Obtaining a Grammar from SM Expressions). *We inductively define the function φ_{SM}, taking an SM expression M and producing a context-free grammar G, generating the language $\mathcal{L}(M)$.*

The rules for the starting symbol of the grammars obtained below have the right-hand side consisting of three non-terminal symbols (for prefix, inner, and suffix expressions).

1. *Given a Regular Expression R_1 and a grammar $G_1 = \{S_1 \rightarrow \alpha, \ldots\}$, such that $\mathcal{L}(G_1) = \mathcal{L}_{RE}(R_1)$, then*

$$\frac{\varphi_R(R_1) = \{S_1 \rightarrow \alpha, \ldots\}, \qquad S, X, Y \text{ are new} \\ G = \{S \rightarrow XS_1Y, \ X \rightarrow \varepsilon, \ Y \rightarrow \varepsilon\} \cup G_1}{\varphi_{SM}(R_1) = G}$$

2. *The grammar defined for the combinations of SM Expressions can be obtained by the rule:*

$$\frac{\varphi_{SM}(E) = \{S_1 \rightarrow \alpha, \ldots\}, \qquad \varphi_{SM}(E') = \{S_2 \rightarrow \beta, \ldots\}, \\ \varphi_{SM}(E_0) = \{S_0 \rightarrow \gamma, \ldots\} \qquad S \text{ is new} \\ G = \{S \rightarrow S_1 S_0 S_2\} \cup \varphi_{SM}(E) \cup \varphi_{SM}(E') \cup \varphi_{SM}(E_0)}{\varphi_{SM}(\texttt{<}E\texttt{>}E_0\texttt{<}E'\texttt{>}) = G}$$

3. *Translation for SM Expressions containing sequences of nested parentheses:*

$$\frac{\begin{array}{c} G_1 = \varphi_{SM}(\texttt{<}P_1\texttt{>}E_0\texttt{<}P_1'\texttt{>}) = \{S_1 \rightarrow XS_0Y, \ldots\} \\ G_2 = \varphi_{SM}(\texttt{<}P_2\texttt{>}E_0\texttt{<}P_2'\texttt{>}) = \{S_2 \rightarrow WS_0Z, \ldots\} \\ G_1' = G_1 - \{w \mid w = S_1 \rightarrow \delta \in G_1\} \\ G_2' = G_2 - \{w \mid w = S_2 \rightarrow \theta \in G_2\} \qquad S, S_1', S_2' \text{ are new} \\ G_3 = \{S_1' \rightarrow WX \mid S_1 \rightarrow XS_0Y \in G_1, \ S_2 \rightarrow WS_0Z \in G_2\} \\ \cup \{S_2' \rightarrow YZ \mid S_1 \rightarrow XS_0Y \in G_1, \ S_2 \rightarrow WS_0Z \in G_2\} \\ G = \{S \rightarrow S_1' S_0 S_2'\} \cup G_1' \cup G_2' \cup G_3 \end{array}}{\varphi_{SM}(\texttt{<}P_2.P_1\texttt{>}E_0\texttt{<}P_1'.P_2'\texttt{>}) = G}$$

4. *The grammar for SM Expressions containing the choice operator is:*

$$\frac{\begin{array}{c} G_1 = \varphi_{SM}(\texttt{<}P_1\texttt{>}E_0\texttt{<}P_1'\texttt{>}) = \{S_1 \rightarrow XS_0Y, \ldots\} \\ G_2 = \varphi_{SM}(\texttt{<}P_2\texttt{>}E_0\texttt{<}P_2'\texttt{>}) = \{S_2 \rightarrow WS_0Z, \ldots\} \\ G_1' = G_1 - \{w \mid w = S_1 \rightarrow \delta \in G_1\} \\ G_2' = G_2 - \{w \mid w = S_2 \rightarrow \theta \in G_2\} \qquad S \text{ is new} \\ G_3 = \{S \rightarrow XS_0Y \mid S_1 \rightarrow XS_0Y \in G_1\} \cup \{S \rightarrow WS_0Z \mid S_2 \rightarrow WS_0Z \in G_2\} \\ G = G_1' \cup G_2' \cup G_3 \end{array}}{\varphi_{SM}(\texttt{<}P_2 + P_1\texttt{>}E_0\texttt{<}P_1' + P_2'\texttt{>}) = G}$$

5. *The grammar for SM Expressions containing the closure operator*

$$\frac{\begin{array}{c} G_1 = \varphi_{SM}(\texttt{<}P\texttt{>}E\texttt{<}P'\texttt{>}) = \{S_1 \rightarrow XS_0Y, \ldots\} \\ G_1' = G_1 - \{w \mid w = S_1 \rightarrow \delta \in G_1\}, \qquad S, W, Z \text{ are new} \\ G_2 = G_1' \cup \{S \rightarrow WS_0Z\} \cup \{W \rightarrow \varepsilon, Z \rightarrow \varepsilon\} \\ G = G_2 \cup \{S \rightarrow XS\,Y \mid S_1 \rightarrow XS_0Y \in G_1\} \end{array}}{\varphi_{SM}(\texttt{<}: P :\texttt{>}E\texttt{<}: P' : \texttt{>}) = G}$$

We now need to prove that the class of SM expressions (and of matching choice languages in [14]) is that of the languages defined by our grammars.

Property 3 (Standard Matching Choice Grammars and Languages). Given an SM Expression M, $\mathcal{L}(M) = \mathcal{L}(\varphi_{SM}(M))$.

Proof. By Rule Induction on M. ∎

4   Defining Query Patterns with SM Expressions

In this section, we show how SM Expressions can be used to build non-regular property paths. We adapt rcfSPARQL [7] to support SM expressions without using a context-free grammar in the query.

Let us present the language by means of examples. The following example (adapted from [7]) illustrates the use of SM expressions in a query.

Same-Generation Queries. This kind of query [1] looks for nodes that are *(i)* equidistant to a common ancestor, and *(ii)* have a given property.

In Fig. 1, we depict a database D containing data about a company's employees. In the following we rewrite the query in [7] by using SM Expressions. This query selects employees having the same job, but different salaries:

```
1  SELECT ?job, ?emp1, ?sal1, ?emp2, ?sal2
2  FROM D
3  WHERE {
4      ?emp1 <: boss :><: boss :> ?emp2 .
5      ?emp1 job ?job .
6      ?emp2 job ?job .
7      ?emp1 salary ?sal1 .
8      ?emp2 salary ?sal2 .
9      FILTER (?sal1 > ?sal2)
10 }
```

This query defines a relation formed by 5-tuples. The variables at line (1) define the attributes of this relation. The property path at line (4) defines a path between employees (?emp1 and ?emp2). Notice that this path uses an SM Expression to look for paths between employees at the same level of the hierarchy. These paths will be formed by nested boss and $\overline{boss}$ edges. Notice that we use the $\overline{x}$ notation to express the inversion of an edge.

Lines (5–8) of the query look, respectively, for the jobs and salaries of each pair of employees identified in line (4). Notice that there is just one variable ?job, denoting that both employees have the same position in the company. Line (9) filters the pairs of employees that have the same job but different salaries.

The SM Expression at line 4 of the query defines paths in the graph initiating with employee nodes. Departing from these nodes, the answer is defined as those employee nodes connected to the initial one by a sequence of boss edges followed by a sequence of equal number of $\overline{boss}$, thus linking employees at the same level in the job hierarchy. We can verify that the SM Expression <: boss :><: $\overline{boss}$:> denotes the set $(\{boss\} \times \{\overline{boss}\}) * \circ \{\varepsilon\} = \{boss^n \ \overline{boss}^n \mid n \geq 0\}$. It is easy to check that the language generated by the SM Expression is the same as the one generated by the grammar $\{S \to boss \ S \ \overline{boss}, S \to \epsilon\}$.

Equal number of a's and b's. The language $\{\alpha \in \{a, b\}^* \mid \#_a(\alpha) = \#_b(\alpha)\}$ contains strings from an alphabet Σ such that the number of symbols a's is equal to the number of b's.

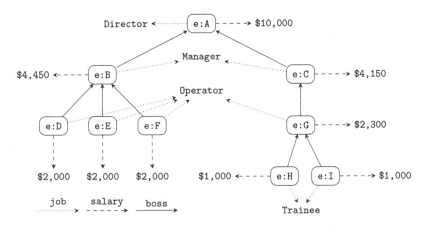

Fig. 1. Example of hierarchy database D [7].

The equivalent SM expression for this language is (<:a + b:><:a + b:>)*. Notice here the use of the choice operator, which ensures that for every a, there will be a b in some subsequent position of the input string or trace, and vice-versa. The Matching-Choice set describing this language is $(\{(a,b),(b,a)\} * \circ\{\epsilon\})^*$.

Double-Length Right Parentheses. The classic example of non-regular context-free language is the one presented in Example 4. A variation of that language is $b^n a b^{2n}$. This language is given by the Matching-Choice set $\{(b,bb)\} * \circ\{a\}$. The equivalent SM expression is <:b:>a<:bb:>.

Class/Type Hierarchy. The RDFS vocabulary is used to define the schema of an RDF database. The terms *type* and *subClassOf* from that vocabulary state, respectively, the type of a resource and a sub-class relationship between two types. The language of balanced pairs of *subClassOf* and *type* edges was defined in a previous work [16]. It retrieves concepts on the same level of a class/type hierarchy. This language is given by the Matching-Choice set $\{(sc,\overline{sc}),(t,\overline{t})\} * \circ\{sc\ \overline{sc}, t\ \overline{t}\}$. In terms of SM expressions, the same language can be expressed as <:sc + t:>(sc sc-)|(t t-)<:t- + sc-:>.

A similar language was also defined in that work. The language of balanced *subClassOf* edges, followed by an extra *subClassOf⁻* retrieves concepts on adjacent levels of the class hierarchy. This language is the same as the Matching-Choice set $(\{(sc,\overline{sc})\} * \circ\{\varepsilon\})\ \overline{sc}$. The equivalent SM expression is <:sc:><:sc-:>sc-.

Queries Over Synthetic Networks. The language $a^{n+1}b^m c^m d^{n+1}$ was defined by [5]. It was used to query synthetic social networks. It corresponds to the Matching-Choice set $a\{(a,d) * \circ(\{(b,c)\} * \circ\{\varepsilon\})\}d$. The equivalent SM Expression is a<:a:>(<:b:><:c:>)<:d:>d.

Recursion Inside Matching Parentheses. A more complicated language, presented in [6], is $(a^n \ b \ c^n)^k \ d \ (e^m \ f^m)^k$. It involves recursion inside the external matching parentheses. The Matching-Choice set $(\{(a,c)\}*\circ\{b\} \times \{(e,f)\}*\circ\{\varepsilon\})*$ $\circ\{d\}$ corresponds to that language. In terms of SM expressions, we can define that language as `<:(<:a:>b<:c:>):>d<:(<:e:><:f:>):>`.

5   Related Work

The design of syntactic expressions to define non-regular languages is a topic that has been studied for decades. Differently from the case of regular expressions and languages, there is no consensus about a notation that may be used to describe context-free languages or, at least, a subset of them. We briefly analyze some of the proposals that inspired the one in this paper.

Cap Expressions [15] define context-free languages by building expressions that contain placeholder symbols that may be replaced by the expression itself. If $P(\alpha)$ is an expression describing a language, such that it does not contain the symbol $\hat{a}$, the expression $\langle P(\hat{a})\rangle_\alpha$ is a cap expression that describes a language. The language described by $\langle P(\hat{a})\rangle_\alpha$ is formed by all the words obtained *(i)* from words of $P(\alpha)$ by any finite number of substitutions of α for words in $P(\alpha)$ and *(ii)* from words obtained in previous substitutions, also performing any number of substitutions of α for words in $P(\alpha)$.

Any context-free language can be defined by expressions using concatenation, union and Cap operations [15]. Even if Cap Expressions can be seen as a step towards a simple notation to describe context-free languages, their use is not popularized. This may be explained by the complicated syntax of the expressions and by their semantics obtained in terms of successive substitutions, which is difficult to describe due to the many possible results at each step in developing the result.

Another notation for the definition of non-regular languages are *Linear Expressions* [10]. This class of expressions use the $_L$ and $_R$ annotations on the terminal symbols of a regular expression, to indicate the appearance of that symbol in the left or right-hand side repeated portion of a string belonging to a context-free language. For instance, the linear expression $(a_L b_R b_R)^*$ describes the language $\{a^i b^{2i} \mid i \geq 0\}$. The language described by the linear expression $(a_L a_R + b_L b_R)^*$ is $\{ww^r \mid w \in \mathcal{L}(\ (a+b)* \ )\}$ [10].

Linear expressions are capable of expressing linear languages, a proper subset of context-free languages. To the extent of our knowledge, these expressions have not been used in query or programming languages.

Nested Regular Expressions [8] are an extension of regular expressions that define query paths in nSPARQL, an extension of the SPARQL graph database query language. Nested regular expressions include a "[_]" operator to define branching. For instance, the expression $a[b*]c$ describes all paths $a \ c$ in the graph, such that there exists a path $b*$ departing from the node of the graph that is reached after the prefix path a. In this way, nSPARQL property paths are defined by nested regular expressions and by the use of *navigational axis*, similar to those in XPATH [11].

Extended Regular Expressions [3] are an extension of regular expressions that use variables to refer to parts of the expression itself. These variables are also called backreferences. The notation describes a proper subset of context-sensitive languages that are not comparable with context-free languages. Despite their use in different implementations, Extended Regular Expressions are problematic, since there are proof of undecidability of some decision problems when they are used, even when they have just one variable.

A previous work [6] proposes *Recursive Expressions*, which were intended to describe languages with matching parentheses. These expressions are regular expressions, extended with a ternary operator to define pairs of matching parentheses around a core set of strings. This ternary operator is similar to the double colon operator ": _ :" defined in SM Expressions and represents the synchronized repetition of matching parentheses. Recursive Expressions define a class of non-regular languages that are a proper subset of context-free languages. Although these expressions are more expressive than regular expressions, they lack the capability of describing some useful languages. They also showed poor pragmatic properties, being difficult to read in case of large examples. Some of the problems of Recursive Expressions are solved by using SM Expressions.

To the extent of our knowledge, there is no previous attempt at the proposal of non-regular expressions based on Yntema's Matching-Choice sets.

6   Conclusions and Future Work

This paper presents SM Expressions, a notation to specify languages of the family of Standard Matching-Choice Sets [14]. This family of languages is an important subset of context-free languages, being built around the notion of parenthesizing strings. SM Expressions can define context-free path queries as part of query languages for graph databases. Several languages have been proposed to express non-regular path queries, but usually they rely upon rules of context-free grammars. Our proposal uses SM Expressions to easily express non-regular path queries to be translated into rules of context-free grammars.

In this paper we provide an overview of the operators defined in [14] for Matching-Choice Sets. These operators pave the way for the definition of a hierarchy of classes of languages, including Standard Matching-Choice Sets. After that, we introduce the syntax of SM Expressions and present the semantics of their constructs in terms of Standard Matching-Choice Sets. We demonstrate how to obtain context-free grammars from SM Expressions and illustrate the usage of SM Expressions in the context of the work proposed in [6]. Our proposal can be implemented by the front-end of a query language processor. Questions of efficiency and quality depends on the query engine used as back-end.

Our work can be extended by considering different perspectives: (*i*) analyzing the asymptotic costs associated to the translation of SM Expressions into context-free grammars; (*ii*) identifying the class of context-free path queries that cannot be expressed with SM Expressions; (*iii*) comparing the expressiveness of

SM Expressions with other proposals of non-regular expressions; and (*iv*) investigating the usage of SM Expressions as part of other non-regular specification languages.

References

1. Abiteboul, S., Hull, R., Vianu, V.: Foundations of Databases. Addison-Wesley, Boston (1995)
2. Francis, N., et al.: Cypher: an evolving query language for property graphs. In: Proceedings of SIGMOD'2018, pp. 1433–1445. ACM, New York (2018). https://doi.org/10.1145/3183713.3190657
3. Freydenberger, D.: Extended regular expressions: succinctness and decidability. Theory Comput. Syst. **53**(2), 159–193 (2012). https://doi.org/10.1007/s00224-012-9389-0
4. Hellings, J.: Conjunctive context-free path queries. In: Schweikardt, N., Christophides, V., Leroy, V. (eds.) Proc. 17th International Conference on Database Theory (ICDT), Athens, Greece, 24–28 March 2014, pp. 119–130. OpenProceedings.org (2014). https://doi.org/10.5441/002/icdt.2014.15
5. Kuijpers, J., Fletcher, G., Yakovets, N., Lindaaker, T.: An experimental study of context-free path query evaluation methods. In: Proceedings of the SSDBM'2019, pp. 121–132. ACM (2019)
6. Medeiros, C., Costa, U., Grigorev, S., Musicante, M.A.: Recursive expressions for SPARQL property paths. In: Bellatreche, L., et al. (eds.) TPDL/ADBIS -2020. CCIS, vol. 1260, pp. 72–84. Springer, Cham (2020). https://doi.org/10.1007/978-3-030-55814-7_6
7. Medeiros, C.M., Musicante, M.A., Costa, U.S.: LL-based query answering over RDF databases. J. Comput. Lang. **51**, 75–87 (2019). https://doi.org/10.1016/j.cola.2019.02.002
8. Pérez, J., Arenas, M., Gutierrez, C.: nSPARQL: a navigational language for RDF. In: Sheth, A., et al. (eds.) ISWC 2008. LNCS, vol. 5318, pp. 66–81. Springer, Heidelberg (2008). https://doi.org/10.1007/978-3-540-88564-1_5
9. Prud'hommeaux, E., Seaborne, A.: SPARQL query language for RDF (2008). Latest version available as http://www.w3.org/TR/rdf-sparql-query/
10. Sempere, J.: On a class of regular-like expressions for linear languages. J. Automata Lang. Comb. **5**, 343–354 (2000)
11. Spiegel, J., Dyck, M., Robie, J.: XML path language (XPath) 3.1. W3C recommendation, W3C (2017). https://www.w3.org/TR/2017/REC-xpath-31-20170321/
12. W3C: SPARQL 1.1 query language (2012). https://www.w3.org/TR/2012/PR-sparql11-query-20121108/
13. W3C: RDF - semantics web standards (2014). https://www.w3.org/RDF/
14. Yntema, M.: Inclusion relations among families of context-free languages. Inf. Control **10**(6), 572–597 (1967). https://doi.org/10.1016/S0019-9958(67)91032-7
15. Yntema, M.: Cap expressions for context-free languages. Inf. Control **18**(4), 311–318 (1971). https://doi.org/10.1016/S0019-9958(71)90419-0
16. Zhang, X., Feng, Z., Wang, X., Rao, G., Wu, W.: Context-free path queries on RDF graphs. In: Groth, P., et al. (eds.) ISWC 2016. LNCS, vol. 9981, pp. 632–648. Springer, Cham (2016). https://doi.org/10.1007/978-3-319-46523-4_38

COVID-19 Portal: Machine Learning Techniques Applied to the Analysis of Judicial Processes Related to the Pandemic

Ana Sodré[1]($\boxtimes$), Dimmy Magalhães[1]($\boxtimes$), Luis Floriano[1]($\boxtimes$), Aurora Pozo[1]($\boxtimes$), Carmem Hara[1]($\boxtimes$), and Sidnei Machado[2]($\boxtimes$)

[1] Departamento de Ciência da Computação, Universidade Federal do Paraná, Curitiba, PR, Brazil
{apas19,dksmagalhaes,lemf19,aurora,carmem}@inf.ufpr.br
[2] Departamento de Direito, Universidade Federal do Paraná, Curitiba, PR, Brazil
sidneimachado@ufpr.br

Abstract. The COVID-19 pandemic created new demands, not only for health services, but also for services in other domains such as the judicial system. New tools that assist in the analysis of the judicial process may help in this problem. In particular, artificial intelligence (AI) techniques may be applied to provide a qualitative analysis of legal documents. Although there exist a number of works that apply AI in the judicial domain, few target the pandemic or publicly provide the information extracted from the texts. Following the suggestions and needs of a legal expert, we have developed the COVID-19 Portal. It extracts documents from the Supreme Federal Court in Brazil, and applies AI technologies to obtain fine-grained quantitative and qualitative information on words used in the texts. This information is made available on a website and can help lawyers identify trends and develop arguments for judicial processes related to the pandemic.

1 Introduction

The COVID-19 pandemic deeply affected all sectors of society. In addition to hugely increase the demand for health services in all countries, it has caused a global economic downturn. This panorama is no different in the judicial area, since the number of legal proceedings has grown due to three main facts: 1) the judiciary is being more demanded by the population; 2) new computational tools create easier access to the judiciary; and 3) a new object of demand: cases directly related to the pandemic. The growth in demand from the Judiciary (mainly referring to Covid) requires the creation of new tools, and in this perspective, the use of artificial intelligence (AI) techniques can help solve this problem.

There have been a number of initiatives to apply AI in the judicial domain [8,17,19], but few target the pandemic. Most of the works that consider the

L. Bellatreche et al. (Eds.): ADBIS 2021 Short Papers, Workshops and Doctoral Consortium, CCIS 1450, pp. 109–120, 2021.
https://doi.org/10.1007/978-3-030-85082-1_10

COVID in particular, provide a quantitative analysis, such as the Portal of the World Health Organization (WHO). Others focus on health related areas [3] or use social networks as their source of data [4, 16]. In contrast, our work follows the suggestions and needs of a legal expert, and proposes the application of AI techniques for judicial texts specifically related to the pandemic. We built the COVID-19 Portal[1] that gathers texts of decisions or lawsuits related to the pandemic addressed to the Federal Supreme Court (STF) in Brazil. STF is the last instance of the Judiciary and Constitutional Court. Accompanying the decisions and precedents of the Court's jurisprudence has enormous relevance to guarantee access to fundamental rights. The Covid-19 Portal is a tool that automates the selection of a set of STF decisions related to labor rights and the Covid-19. Moreover, the system shows the network of words on which the decisions are based, which allows the visualization of the Court's jurisprudence in terms of the connection between decisions and patterns of citation of precedents. Another important functionality of the system is its ability to filter and represent the arguments of each minister of the Court. The result of the analysis for a specialist (lawyer or judge) is the instantaneous identification of jurisprudence and the precedent cases to which it relates by reiterating the argumentative construction of guiding standards of judicial actions.

To build the Portal, we collected the judicial processes from the STF portal[2] that contain the terms: *'pandemia'* and *'covid'*. For each process, we collected all associated documents: opinions and decisions. Then, all documents are processed by AI techniques to perform quantitative, qualitative, and semantic analysis of the data. We apply mass word processing techniques, such as NLTK [1], Word2Vec [11], Sentence-BERT [14] and K-Means [18]. Besides, we use a word embedding model based on Word2Vec applied to lawsuits called Lex2Vec[3]. The application of artificial intelligence on the COVID-19 Portal is extremely relevant due to its ability to build representations, cognitive models, and computational models for automated reasoning.

The COVID-19 Portal provides as output: 1) a quantitative analysis, containing the most frequent words, a process count by class, and word count by field of law and legal class; 2) a context graph showing the proximity among words; and 3) a general document clustering and a clustering based on ministers of the STF. The user may choose to obtain the information concerning the entire country, or a specific state, or a particular field of law. The ultimate goal of the Portal is to assist the user in the process of constructing legal arguments and in decision making.

This work is organized as follows. Section 2 covers previous approaches applied to text classification and that combine artificial intelligence and the domain of law. The Portal architecture is presented in Sect. 3 with details about the components that provide each of the system's functionalities. Next, Sect. 4 describes the techniques applied for developing the Portal. Finally, Sect. 5 concludes the paper with a discussion on the results and future directions.

---

[1] http://portalcovid-cbio-cd.herokuapp.com/.

[2] http://portal.stf.jus.br/.

[3] https://github.com/thefonseca/lex2vec.

2   Related Work

Even though research on natural language processing is advanced in many aspects and languages, there are few published works related to the semantic analysis of Brazilian judicial processes.

In the international context, researchers use different natural language processing (NLP) techniques for analyzing judicial documents. [10] proposes a formal model, *quasi-logical form* (QLF) for semantic feature extraction from legal documents. [5] shows the JUMAS project with Automatic Speech Recognition, Emotion Recognition, and Legal Information Recover. [8] proposes a model using segmentation of legal text topics and then applies clustering techniques and SVM to merge groups of texts from the perspective of their topics. [17] highlights text mining techniques to create relevant topics about legal proceedings. [19] proposes a model based on rules and semantic relationships. Although all these works focus on the analysis of legal documents, they do not provide a Web portal to visualize their results and they have not been proposed to manage processes directly related to the pandemic.

Among the sources that provide data and quantitative analysis related to the pandemic, we mention two. The first is the Coronavirus Resource Center from Johns Hopkins University[4]. It presents a quantitative analysis of COVID-19 cases around the world. The second is the World Health Organization (WHO), which provides daily reports in a portal[5]. Both sources provide general data, such as the number of cases, reported by country, state or region, the number of deaths, and number of tests. However, there are no cross references of these information to other sources of information and, in particular, with judicial documents.

In [3], the authors present a perspective on how NLP can be used to assist with assessment and rehabilitation for acute and chronic conditions related to COVID-19. In our case, we seek to present AI and NLP as techniques used for qualitative analysis of pandemic data in the legal context. Social networks have also been considered as data sources related to the pandemia. [4] uses NLP techniques for massive data analysis on social networks. Similar to our work, the words in the text corpus are represented as numerical vectors in order to use Word Embedding methods to identify the similarity between lexical contexts. But, for clustering, they apply Partitioning Around Medoids, while our Portal uses K-means, as explained in Sect. 4.3.

[16] also collects data related to the pandemia from social networks, in particular, from Twitter. It provides features similar to the ones present in our Portal, such as the listing of the 10 most frequent words and a word cloud. However, their focus is to provide a quantitative analysis of the tweets in order to identify a spatio-temporal relationship between Twitter and cases of COVID-19. Our work differs on its purpose. We aim to provide a qualitative data analysis in the Brazilian juridical context.

---

[4] https://coronavirus.jhu.edu/map.html.
[5] https://covid19.who.int/.

The work most closely related to ours is the Datalawyer Insights[6]. Similar to our work, it focus on judicial processes related to the pandemia in Brazil. However, they use as their source processes from the Labor Justice Court, while we obtain the data from the Federal Supreme Court. Our work also distinguishes from theirs on the type of information provided. Datalawyer reports the number of processes and amount of monetary damages accounted by the type of the process and economic activity. Our Portal, on the other hand, processes the documents in a finer-grained level in order to compute the proximity of words and categorize them.

3   COVID-19 Portal

In this section, we describe the general architecture of the COVID-19 portal. We divide the architecture into three major modules: NLP Module, API, and Front-End. Figure 1 shows the relationship between the modules and their components. The NLP Module is an offline data processing module. Once the crawler obtains the judicial document data, this module executes the following steps: 1) data preprocessing, by cleaning, tokenizing, and removing stop words defined in NLTK[7]; 2) word counting; 3) frequency analysis; 4) word context analysis; 5) word training and document embedding; and 6) document clustering. The results of this process are stored in 3 datasets: Frequency Data, Graph Data, and Cluster Data.

These datasets are used by the API Module, shown in yellow. This module is responsible for creating micro-services that search the datasets and return information in JSON format, based on the user-defined search criteria, such as: the state, field of law, and minister name. The results retrieved from these micro-services are given as input to the Front-end Module, which provides a user-friendly interface. This module contains one component for each type of output in the Portal. These components render charts, word cloud, and context graphs. Currently, the Front-end module has filters by state, field and minister.

The Portal provides information based on three main flows: Quantitative Analysis, Context Graph, and Clustering. These flows are described in the following sections.

3.1   Quantitative Analysis

The quantitative analysis of court documents (shown in green in Fig. 1) is responsible for word counting and frequency analysis. The goal is to determine the set of most relevant words, based on a statistical analysis of the set of documents considered as input. To do so, words are analyzed according to their importance throughout the context of the documents. Thus, more discriminating words have higher measures, highlighting their relevance.

---

[6] https://www.datalawyer.com.br/dados-covid-19-justica-trabalhista.
[7] http://www.nltk.org/.

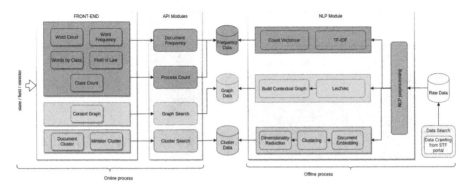

Fig. 1. COVID-19 portal architecture

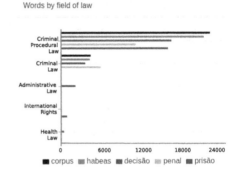

Fig. 2. Word cloud.

Fig. 3. Words by field of law.

In the system architecture of Fig. 1, the quantitative analysis flow consists of the *CountVectorizer* and *TFIDF* components of the NLP Module, the *Document Frequency* and *Process Count* components of the API Module, and the *Word Cloud, Word Frequency, Word by Class, Field of Law* and *Class Count* components of the Front-end Module.

Figure 2 shows the output of the *Word Cloud* component. It contains around 50 words and adjusts their size according to their relevance. The frequency analysis allows the user to observe trends among legal processes. As an example, Fig. 2 shows a predominance of criminal proceedings related to *habeas corpus*, over *prisão* and *cassação*.

From the juridical point of view, it is also relevant to visualize quantitative data about the processes. It is essential to know the most frequently words in each field of law and each legal class, as well as the number of processes in each class. An example of the *Words by Field* output is presented in Fig. 3. Since there may be a large number of fields of law, the chart shows five with the largest number of relevant words. Then, among these fields, the five most frequently used relevant words are chosen to be presented. A similar process is used to determine the

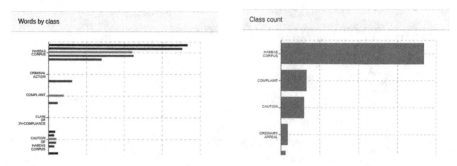

Fig. 4. Words by class. The 10 classes with the highest number of words are selected

Fig. 5. Class count. Number of process by legal class

ten legal classes and five words presented by the output of component *Words by Class*, shown in Fig. 4. The *Class Count* component presents the ten legal classes with largest number of processes. An example is presented in Fig. 5.

3.2   Context Graph

The semantic analysis allows the association of a meaning to a syntactic structure. The interpretation of an expression is one of the challenges of NLP since the sentence understanding can be ambiguous or wrong. Through this context analysis, users can make better-informed decisions and recommendations. We apply semantic analysis to build the relationship graph between words, where metrics such as the frequency of words, context windows, and frequency of words for context windows are considered. The goal is to determine the proximity among words, that is, which words are most often used together in the same context.

In the system architecture of Fig. 1, the context graph flow (shown in purple) is composed of the *Lex2Vec* and *Build Contextual Graph* of the NLP Module, the Graph Search component of the API Module, and the *Context Graph* of the Front-End Module.

An example of the output is given in Fig. 6. The graph shows the relationship between words. We start with the ten most relevant words, and based on them, we follow the five closest words and represent them in the graph. In order to compute the proximity among words, we use Lex2Vec, a trained Word2Vec model of the Brazilian federal legislation. Details on the model are presented in Sect. 4.2.

The motivation for showing the closest words of the most frequent ones in existing processes is to help counselors develop arguments for new processes. Granular graph analysis can give the user a perspective on how a particular topic is being addressed in legal documents.

3.3   Clustering

Clustering legal documents can help the user find semantic patterns among the documents and their authors. The goal is to group the legal documents according to their semantic features, i.e., similar texts must be close in the vector space.

In our architecture, the clustering flow (shown in orange in Fig. 1) is composed of the *Document Embedding*, *Clustering* and *Dimensionality Reduction* components of the NLP Module, the *Cluster Search* of the API Module, and the *Document Cluster* and *Minister Cluster* of the Front-end Module. The *Document Embedding* module is responsible for converting the text into a vector. The conversion must carry with it the text's semantic characteristics, and thus semantically close texts will have similar vectors (according to some vector distance metric). The *Clustering* component then groups the legal documents based on their context and according to their author, while the *Dimensionality Reduction* component prepares the data for meaningful visual representations of the groups.

An example of *Document Cluster* output is shown in Fig. 7. The *Minister Cluster* output, presented in Fig. 8, shows the same information, but makes a distinction among the ministers, members of the STF, responsible for the judicial document. The clustering of judicial documents allows the user to understand how the court documents are related to each other. Besides, it allows verifying the legal guidelines of each STF minister, for example, highlighting possible unanimous votes given a particular theme.

Fig. 6. Context graph. Each node is a significant word and the edges are calculated according to the cosines distance

4   Development of the COVID-19 Portal

In this section, we describe the main languages, databases, techniques, and frameworks used in the development of the COVID-19 Portal.

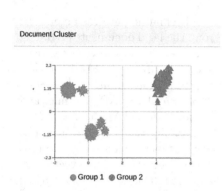

Fig. 7. Document cluster. The documents were computationally grouped into 2 groups using the K-Means algorithm

Fig. 8. Minister cluster. The documents were computationally grouped into 2 groups using the K-Means algorithm and highlight STF minister

4.1 Languages and Database

The COVID-19 portal was built in three stages, following the three-modules system architecture illustrated in Fig. 1. Each module was built independently, and the communication among components is mostly based on requests and responses in the JSON format[8].

Components of the NLP module were developed in Python[9] version 3.7, with the Sklearn framework [13]. All components are executed off-line and periodically. More specifically, legal documents that contain the words "pandemia" and "covid" are retrieved from the STF Portal. These documents are given as input to the components, that generate JSON data to update the *Frequency*, *Graph* and *Cluster* datasets, kept in Mongo DB[10].

To develop the API module, we used Node.js[11]. This framework is an asynchronous event-driven JavaScript runtime and it is designed to build scalable network applications. Upon each connection, the callback is fired, but if there is no work to be done, Node.js will sleep. We chose this framework for its ease of implementation of micro-services and the high performance presented in our tests. This choice allows us to envision different ways of viewing the data in the future, such as in a mobile interface.

Finally, the Front-End module was developed using the ReactJS[12] framework. ReactJS is a component-based framework, i.e., each output in the portal is an encapsulated component that manages its state, which is then used to

---

[8] https://www.json.org/.
[9] https://www.python.org/downloads/release/python-370/.
[10] https://cloud.mongodb.com/.
[11] https://nodejs.org/.
[12] https://reactjs.org/.

compose the user interface. This feature allows us to treat each output as an independent element, speeding up the delivery of information to the user.

4.2   Word2Vec and Lex2vec

One way to analyze the context of a given sentence is by calculating the distance among its words. Thus, it is possible to represent each word mathematically as a vector to compute their similarity.

In this perspective, Word2Vec [11] is a statistical method to perform tasks related to natural language processing, e.g., context analysis. Developed by Tomas Mikolov and published in 2013, it uses a neural network model to learn word embedding from a text corpus. As the name suggests, Word2Vec represents each word as a numeric vector, where each position refers to a dimension. Thus, the distance between words is calculated through the similarity of cosines between the vectors.

For the COVID-19 Portal, we used a previously trained model called Lex2Vec. It is a trained Word2Vec model with vector representations of words present in the corpus of Brazilian federal legislation. We have extended Lex2Vec, by incrementally training the model with the set of documents extracted from the STF. Although the result of the combined training deserves further analysis, we have noticed that the final group of node and edges of the *Context Graph* differs from our previous implementation, that adopted Pointwise Mutual Information (PMI) and cosine distance. For example, the edge between the words *habeas* and *impetração* exists in the graph created based on Lex2Vec, but does not exist using the PMI metric.

4.3   K-Means

Clustering is a process that partitions a data set into groups such that elements in the same group are more similar to each other than to elements in other groups. For the development of the *Clustering* component of the NLP Module, we executed experiments with several algorithms, such as: K-Means [9], DBScan [15], Spectral clustering [7] and Agglomerative Clustering [12]. Since all the algorithms generated similar results, we opted for K-Means. The K-Means algorithm requires as input the number of clusters (K), besides the dataset to be partitioned. For this work, a publicly available implementation of the algorithm[13] was used.

4.4   Document Embedding

Before running the clustering algorithm, texts must be converted into float vectors that represent their semantics. This step is called *Document Embedding*. In this work, we compared three frameworks: Word2Vec [11], Doc2Vec [6] and

---

[13] https://scikit-learn.org/stable/modules/generated/sklearn.cluster.KMeans.html.

Sentence-BERT [14]. After running each of the algorithms, we execute the K-Means algorithm several times, varying the value of K, which determines the number of groups. We collect the silhouette metric value[14]. The representation with the best silhouette value was considered the best representation.

For the first option for document embedding, we used the Word2Vec model based on Lex2Vec to compute the vectors for each word in the text. Then, an average vector between those words was computed, generating one vector for each judicial document. We call this representation as $Word2Vec_{avg}$. As the second alternative, we used Doc2Vec to generate the representation of the documents. This method learns paragraph and document embeddings via the distributed memory and distributed bag of word models and generates one vector for each judicial document. Finally, we used Sentence-BERT, a modification of the pre-trained BERT network that uses siamese and triplet network structures to derive semantically meaningful sentence embeddings that can be compared using cosine-similarity [14]. Table 1 shows the silhouette value for each representation model according to the K-value of the K-Means algorithm. Given that Sentence-BERT produced the highest silhouette values, it was chosen to implement the *Document Embedding* component.

Table 1. Document embedding evaluation according to silhouette metric

Model	K-value		
	2	3	4
$Word2Vec_{avg}$	0.1277	0.0799	0.0031
Doc2Vec	0.2348	0.2001	0.0039
Sentence-BERT	0.6189	0.5744	0.5298

4.5   Dimensionality Reduction

BERT uses a vector of 768 dimensions (float values) to represent each document. The result of document clustering using the K-means algorithm is a set of K groups of these documents. However, in order to graphically present the grouped documents in a useful and meaningful way, it is impractical to have visual representations for all the dimensions. Thus, it is important to determine the set of principal or representative dimensions. Usually, groups can be easily represented graphically in two or three dimensions. In this work, we chose to represent both the *Document Cluster* and *Minister Cluster* in two dimensions. Thus, we have applied a PCA-based dimensionality reduction [20] to generate two-dimension descriptions for the documents. They are used to represent them in a graph as points, using different colors for each group.

---

[14] In clustering evaluation using silhouette metric, the best value is 1, and the worst value is −1. Values near 0 indicate overlapping clusters. Negative values generally indicate that a sample has been assigned to the wrong cluster, as a different cluster is more similar.

5   Conclusion

This study describes the Covid-19 Portal that contains quantitative and qualitative information related to judicial documents addressed to the STF. The Portal has components that show the most frequent words and terms related to the topic. Also, it exhibits the semantic relationship between documents of different supreme court ministers. The data on the Portal has been processed using several AI techniques, such as clustering, word and document embedding, term frequency analysis, among others. The Portal was built with techniques of Web development that aim at scalability and performance. The information presented in the Portal allow lawyers to contextualize the judicial processes, checking the most common terms and fields and how they are being addressed in other cases. Furthermore, it allows the user to find patterns among court ministers. The semantic analysis can also assist lawyers by indicating possible legal argumentation.

Some interesting future directions of this work include extending the functionalities of the COVID-19 portal, among which we highlight: 1) provide more interactive components, in addition to enable refinements of query results; 2) add multiple selection query filters; 3) generation and comparison of context graphs based on legal documents written by different STF ministers; 4) extraction and representation of the logic used in legal arguments for reaching a decision; and 5) application of clustering algorithms on legal arguments; 6) use pattern analysis, such as LDA [2], to analyze forensic topics.

References

1. Bird, S., Klein, E., Loper, E.: Natural Language Processing with Python: Analyzing Text with the Natural Language Toolkit. O'Reilly Media Inc., Sebastopol (2009)
2. Blei, D.M., Ng, A.Y., Jordan, M.I.: Latent dirichlet allocation. J. Mach. Learn. Res. **3**, 993–1022 (2003)
3. Carriere, J., et al.: Case report: utilizing AI and NLP to assist with healthcare and rehabilitation during the COVID-19 pandemic. Front. Artif. Intell. **4** (2021)
4. Cinelli, M., et al.: The COVID-19 social media infodemic. arXiv preprint arXiv:2003.05004 (2020)
5. Fersini, E., Messina, E., Archetti, F., Cislaghi, M.: Semantics and machine learning: a new generation of court management systems. In: Fred, A., Dietz, J.L.G., Liu, K., Filipe, J. (eds.) Knowledge Discovery, Knowledge Engineering, and Knowledge Management, vol. 272, pp. 382–398. Springer, Heidelberg (2010). https://doi.org/10.1007/978-3-642-29764-9_26
6. Le, Q., Mikolov, T.: Distributed representations of sentences and documents. In: International Conference on Machine Learning, pp. 1188–1196 (2014)
7. Liu, J., Han, J., Aggarwal, C., Reddy, C.: Spectral clustering (2013)
8. Lu, Q., Conrad, J.G., Al-Kofahi, K., Keenan, W.: Legal document clustering with built-in topic segmentation. In: Proceedings of the 20th ACM International Conference on Information and Knowledge Management, pp. 383–392. ACM (2011)
9. MacQueen, J., et al.: Some methods for classification and analysis of multivariate observations. In: Proceedings of the Fifth Berkeley Symposium on Mathematical Statistics and Probability, vol. 1, no. 14, pp. 281–297 (1967)

10. McCarty, L.T.: Deep semantic interpretations of legal texts. In: Proceedings of the 11th International Conference on Artificial Intelligence and Law, pp. 217–224. ACM (2007)
11. Mikolov, T., Chen, K., Corrado, G., Dean, J.: Efficient estimation of word representations in vector space. In: 1st International Conference on Learning Representations, ICLR 2013, Scottsdale, Arizona, USA, 2–4 May 2013, Workshop Track Proceedings (2013). arXiv:1301.3781
12. Murtagh, F., Legendre, P.: Ward's hierarchical agglomerative clustering method: which algorithms implement ward's criterion? J. Classif. **31**(3), 274–295 (2014). https://doi.org/10.1007/s00357-014-9161-z
13. Pedregosa, F., et al.: Scikit-learn: machine learning in Python. J. Mach. Learn. Res. **12**(85), 2825–2830 (2011). http://jmlr.org/papers/v12/pedregosa11a.html
14. Reimers, N., Gurevych, I.: Making monolingual sentence embeddings multilingual using knowledge distillation. arXiv preprint arXiv:2004.09813 (2020)
15. Schubert, E., Sander, J., Ester, M., Kriegel, H.P., Xu, X.: DBSCAN revisited, revisited: why and how you should (still) use DBSCAN. ACM Trans. Database Syst. (TODS) **42**(3), 1–21 (2017)
16. Singh, L., et al.: A first look at COVID-19 information and misinformation sharing on Twitter. arXiv preprint arXiv:2003.13907 (2020)
17. Wagh, R.S.: Knowledge discovery from legal documents dataset using text mining techniques. Int. J. Comput. Appl. **66**(23), 32–34 (2013)
18. Wagstaff, K., Cardie, C., Rogers, S., Schroedl, S., et al.: Constrained k-means clustering with background knowledge. In: ICML, vol. 1, pp. 577–584 (2001)
19. Walker, V.R., Han, J.H., Ni, X., Yoseda, K.: Semantic types for computational legal reasoning: propositional connectives and sentence roles in the veterans' claims dataset. In: Proceedings of the 16th edition of the International Conference on Artificial Intelligence and Law, ICAIL 2017, London, United Kingdom, 12–16 June 2017, pp. 217–226 (2017). https://doi.org/10.1145/3086512.3086535
20. Yang, J., Zhang, D., Frangi, A.F., Yang, J.Y.: Two-dimensional PCA: a new approach to appearance-based face representation and recognition. IEEE Trans. Pattern Anal. Mach. Intell. **26**(1), 131–137 (2004)

LACLICHEV: Exploring the History of Climate Change in Latin America Within Newspapers Digital Collections

Genoveva Vargas-Solar[1(✉)], José-Luis Zechinelli-Martini[2],
Javier A. Espinosa-Oviedo[3], and Luis M. Vilches-Blázquez[4]

[1] CNRS, LIRIS Campus de la Doua, 69622 Villeurbanne, France
genoveva.vargas-solar@liris.cnrs.fr
[2] Fundación Universidad de las Américas Puebla, 72820 San Andrés Cholula, Mexico
joseluis.zechinelli@udlap.mx
[3] Université Lumière Lyon 2, ERIC, Lyon, France
javier.espinosa-oviedo@univ-lyon2.fr
[4] Centro de Investigación en Computación, IPN, 07738 Ciudad de México, Mexico

Abstract. This paper introduces LACLICHEV (Latin American Climate Change Evolution platform), a data collections exploration environment for exploring historical newspapers searching for articles reporting meteorological events. LACLICHEV is based on data collections' exploration techniques combined with information retrieval, data analytics, and geographic querying and visualization. This environment provides tools for curating, exploring and analyzing historical newspapers articles, their description and location, and the vocabularies used for referring to meteorological events. The objective being to understand the content of newspapers and identifying possible patterns and models that can build a view of the history of climate change in the Latin American region.

Keywords: Data curation · Metadata extraction · Data collections exploration · Data analytics

1 Introduction

Historical analysis of climate behaviour can provide conclusions about climatic phenomena and Earth climate behaviour evolution. Even if there is an increasing interest in analysing digital data collections for performing historical studies on climatic events [1,2], the history of climate behaviour is still an open issue that has not revealed missing knowledge. Digital data collections make it possible to have an analytic vision of the evolution of environmental, administrative,

We thank the master student Santiago Ruiz Angulo of the Universidad Autónoma de Guadalajara who implemented the first version of LACLICHEV during his internship at the Barcelona Supercomputing Centre, Spain, funded by the CONACYT "beca mixta" fellowship program of the Mexican government.

L. Bellatreche et al. (Eds.): ADBIS 2021 Short Papers, Workshops and Doctoral Consortium, CCIS 1450, pp. 121–132, 2021.
https://doi.org/10.1007/978-3-030-85082-1_11

economic, and social phenomena. In this context, our work deals with data collections that report the emergence of meteorological events (e.g., temperature changes, avalanches, river flow growth, volcano eruptions).

This paper introduces a Latin American Climate Change Evolution platform called LACLICHEV. The objective is to expose and study the history of climate change in Latin America. The hypothesis is that Latin America history is contained in newspapers articles lying in digital collections available in the national libraries of Mexico, Colombia, Ecuador, and Uruguay.

LACLICHEV addresses three issues. First, newspaper archaeology by chasing articles talking about climatic events using specific vocabulary to discover as many articles as possible. The challenge is choosing the adequate vocabulary to increase chances to get articles actually talking about climatic events. Second, once an article talks about a climatic event, it is Geo-Temporal tagged with metadata specifying what happened, where and when it happened, and its duration and geographical extent. The objective here is to build a climatic event history. Finally, on top of this history, the objective is to run analytics questions and visualize results in maps, given that the content is highly spatial.

The remainder of this paper is organized as follows. Section 2 introduces the curation process that we propose for historical newspapers articles potentially reporting on climatic events. Section 3 describes the general architecture of LACLICHEV and the experiments we conducted for evaluating it. Section 4 studies approaches that promote datasets exploration for defining the type of analysis possible on top of them. Finally, Sect. 5 concludes the paper underlying the contribution and discussing future work.

2   Curating Historical Newspapers Articles

The objective of curating historical newspapers articles is to build a dataset of documents reporting meteorological events and associating them with metadata, providing as much information as possible about the reported event. The newspapers curation process is a semi-automatic process devoted to: (i) find articles reporting climatic events within newspapers articles collections; and (ii) geo-tag and store those articles that report such events for building a database containing the climatic event history.

Curation tasks are performed on a collection of textual digital documents provided by digital libraries that own the collections. Each library adopts a meta-data schema specifying the newspaper's name, the country, the day and number, the number of pages, the window time in which it circulated. The curation process generates data structures that provide an abstract representation of the content of each article describing an event. We propose a meteorological event knowledge model (see Fig. 1) to represent climatic event reports in digital documents. The model describes events from different perspectives using the information from the articles and newspapers. As shown in Fig. 1, an event is associated with the newspaper article(s) that describe it (reading from right to left). Each article is annotated with linguistic labels that curate it. Curation tasks can be recurrent and include a human-in-the-loop strategy for validating and adjusting the results. For example, suppose an event is geo-tagged to associate it to a geographic location and the event is described in an article about Montevideo news from Uruguayan newspapers collections. In that case, a human will verify that the geographic location refers to Montevideo in Uruguay and not in the US.

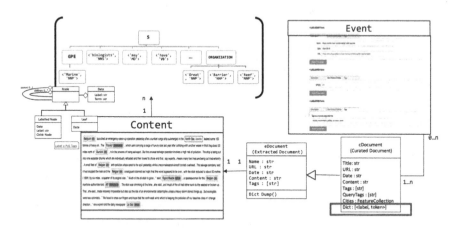

Fig. 1. Event data model

The event knowledge model provides concepts for representing a *meteorological event* according to descriptive, meteorologic, linguistic, and knowledge domain perspectives.

Descriptive Perspective. Meteorological events are described in different ways in historical newspapers articles depending on the author. However, we can often collect information related to location, date, duration, scope, and damages. Meteorological features (like millimetres of precipitations, wind speed, temperature, pressure, etc.) can be explicitly described in articles or deduced according to the description of the event. For example, an event reported in Montevideo and describing an overflow of the river implies winds higher than 100 km/h and rain of more than 10 ml/hour according to the knowledge provided by meteorologists. This knowledge domain is used for completing the meteorologic features describing the reported event.

Linguistic perspective gathers the terms used for describing an event in one or several articles belonging to a given newspaper. We propose a tree-based data structure named *content tree* for representing the content of a historical newspaper article (see UML class diagram in Fig. 2). The tree corresponds to the grammatical analysis of each sentence in the textual content of the article commonly used in Natural Language Processing (NLP) techniques. The **content tree**, as shown in the diagram, consists of a set of sentences. A **sentence** is defined as a set of nodes representing grammatical elements of a sentence and leaves representing the terms composing a sentence in a specific article.

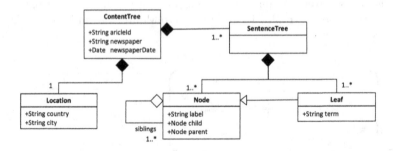

Fig. 2. UML class diagram representing the general structure of a content tree

As shown in Fig. 2, a **node** represents a type of grammatical element given in a specific linguistic model defined for a specific language. It is labelled adopting the entity labels produced by classic natural language processing tools known as Part Of Speech (POS) tags. For instance, *noun, proper singular* (NNP), *noun, plural* (NNS), *verb, modal auxiliary* (MD), *Geopolitical entity* (GPE), or Organization. In the case of subjects (NNP), they can be grouped into more general entities that identify geographic locations (GPE), places, names, and organization.[1] A **node** has a child, which is also a Node or a Leaf, and a set of siblings, which are other nodes. A **leaf** specializes a node and it represents a term contained in the article. A term is a string and has a parent, a **node** represents a POS tag.

Every article in a newspaper is associated with its content tree. Thereby, a data scientist or expert domain can explore the articles by navigating their content trees without reading the full content. For example, *retrieve articles reporting heavy storms in Uruguay in December 1810.*

Meteorological perspective characterizes the climate event (see Fig. 3) with attributes that describe it in one or several newspaper articles. Nevertheless, not all the attributes can necessarily have a value since there might not be any evidence within the articles that report it. Attributes, like the date of the event, its geographical scope, or the location of the damaged regions, are computed by navigating through the *content tree* of every article reporting the event.

Knowledge domain perspective describes meteorological events using knowledge domain statements created by experts of the National Library of Uruguay. This knowledge has been associated with events through manual analysis of newspapers collections and interacting with meteorologists. This knowledge can help interpret the empirical information reported in the articles and complete the information associated with the event description. For example, if the river

---

[1] A full list of POS tags can be found in https://www.cms.gov/.

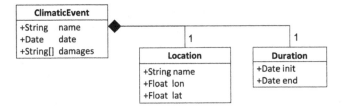

Fig. 3. UML class diagram representing a climatic event

was flooding due to a storm, it is possible to estimate the wind speed and the approximate litters of rain. The knowledge domain perspective is modelled as a glossary.

2.1 Building a Climatic Event History

Given a documents' collection and its associated data structures describing the content of its articles, the data scientist can explore articles to determine whether they report climatic events. This phase integrates the human-in-the-loop. The reason is that newspaper articles use of colloquial terms can be tricky and refer to metaphors that might not denote a climatic event. Language subtilities are not easy to handle manually, mainly because we are dealing with a language used some centuries ago, which increases the challenge of classifying the content of the articles. During this phase, articles referring to climatic events are geo-temporally tagged to associate them with the region and/or time window in which they happened. Tags are validated by the data scientist. Since the result can contain a significant number of articles, the user can use three tools for understanding the content of the result. The tools let her manipulate a terms frequency matrix and a terms heat map. She can also explore the content of the article text using a view that provides information about the context in which the terms potentially describing an event appear in the document. For example, the name of geographic locations in the document might refer to the location of the event and the region that it touched, and a list of geopolitical entities (e.g., school, public building, etc.) to determine the damages caused by the event.

The data scientist can perform the following actions:

- Correct the terms associated with climatic events that might not be used in such a sense in the text. Indeed, some social and political demonstrations are often described as climatic events. For a classic automatic text analysis process, this can be not easy to identify and filter. For example, an article entitled *Stormy weather within the ails of the senate in Ecuador* has nothing to do with a climatic event but with a political one.
- Determine whether personal names correspond to the event's name (e.g. hurricane or storm name). If that is the case, this information will be used for inserting the event in the history.

- Verify whether the names of cities, regions, and countries correspond to geographic entities. The system underlines the names of patronyms and the data scientist can see the location of the possible geographic entities. Thus, the user can also confirm whether the article refers to the geographic zone that she is searching for. For instance, if "Santa Clara" is underlined, it can refer to a place, city, or village.
- Determine the date of the event and its characteristics. The temporal terms and adjectives are also underlined to let the data scientist click on those that describe the event.
- Determine the type of damages caused by the event by exploring those terms that describe such kind of information.

The previous actions are used to adjust the representation of the articles' content and identify meteorological events more accurately since the data scientist or domain expert knowledge is used (see Fig. 4 showing LACLICHEV interfaces for curation).

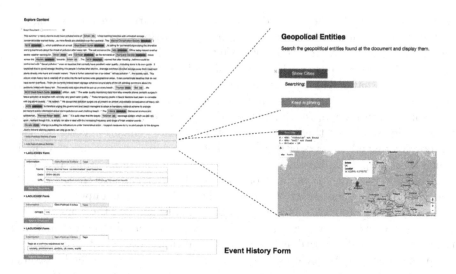

Fig. 4. Event curation process interface for tagging events

2.2   Climatic Events History

Once the events have been validated, the data scientist can use a form to define and store the event. Meta-data is stored in persistence support, a key-value or a document store depending on the technology adopted by each library, whereas the raw documents remain archived in a different server or the same store. The climatic event's history provides an interface for performing querying and analytics tasks on top of it. For example, locate events that happened during the XIX century, enumerate and locate the most famous climatic events in the region, or create a heat map of the events in Latin America that happened in the last ten years of the XIX century.

2.3   Exploring the Collections of Digital Newspapers

Newspaper articles are explored by conjunctive or disjunctive keyword queries, where keywords can belong to several vocabularies. For example, search articles reporting heavy storms and rivers flooding. The query expressed by a data scientist is automatically completed by using rewriting techniques that consider synonyms, more specific or more general concepts [4]. Thus, three tools can be used for exploring meteorological events depending on expert knowledge of what she/he is looking for:

- *Filtering.* Retrieving factual information. For example, filtering events by region or country or by year.
- *Term frequency.* Understanding the content of digital newspaper collection through the vocabulary used in its articles. Therefore, LACLICHEV exposes the terms frequency matrix and a terms heat map under an interactive interface. The domain expert can see which are, statistically, the terms most used in the articles, group documents according to the terms used, choose articles using a specific term.
- *Additional information.* Exploring the content of a specific article using a view that provides information about the name of geographic locations in the document that might refer to the location of the event and the region that it touched, and a list of geospatial features (e.g., school, public building, etc.) to determine, for example, the damages caused by the event.

3   LACLICHEV in Action

We have implemented LACLICHEV on top of the Jupyter platform[2] for executing the human-in-the-loop tasks that implement the data collection exploration process. Figure 5 shows the general architecture of LACLICHEV organised into three layers: (i) a frontend giving access to the events history containing curated articles reporting climatic events and providing tools for curating articles and creating events descriptions; (ii) a backend with the climatic event history and tools for pre-processing newspaper articles; and (iii) an external layer connecting to documents providers that are available through servers accessible in the web and through APIs exported by libraries.[3]

3.1   Building a Latin American Climatic Event History

We have worked with the national libraries of Mexico, Colombia, Ecuador and Uruguay to access their newspapers digital collections. For our experiments, we worked with the collections of the XVIII and XIX centuries of newspapers written in Spanish with the linguistic variations of these four countries. We curated

---

[2] https://jupyter.org.

[3] A demo version of LACLICHEV capable of exploring articles from TheGuardian is accessible on Github: http://github.com/javieraespinosa/ipgh-lab.

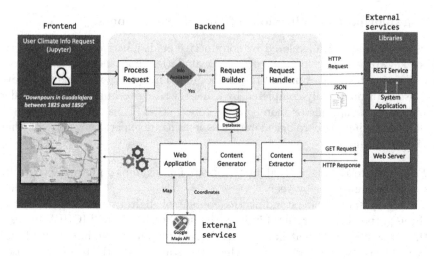

Fig. 5. General functional architecture of LACLICHEV

collections and generated the vocabulary used on articles identified as reporting a meteorological event. Digital newspaper collections remain in the initial repositories that belong to the libraries. Then, terms and links to the OCR (Optical Character Recognition) archives containing documents with articles reporting meteorological events were stored in distributed histories managed in each country. The process consists of five steps usually used in natural language processing techniques: sentence segmentation, tokenization, speech tagging, entity, and relation detection. LACLICHEV implements these phases on Python, relying on the NLTK library.

Besides curating the data collections content, we wanted to discover linguistic variations in different Latin American countries to describe meteorological events. People's language and variations can picture civilians' perception of these events, consequences, and associated explanations. Thus, local vocabularies were created out of the terms used in newspapers' articles[4] (see Fig. 6). Then we updated and enriched through queries, exploration and analytic activities these vocabularies through human-in-the-loop actions. Data scientists tagged "colloquial" terms used to describe climatic events and associated them with more scientific terms. These terms can be then used for defining keyword queries for exploring newspapers datasets.

3.2   Querying, Exploring and Curating Data Collections

LACLICHEV proposes a frontend for data scientists to:

– Query the events history of already tagged events. The queries can be keyword oriented (e.g. locate the most famous events in Mexico during the XVIII

---

[4] In Mexico a storm is called a "chaparrón" and in Uruguay it is called a "chubasco".

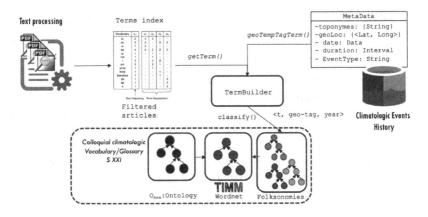

Fig. 6. Collecting colloquial vocabulary

century). Users decide to use some terms that can belong to any of the vocabularies generated in the pre-processing phase. LACLICHEV applies then query rewriting techniques to extended user expressed queries with synonyms, subsuming, and general terms. The particular characteristic of this task is that the user (i.e., data scientist) can interact and guide the process according to her/his knowledge and expectations about what she/he expects to explore and search.

- Perform analytics operations and analyse results generally presented within maps (e.g. where and how did rainy periods evolved in the region?). In the current version of LACLICHEV analytics queries cannot be expressed in the frontend. They are implemented manually through notebooks running on top of the events history.
- Exploring data collections for curating them and building the history of climatic events.
- Manage vocabularies, adding terms, guiding their classification, and studying the linguistic connections between the terms used in the different countries.

Data scientists can express queries that can potentially explore historical newspapers content to find articles describing meteorological events. The aim is to have a good balance between precision and recall despite the ambiguity of the language (Spanish variations in naming climatic events). The domain experts must express "clever" queries that can exploit the collections to achieve this goal.

Queries can be initial conjunctive and disjunctive expression combining terms chosen from the built-in vocabularies or not. Then, queries are rewritten in an expression tree where nodes are conjunction and disjunction operators and leaves are terms, according to an input query expressed as a conjunction and disjunction of terms potentially belonging to a meteorological vocabulary. Our approach for rewriting queries is based on a queries-as-answers process. This technique rewrites user queries into queries that can produce more precise results according to the explored dataset content. Queries-as-answers proposed by LACLICHEV consists of a list of frequently used queries. Thus, we focus on the following aspects:

Extending Query Alternatives Using Hypernyms and Synonymes. Given an initial conjunctive/disjunctive query, the query is rewritten by extending it with general and more specific terms, synonyms, etc. The terms used in the query belong to colloquial vocabulary for denoting meteorological events. The rewriting process can be automatic or interactive, in which case the system proposes alternatives and the user can validate the proposed terms (e.g., if the query is *"heavy storms"*, the query can be extended by adding *"heavy storm dust"*). We use Wordnet[5] for looking for associated terms and synonyms that help address concepts used in different Spanish speaking countries. We do not translate the query terms to other languages because our digital data collections consist of newspapers written in Spanish. LACLICHEV allows equivalent terms searching to morph a query. For a new term, LACLICHEV generates a node with the operator and then connects the initial term with the equivalent terms in a disjunctive expression subtree. Thereby, more general terms are collected and connected to the initial term with these terms in a conjunctive expression subtree. The result is a new expression tree corresponding to an extended query Q_{ExT}. The query morphing algorithm behind LACLICHEV is described in [8].

Extending Query Alternatives Using Cultural Terms. We use local vocabularies ("folksonomies" in a metaphorical way) for generating new query expression trees that substitute the terms used in Q'_{ExTi} with equivalent terms used in a target country (e.g., blizzard instead of a heavy storm). This will result in transformed expression trees each one using the terms of a country ($Q''_{ExT1} \ldots Q''_{ExTj}$) [9]. We create and feed each vocabulary according to the country of origin of the processed newspaper article. This lets us extract the vocabulary used during the XVIII and XIX centuries for describing meteorological events in different Latin American countries (i.e. Mexico, Colombia, Ecuador, and Uruguay). Using this information LACLICHEV can answer the following queries: *How terms used to describe climatic events have changed between XIX–XX c.? Which are standard terms used to describe climatic events across Latin American countries? Which are the most popular terms used in XIX c. for describing climatic events?*

Defining Filters Using Knowledge Domain. We also use domain knowledge for rewriting the queries. We have a knowledge base provided by domain experts that contains some meteorological event rules. For example, rules state that in the presence of a heavy storm: the wind speed is higher that 118 km/hr (R1); the rivers can grow and produce big waves (R2); there are rains between 2,5 and 7,5 mm/hr (R3); the range of surface that can be reached by a 100 km wind speed storm is of 1000 km (R4). Our approach uses this information for generating possible queries that help the domain expert better precise her/his query or

---

[5] http://timm.ujaen.es/recursos/spanish-wordnet-3-0/.

define several queries that can be representative of what she/he is looking for. For example, the previous initial query "Q_1: heavy storm" is rewritten into new additional queries: "Q_{11}: heavy storm *or storm with wind speed > 100 km*" (using R1); "Q_{12}: storms with 100 km speed that reached Mexico City" (using R2 and knowing the initial point and geographic information); "Q_{13}: storms touching villages 500 km around Mexico city happening in the same period" (using R4). Instead of having a long query expression, our approach proposes sets of queries that the domain expert can choose and combine.

Analytics Queries. The climatic events history provided and maintained by LACLICHEV can visualize information and perform analytical tasks. For example, LACLICHEV can answer spatio-temporal queries like: *locate climatic events in the XVII c.*; *enumerate & locate the most famous events in the region*; *create a heat map of events in Latin America in the last years of the XIX c.* The objective is to answer analytics queries that imply aggregating information stored in the events history. For example, *how did rainy time evolve in time in the region? In which way was climate different between XVII and XIX c.? How did vocabulary evolve from colloquial to scientific and standardized in the XX c.?* In future versions, LACLICHEV is willing to answer prediction queries like *could it have been possible to predict the evolution of climate behaviour from the data in XVIII and XIX c.?* This type of queries requires the collection, curation, and preparation of more newspapers articles and other complementary data. This concerns future work.

4   Related Work

Despite the availability of datasets, very often, users are unsure which patterns they want to find. Data exploration [6] proposes strategies for going into the whole or samples of datasets to understand their content and determine the type of questions that can be answered. Some of the techniques are data grooming, multi-scale queries, result set post-processing, query morphing, and queries as answers. *Data grooming* denotes the process of transforming raw data into analysable data with various data structures. *Multi-scale queries* propose to split a query into multiple queries executed on different fragments of the database and then perform a union of those queries. *Result set* post-processing assumes an array of simple statistical information such as min, max, and mean to be more helpful, especially on massive data sets. *Query morphing* and *queries as an answer* are rewriting techniques that compute alternative queries (e.g. adding terms) that can potentially better explore a dataset than an initial query. Approaches such as *interactive query expansion* (IQE) [3,5,7] have shown the importance of data consumers in the data exploration process. Users' intention helps to navigate through the unknown data, formulate queries, and find the desired information. In most of the occurrences, user feedback acts as vital relevance criteria for next query search iteration.

Existing solutions are not delivered in integrated environments that data scientists can comfortably use to explore data collections. The technical effort is still necessary to combine several tools to explore and process datasets and go from raw independent data sets to knowledge, for example, on climate change. Therefore, LACLICHEV aimed to tailor a data exploration environment that could help explore digital datasets using a human-in-the-loop approach.

5 Conclusion and Future Work

This paper presented our approach for exploring newspaper digital collections for building knowledge about the history of climate change in Latin America. Using well-known information retrieval and analytics techniques, we developed a data science exploration environment that provides tools for understanding the content of collections. We used digital newspapers collections for applying such techniques for building and analyzing the history of climate change in Mexico, Colombia, Ecuador, and Uruguay. The work reported here is the first step towards this ambitious challenge. We continue enriching data collections, developing and testing solutions for generating and sharing step-by-step this history.

References

1. Climate of the past: an interactive open-access journal of the European geosciences union. European Geosciences Union. http://www.climate-of-the-past.net. Accessed 23 Apr 2021
2. Historical climate data. Government of Canada. http://climate.weather.gc.ca. Accessed 23 Apr 2021
3. Belkin, N.J.: Some (what) grand challenges for information retrieval. In: ACM SIGIR Forum, vol. 42, pp. 47–54. ACM, New York (2008)
4. Carvalho, D.A.S., Souza Neto, P.A., Ghedira-Guegan, C., Bennani, N., Vargas-Solar, G.: *Rhone*: a quality-based query rewriting algorithm for data integration. In: Ivanović, M., et al. (eds.) ADBIS 2016. CCIS, vol. 637, pp. 80–87. Springer, Cham (2016). https://doi.org/10.1007/978-3-319-44066-8_9
5. Goswami, P., Gaussier, E., Amini, M.R.: Exploring the space of information retrieval term scoring functions. Inf. Process. Manag. **53**(2), 454–472 (2017)
6. Kersten, M.L., Idreos, S., Manegold, S., Liarou, E., et al.: The researcher's guide to the data deluge: querying a scientific database in just a few seconds. PVLDB Chall. Vis. **4**, 1474–1477 (2011)
7. Ruthven, I.: Re-examining the potential effectiveness of interactive query expansion. In: Proceedings of the 26th Annual International ACM SIGIR Conference on Research and Development in Information Retrieval, pp. 213–220 (2003)
8. Vargas-Solar, G., Farokhnejad, M., Espinosa-Oviedo, J.: Towards human-in-the-loop based query rewriting for exploring datasets. In: Proceedings of the Workshops of the EDBT/ICDT 2021 Joint Conference (2021)
9. Vargas-Solar, G., Zechinelli-Martini, J.L., Espinosa-Oviedo, J.A.: Computing query sets for better exploring raw data collections. In: 2018 13th International Workshop on Semantic and Social Media Adaptation and Personalization (SMAP), pp. 99–104. IEEE (2018)

Public Health Units - Exploratory Analysis for Decision Support

Tatiane Lautert[1], Nádia P. Kozievitch[1(✉)], Ismael Villanueva-Miranda[2], and Monika Akbar[2]

[1] Federal University of Technology – Paraná (UTFPR), Curitiba, PR, Brazil
tatianelautert@alunos.utfpr.edu.br, nadiap@utfpr.edu.br
[2] University of Texas at El Paso, 500 West University Avenue,
El Paso, TX 79968, USA
ivillanueva5@miners.utep.edu, makbar@utep.edu

Abstract. Guaranteeing adequate health services to the population is a challenge, especially in developing countries where limited resources must be optimized in order to reach a larger percentage of the population. To properly assess health services and prioritize new investments, it is important to collect, integrate, and analyze a large amount of relevant data. The paper presents a prototype for data visualization, along with challenges related to meaningful presentations of results.

Keywords: Open data · Health data · Exploratory analysis · Visualization

1 Introduction

Cities influence people's health and well-being through policies and interventions, including those addressing social inclusion and social support; support for healthy and active lifestyles (e.g., cycling lanes); safety and environmental regulations supporting children and elderly population; working conditions; climate change preparedness; among others[1]. According to the World Health Organization (WHO)[2], today's cities are facing a triple health burden: infectious diseases (such as pneumonia, dengue, HIV/AIDS, tuberculosis, pneumonia); noncommunicable diseases (such as heart disease, stroke, asthma) and other respiratory illnesses, cancers, diabetes and depression; and violence and injuries, including road traffic injuries.

Several health challenges can be listed as urgent for the next decade: (i) harnessing new technologies, (ii) stopping infectious diseases, (iii) elevating health in

---

[1] https://www.euro.who.int/en/health-topics/environment-and-health/urban-health/publications/2019/implementation-framework-for-phase-vii-20192024-of-the-who-european-healthy-cities-network-goals-requirements-and-strategic-approaches-2019, last accessed 22-Jul-2020.
[2] https://www.who.int/health-topics/urban-health, last accessed 22-Jul-2020.

© Springer Nature Switzerland AG 2021
L. Bellatreche et al. (Eds.): ADBIS 2021 Short Papers, Workshops and Doctoral Consortium, CCIS 1450, pp. 133–138, 2021.
https://doi.org/10.1007/978-3-030-85082-1_12

the climate debate, (iv) delivering health in conflict and crisis, among others [7]. In parallel, governments from several countries are taking initiatives to provide open health data to support planning and policy making (such as Open Cities Project[3] and European Data Portal[4]) to support primary health care and local services[5]. Curitiba, the eighth most populous city of Brazil, has been participating in the open data initiatives along with several government stakeholders, such as *Instituto de Planejamento de Curitiba* (IPPUC)[6] and the Municipality of Curitiba[7].

The Global Research Data Alliance Community (RDA) [6] created a set of guidelines and recommendations[8] for data sharing of health data under the present COVID-19 pandemic circumstances, including legal and ethical considerations, research software, community participation and indigenous data.

In this direction, a disease outbreak is the occurrence of disease cases in excess of normal expectancy[9]. The number of cases varies according to the disease-causing agent, and the size and type of previous and existing exposure to the agent, and the definition may vary for countries, regions and even cities. In Brazil, this information can be found in Epidemiological Bulletins, which contains detailed data and analysis[10]. In USA the information is provided by the Center for Disease Control and Prevention (CDC)[11]. The WHO outbreak definition [8] states that for a defined area, the average number of cases from previous years can be taken as a threshold. All observations above that threshold should be considered as an outbreak. In order to detect the outbreaks of these diseases, this paper uses a modified version of the World Health Organization (WHO) outbreak definition. Instead of using the raw disease count, we set the threshold for outbreak at two standard deviations in excess of the endemic channel (i.e., average) [1].

This article presents a prototype (using data from the public Health Units of Curitiba) for data visualization, along with challenges related to meaningful presentation of results.

---

[3] https://www.worldbank.org/en/region/sar/publication/planning-open-cities-mapping-project, last accessed 13-Sep-2020.

[4] https://www.europeandataportal.eu/en/highlights/open-health-data-european-data-portal, last accessed 22-Feb-2020.

[5] https://www.euro.who.int/__data/assets/pdf_file/0003/376833/almaty-acclamation-mayors-eng.pdf, last accessed 22-Jul-2020.

[6] http://ippuc.org.br/, last accessed 22-Feb-2020.

[7] http://www.curitiba.pr.gov.br/DADOSABERTOS/, last accessed 22-Feb-2020.

[8] https://www.rd-alliance.org/global-research-data-alliance-community-response-global-covid-19-pandemic, last accessed 06-May-2020.

[9] https://www.who.int/environmental_health_emergencies/disease_outbreaks/en/, last accessed 19-July-2020.

[10] https://www.saude.gov.br/boletins-epidemiologicos.

[11] https://wwwnc.cdc.gov/travel/destinations/traveler/none/brazil, last accessed 19-July-2020.

2   Prototype

The main datasets used in this paper come from IPPUC[12] and the Municipality of Curitiba's[13]. Additionally, this paper also uses socioeconomic and census data[14], and population density published by IBGE[15]. A preliminary exploratory analysis can be found in [4].

The prototype uses only open-source technologies to avoid any kind of spending on software acquisition and/or contracting. From the database server, it was used a Linux server with Debian 9 distribution with two cores of AMD EPYC 7401 processors and 1.5G of memory,with PostgreSQL 9.5 x64. From the application server, it was used Linux server with 12 Cores, 16 GB of RAM and Ubuntu distribution, with Apache Server and PHP. For maps and libraries, we used Open Street Map combined with Leaflet library and Heatmap puglin have been used for Mapping and Data visualization. Other Libraries such as: JQuery v. 3.3.1, Bootstrap v. 3.3.7, and Chart.js v. 2.7.3. Figure 1 shows the technologies used.

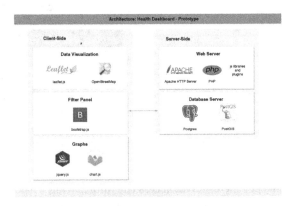

Fig. 1. Health dashboard architecture.

The application used the following tables: (i) Health Data Source Table: Similar to the data used during the exploratory data analysis phase, only the Medical appointments data has been used, from Basic Health Units (BHU) and Emergency Care Units (ECU). A total of 14,382,414 medical records from 2016 to 2020 have been inserted into this database table; and (ii) Geodata Tables: Database table limites_legais.divisa_de_bairros stores the geo-referenced data of Curitiba neighborhoods. Table limites_legais.divisa_de_regionais stores the geo-referenced data of Curitiba macro regions. Table saude.unidade_saude stores the geo-referenced data of Curitiba Health Units.

---

[12] http://ippuc.org.br, last accessed Nov 24, 2017.

[13] http://www.curitiba.pr.gov.br/DADOSABERTOS/, last accessed Nov 24, 2017.

[14] http://www.agenciacuritiba.com.br/, last accessed 16-May-2020.

[15] https://www.ibge.gov.br/, last accessed 16-May-2020.

In the prototype, all interaction takes place on a single screen. As shown in Fig. 2, the web interface is divided vertically in two: an interactive map with a search panel and map layer options is available on the right; on the left, the search results are available in the form of dynamic graphs. Through the search panel, users can filter the data according to their needs (year, health unit, day of the week, infectious disease), and through the map layer option users can select the maps layers they wish to see on the screen.

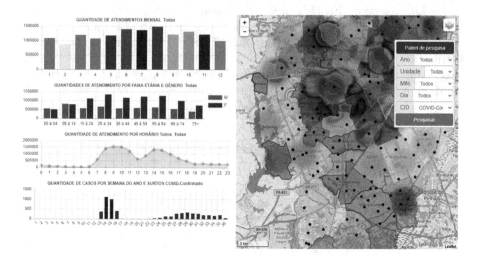

Fig. 2. Interface of the health dashboard prototype.

With the prototype, the user can access the information without any previous knowledge, and data are dynamically displayed according to the user's preference and needs.

Note that Fig. 2 also shows the number of medical appointments by age range and gender. The data are displayed according to the filter criteria defined by the user.

An experiment was conducted between 10-Oct-2020 to 23-Oct-2020 with seven participants to perform usability test and evaluation, 1 participant from the Health Science area, 1 from Architecture, and 5 from Computer Science area (1 IT Architect, 1 IT Quality Engineer, 1 IT Manager, 1 System Analyst, 1 Post Doctoral Student). It is important to mention that given the current COVID-19 pandemic scenario, it was not possible to perform the usability test scenario in deeper detail or in person due to the current social distancing guidelines.

The participants were asked to execute 1 low, 1 medium, and 1 high complexity usability testing scenarios. For each scenario, participants were asked to provide their responses based on the results they could see either on the Dashboard Graphs or on the Dashboard Map. The purpose of this usability test is to measure the user success rate, which is defined by the percentage of tasks that users complete correctly, as explained by Nilsen Normam Group[16]. After executing the usability testing scenarios, participants were asked to respond an evaluation questionnaire to assess their perception in the following aspects (i) easiness of use, (ii) relevance of filters available, (iii) complexity in understanding the graphs, (iv) suggestion for improvements, in case of any.

The results shows that all 7 users executed the usability testing scenario and provided the correct answer to the question. All 7 users evaluated the prototype as easy to use, 6 uses evaluated the complexity in understanding the graph as easy and 1 user evaluated as difficult. All 7 users evaluated the filter options as relevant.

Within the challenges in analyzing health data, we can mention: (i) **Privacy concerns:** Technologies helps improve the quality of health care [5], but data access, storage, and integrity are key challenges when it comes to electronic patient records. (ii) **Differences in data definitions and measurement methods:** Although guidelines are available[17], countries may have multiple sources of data for the same year, it is more usual for data not to be available for every population or year[18]. (iii) **Data Integrity and consistency:** When correlating multiple data sources, integrity and consistency among the different sources of data is a challenge, specially in terms of inconsistencies in data vocabulary, lack of common identifier across different data bases, and missing data. (iv) **Big data:** The challenge is how to make sense of such large amount of data, since the data is complex, and might contain geographical and temporal components [2], along with other issues (such as data quality). (v) **Prior knowledge required to make use of data:** There are several limitations in the use of open data [3], which includes technologies, metadata and standardization.

In parallel, several factors need to be considered when designing a health dashboard prototype such as: (i) Analysis of the technology to be used in terms of cost and libraries compatibility; (ii) Previous knowledge of the data is required; (iii) Identification of relevant filters for end-users; (iv) Analysis of what are the best ways to display the data and what statistical method to apply; (v) The prototype must provide a good user experience, functionalities need to be intuitive and simple; (vi) Use of techniques to enhance the system performance in fetching and retrieving the data from the database, (vii) Use of data security measures to avoid malicious SQL injections; (viii) The Public health data must always be anonymized to be displayed in this kind of dashboard; (ix) Consideration of user feedback for prototype enhancement and (x) it would be helpful if there were standards defined for data sharing and use of technologies by governing bodies.

---

[16] https://www.nngroup.com/articles/success-rate-the-simplest-usability-metric/, last accessed in 01/Nov/2020).

[17] https://www.cdc.gov/nchs/data/series/sr_02/sr02_141.pdf, last accessed 09-Jun-2020.

[18] https://www.who.int/gho/publications/world_health_statistics/2018/en/, last accessed 09-Jun-2019.

The source-code of the prototype is available in GitHub[19] and video showing the prototype usage is available in Youtube[20]. Further details about the exploratory analysis using the same data is available in [4].

3 Conclusion

This paper presents a prototype using data for Curitiba over three years, along with the main research challenges. The Health Data Prototype showed that it can facilitate the analysis of the health data aggregated to other sources to either the general public or professionals of the Health Science departments. It also showed that even with the use of a great amount of data the dashboard performance to show the graphs and maps is not affected with the appropriate techniques of data modeling design. As future work, we can mention the enhancement of the Health Data prototype based on the users' inputs, the inclusion of other datasets such as air quality data and how it correlates with diseases of respiratory tract.

References

1. Brady, O.J., Smith, D.L., Scott, T.W., Hay, S.I.: Dengue disease outbreak definitions are implicitly variable. Epidemics **11**, 92–102 (2015)
2. Ferreira, N., Poco, J., Vo, H.T., Freire, J., Silva, C.T.: Visual exploration of big spatio-temporal urban data: a study of New York City taxi trips. IEEE Trans. Vis. Comput. Graph. **19**(12), 2149–2158 (2013). https://doi.org/10.1109/TVCG.2013.226
3. Janssen, M., Charalabidis, Y., Zuiderwijk, A.: Benefits, adoption barriers and myths of open data and open government. Inf. Syst. Manag. **29**, 258–268 (2012). https://doi.org/10.1080/10580530.2012.716740
4. Lautert, T.A.M.: Public health units in Curitiba: an exploratory analysis and a health dashboard prototype for decision support in health management domain. Master's thesis, UTFPR (2020)
5. Meingast, M., Roosta, T., Sastry, S.: Security and privacy issues with health care information technology. In: 2006 International Conference of the IEEE Engineering in Medicine and Biology Society, pp. 5453–5458, August 2006. https://doi.org/10.1109/IEMBS.2006.260060
6. RDA COVID-19 Working Groups: RDA COVID-19 Working Group Recommendations and Guidelines, 1st release. Research Data Alliance (2020). https://doi.org/10.15497/rda00046
7. World Health Organization: Urgent health challenges for the next decade (2020). https://www.who.int/news-room/photo-story/photo-story-detail/urgent-health-challenges-for-the-next-decade
8. World Health Organization and Others: Who guidelines for epidemic preparedness and response to measles outbreaks. Technical report, World Health Organization (1999)

---

[19] https://github.com/TatianeLautert/dadosAbertosSaudeCuritiba, last accessed on 18-Oct-2020.
[20] https://youtu.be/47KMZO-N91A, last accessed on 08-Nov-2020.

The Formal-Language-Constrained Graph Minimization Problem

Ciro Medeiros[1,2]($\boxtimes$), Martin Musicante[1], and Mirian Halfeld-Ferrari[2]

[1] Federal University of Rio Grande do Norte, Natal, Brazil
cirommed@ppgsc.ufrn.br, mam@dimap.ufrn.br
[2] Université d'Orléans, INSA CVL, LIFO, Orléans, France
mirian@univ-orleans.fr

Abstract. We formally define the *Formal-Language-Constrained Graph Minimization* problem. The goal is to minimize the amount of data associated to the edges of a graph database, while preserving its utility, defined as a conjunctive path query. We show that the problem is NP-hard. A solution to this problem contributes to balancing the trade-off between privacy protection and utility when publishing a graph database.

Keywords: Privacy · Graph databases · Path queries

1 Introduction

Many strategies for preserving both privacy and utility in published sensitive data have been developed in the last two decades. Too much protection reduces utility and thus the value of data being published. Too much utility augments the risks of sensitive information exposure. Balancing the trade-off between privacy protection and data utility is, therefore, the main concern. This paper concentrates on the problem of reducing the amount of data disclosed in a graph database respecting a set of user-defined utility queries. Such graph reduction would be a compromise between individual's demand for privacy safety and companies' need for mining personal data. This work concerns *utility preservation* protecting as much privacy as possible, although no guarantees are provided.

Given a graph D and a set of queries Q, we focus on the reduction of D to a minimal subgraph D^- capable of answering all queries in Q. The goal of this paper is to formalize that problem and study its complexity. Reducing the data graph to the smallest subgraph necessary to ensure the answers for selected queries is an important contribution for preserving a useful querying capacity while applying some data protection strategy. The following example illustrates our viewpoint.

Example 1. To detect the spread of a disease transmissible via face-to-face contact, the natural question raised by authorities is "Given the infection cases so far, who else could have been infected by those?". The government requests

© Springer Nature Switzerland AG 2021
L. Bellatreche et al. (Eds.): ADBIS 2021 Short Papers, Workshops and Doctoral Consortium,
CCIS 1450, pp. 139–145, 2021.
https://doi.org/10.1007/978-3-030-85082-1_13

data from third-party companies that keep geolocation information of people to analyze the spread of the disease and take refraining measures. Figure 1 shows an example graph database representing people's daily relevant displacements (consider all nodes and edges, independently of their color). Every person (Alice, Bob and Claire) is linked to a trace (nodes labeled "tx") composed of a sequence of places (nodes labeled "pyx") they visited along the day. Every edge has a non-negative weight (0 if not provided) assigned by the data keeper corresponding to how sensitive is that information. Alice, Bob and Claire started their day at home. Alice went to work, then to a bar, then back home. Bob left his kids at the nursery, went to work, then to the same bar as Alice, then he took his kids back at the nursery, and went back home. Claire spent her whole day at the nursery and then returned home. One can detect possible infection cases by querying for paths (*i.e.* sequences of edge labels), starting infected at people, following the places they visited and then finding people that were at the same place. Such paths belong to the formal language defined by the regular expression: `trace next* place ^place ^next* ^trace`, where ^ denotes the inverse direction of an edge. For simplicity, we consider that recently infected people cannot infect others immediately.

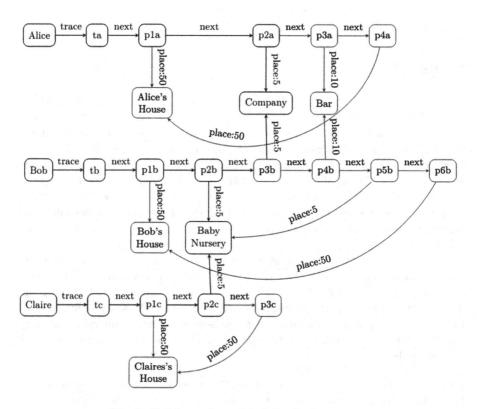

Fig. 1. Database of people's daily displacements.

Suppose Alice is infected, the government wants to analyse the spread of the infection and asks for localization data from a third-party company. To avoid privacy violation, neither the government shares the fact that Alice is infected, nor the third-party company discloses the entire database. The solution we propose relies on the fact that it is possible to reduce the amount of sensitive information disclosed by the third-party company by sharing only a small subset of edges that is sufficient for the government purposes.

The blue sub-graph in Fig. 1 represents the exported database with much less sensitive information than the original one – the sum of weights of the blue part is 20, while it is 345 for the whole graph. Now, even if we are still capable of showing how the infection may spread, we just reveal the information that Bob and Alice are coworkers (and not that they met in a bar). Notice that if another person, say, David, met Alice at the same bar and if there is no other way to reach David from Alice, the situation for Bob and Alice does not change: to preserve utility, we would store the path via Bar connecting Alice and David, without changing the situation for Bob and Alice. Several techniques can be combined to achieve higher levels of privacy protection.   □

Related Work. Our problem differs to the construction of minimum spanning trees since that problem does not require the preservation of query results. Works on privacy and graph databases do not usually consider minimizing graph path data. The Minimum Exposure (ME) problem is introduced in [1] which does not deal with graph databases. The goal is to minimize the information provided by clients to companies, *i.e.*, the company receives only the information needed to satisfy the conditions for that offer. In [2] we find the foundations for publishing graph data with formal, query-based, privacy guarantees against linkage attacks. In [3] we find algorithms to implement privacy and utility queries, but path queries are not supported. In [4], a survey on graph data anonymization, de-anonymization, and de-anonymizability quantification techniques, the authors conclude that it might be impossible to develop some technique that can provide good privacy guarantees while preserving all the data utility. In a graph database, much data can be used to identify individuals and reconstruct sensitive information. Vertex degree, path length, hub score and centrality are some of them. Although the variety of techniques proposed so far, to the best of our knowledge, none of them focuses on minimizing the amount of information disclosed in graph databases considering path queries.

Paper Organization. Section 2 provides theoretical background. Section 3 formally defines the problem and analyzes its complexity. Concluding remarks are offered in Sect. 4.

2   Preliminaries

In our setting, a path query takes as input a graph database, a grammar and a set of start vertices.

Definition 1 (Grammar). *A grammar is a tuple* $G = (N, \Sigma, P, S)$, *where* Σ *and* N *are respectively the set of terminal and non-terminal symbols,* P *is the set of production rules* $\alpha \rightarrow \beta$, *where* $\alpha, \beta \in (N \cup \Sigma)^*, \alpha \neq \epsilon$, *and* S *is the start symbol.*

Given a string $\ldots \alpha \ldots$, one can apply the production rule $\alpha \rightarrow \beta$ by replacing α by β, producing a new string $\ldots \beta \ldots$. If one can successively apply production rules to α generating a string γ, we say that γ is α-derivable.

Definition 2 (Graph). *A graph is a set of triples* $D : V \times E \times V$, *where* V *is the set of vertices and* E *is the set of edge labels.*

Definition 3 (Path). *A path* π *of length* k *is a sequence of triples* t_i *such that* $t_i = (v_{i-1}, e_i, v_i)$, *for* $1 \leq i \leq k$.
The concatenation of edge labels e_i *of a path form a string called* trace.

Definition 4 (Query). *Given a grammar* G *and a graph* D, *a query is a pair* (a, X), *where* $a \in V, X \in N$. *Answers for such query are the destination vertices* b *of paths that start at vertex* a *and form an* X-derivable trace to b.

We can now define a reachability function [5], from a node x by a path α.

Definition 5 (Reachability Function). *Let* G *be a grammar, and* D *be a data graph. Given a vertex* $x \in V$ *and a string* $\alpha \in (\Sigma \cup N)^*$, *the function* $\mathcal{R}_G^D(x, \alpha)$ *defines the set of vertices reachable from* x *by following an* α-derivable *path in* D: $\mathcal{R}_G^D(x, \alpha)$: $V \times (\Sigma \cup N)^* \mapsto \mathcal{P}(V)$. *This function is recursively defined on* α, *as follows:*

1. *For* $\alpha = \varepsilon$, *where* ε *is the empty string, every vertex* x *is reachable from itself:*
   $\mathcal{R}_G^D(x, \varepsilon) = \{x\}$
2. *For* $\alpha = p \in \Sigma$, *the set of vertices reachable from* x *are those connected to* x *via a* p-labeled edge: $\mathcal{R}_G^D(x, p) = \{y \mid (x, p, y) \in D\}$
3. *If* $\alpha \rightarrow \beta \in P$, *the set of vertices reachable from* x *is defined by using the right-hand side of the production rules of* α *in* G:
   $\mathcal{R}_G^D(x, \alpha) = \bigcup_{\alpha \rightarrow \beta \in P} \mathcal{R}_G^D(x, \beta)$
4. *For* $\alpha = \alpha_1 \alpha_2$, *the set of vertices reachable from* x *is given by:*
   $\mathcal{R}_G^D(x, \alpha_1 \alpha_2) = \bigcup_{w \in \mathcal{R}_G^D(x, \alpha_1)} \mathcal{R}_G^D(w, \alpha_2)$.

3 Problem

Data *utility* can be modeled as a set of user-defined path queries that retrieve useful information [2]. To keep the minimum information possible to answer those path queries we propose a function to measure the amount of information being disclosed. Given a graph database D, an exposure function $f(D) = \sum_{t_i \in D} w_i$, where $w_i \in \mathbb{N}$ is the weight[1] assigned to triple t_i, returns a non-negative number corresponding to the amount of information disclosed. Identifying a subgraph that minimizes the amount of information disclosed configures an optimization problem, as formalized in the following definition.

---

[1] Consider $w_i = 1$ as the default value.

Definition 6 (Formal-Language-Constrained   Graph   Minimization Problem). *Given a grammar G, a data graph D, a set of queries Q and a weight function f, the* Formal-Language-Constrained Graph Minimization Problem (FLGM) *consists of identifying a subgraph $D^- \subseteq D$ such that for all queries $(a, X) \in Q$, it holds that $\mathcal{R}_G^{D^-}(a, X) = \mathcal{R}_G^D(a, X)$, and $f(D^-)$ is minimum.*

The FLGM optimization problem is NP-hard, and its decision version is NP-complete for polynomial-time decidable languages. Inspired by Anciaux et al. [6], we demonstrate these statements by reducing an instance of the All Positive Minimum Weighted Boolean Satisfiability (APMWS) [7] problem to an instance of the FLGM problem.

Theorem 1. *The APMWS problem is reducible to the FLGM problem.*

Proof. An instance of the APMWS problem consists of, given a formula in the conjunctive normal form $F = \bigwedge_j (\bigvee_k b_{j,k})$ and a weight function $f(h) = \sum_{b_{j,k}=true \in h} w_{j,k}$, finding an assignment of truth-values to variables $h : b_{j,k} \mapsto \{true, false\}$ such that $h(F) = true$, and $f(h)$ is minimum.

The FLGM problem can be represented by a boolean formula similar to that of the APMWS problem as $F' = \bigwedge_j (\bigvee_k (\bigwedge_m b_{j,k,m}))$, where: (1) the outer conjunction is true if all queries $(a, X) \in Q$ have their answers preserved; (2) the disjunction is true if there exists at least one path from vertex a to each destination b following an X-derivable path; (3) the inner conjunction is the set of triples that form an X-derivable path from a to b; (4) a variable $b_{j,k,m}$ is true if its corresponding triple $t_{j,k,m}$ is in D^-, and false otherwise (it can occur that $t_{j,k,m} = t_{j',k',m'}$ for $j \neq j', k \neq k', m \neq m'$).

Solving this problem means finding an assignment $h' : b_{j,k,m} \mapsto \{true, false\}$ such that $h'(F') = true$, and $f(h')$ is minimum. Considering triple sets of size 1 in the inner conjunction of the FLGM problem, an APMWS variable $b_{j,k}$ can be rewritten as $b_{j,k,1}$. We can then solve an APMWS instance by solving its equivalent FLGM instance.

Corollary 1. *The FLGM optimization problem is NP-hard.*

Proof. From Theorem 1 and from the fact that the APMWS optimization problem is NP-hard [7], it follows that the FLGM problem is NP-hard as well.

Corollary 2. *If input grammar G belongs to a class of polynomial-time decidable grammars, the FLGM decision problem is NP-complete.*

Proof. Given a solution to an instance of the FLGM decision problem, which includes a constant k, we can verify the solution in polynomial time by (1) summing the weight of all edges in D^- and comparing it to k; and (2) running a polynomial-time algorithm to evaluate the path queries Q. Therefore, $FLGM \in NP$. From that, and from Corollary 1, we have that the FLGM decision problem is NP-complete for polynomial-time decidable grammars.

Notice that such restriction on the class of the input grammar is necessary, since the sole task of recognizing a string using a non-polynomial-time decidable grammar is, by definition, not in P. Regular and Context-free grammars are polynomial-time decidable.

4 Concluding Remarks

In this paper, we introduce and formalize graph optimization problem that is relevant to the area of privacy preservation, aiming at minimize the exposure of sensible data. The Formal-Language-Constrained Graph Minimization problem consists of computing a minimal subgraph that preserves utility queries, which are defined by grammars. Our approach may be used in conjunction with Regular and Context-free path queries [8]. Although presented in the context of privacy protection, the FLGM problem emerges from different scenarios such as network minimization, source-code analysis [9], among others.

We are currently working on a heuristic allowing to treat the FLGM problem in polynomial-time. The idea is to iteratively remove edges from the input graph while respecting the paths defined in the utility queries. We are producing a prototype to assert the feasibility of our technique via experiments. Our technique will be applicable along with other privacy protection ones, in a complementary fashion.

Acknowledgements. Work partly supported by INES (CNPq/465614/2014-0), CAPES - Finance Code 001. Participation in the ANR-SendUp project and DOING action (DIAMS, MADICS).

References

1. Anciaux, N., Nguyen, B., Vazirgiannis, M.: Miminum exposure in classification scenarios. Research report (2011)
2. Grau, B.C., Kostylev, E.V.: Logical foundations of privacy-preserving publishing of linked data. In: 30th AAAI Conference on Artificial Intelligence, pp. 52–95 (2016)
3. Delanaux, R., Bonifati, A., Rousset, M.-C., Thion, R.: RDF graph anonymization robust to data linkage. In: Cheng, R., Mamoulis, N., Sun, Y., Huang, X. (eds.) WISE 2020. LNCS, vol. 11881, pp. 491–506. Springer, Cham (2019). https://doi.org/10.1007/978-3-030-34223-4_31
4. Ji, S., Mittal, P., Beyah, R.: Graph data anonymization, de-anonymization attacks, and de-anonymizability quantification: a survey. IEEE Commun. Surv. Tutor. **19**(2), 1305–1326 (2017)
5. Medeiros, C.M., Musicante, M.A., Costa, U.S.: An algorithm for context-free path queries over graph databases. In: Proceedings of the 24th Brazilian Symposium on Context-Oriented Programming and Advanced Modularity, SBLP 2020, New York, NY, USA, pp. 40–47. ACM (2020)
6. Anciaux, N., Nguyen, B., Vazirgiannis, M.: Limiting data collection in online forms. IEEE PST (2012)
7. Alimonti, P., Ausiello, G., Giovaniello, L., Protasi, M.: On the complexity of approximating weighted satisfiability problems (extended abstract) (1997)

8. Medeiros, C.M., Musicante, M.A., Costa, U.S.: LL-based query answering over RDF databases. J. Comput. Lang. **51**, 75–87 (2019)
9. Li, Y., Zhang, Q., Reps, T.: Fast graph simplification for interleaved Dyck-reachability. In: Proceedings of the 41st ACM SIGPLAN Conference on Programming Language Design and Implementation, pp. 780–793 (2020)

Interpreting Decision-Making Process for Multivariate Time Series Classification

Rufat Babayev[✉][iD] and Lena Wiese[iD]

Fraunhofer ITEM, Nikolai-Fuchs-Strasse 1, 30625 Hannover, Germany
{rufat.babayev,lena.wiese}@item.fraunhofer.de

Abstract. Over the last years, several time series classification (TSC) algorithms have been proposed both in traditional machine learning and deep learning domains which have shown remarkable enhancement over the previously published state-of-the-art methods. However, their decision-making processes generally stay as black boxes to the user. Model-agnostic (post-hoc) explainers, such as the state-of-the-art SHAP, are proposed to make the predictions of machine learning models explainable with the presence of well-designed domain mappings. In our paper, we first apply univariate classifiers on the dimensions of multivariate time series data individually. This is a straightforward technique for multivariate time series classification (MTSC). Then, we use state-of-the-art timeXplain framework to interpret the decision making process of the univariate classifiers on the multivariate time series data. With a careful choice of interpretability parameters, we demonstrate that it is possible to obtain explainability for such time series data.

Keywords: Time series analysis · Multivariate time series classification · Interpretability of machine learning models

1 Introduction

Time Series Classification (TSC) is a type of supervised machine learning where attributes of the input vector have ordered and real-valued entries. This makes time series data different from what the traditional algorithms are designed for.

Over the years, the main emphasis has been on univariate time series classification, where each series contains records of a single variable and a class label [9]. However, it is more common to notice MTSC tasks, where time series contains records of multiple variables and associated label. For example, activity recognition, measurements based on EEG, ECG and EHR are all inherently multivariate. Nevertheless, the general focus in TSC field has been on the univariate case. For example, *sktime* [6] which is a popular framework for time series analysis, contains significantly more univariate algorithms than the multivariate ones. The sktime is designed to work with datasets from UEA [3] archive which is a resource for MTSC. Since, the datasets there are of equal length and do not have missing values, comparing algorithms on them becomes straightforward.

© Springer Nature Switzerland AG 2021
L. Bellatreche et al. (Eds.): ADBIS 2021 Short Papers, Workshops and Doctoral Consortium,
CCIS 1450, pp. 146–152, 2021.
https://doi.org/10.1007/978-3-030-85082-1_14

Using sktime, we compare two recent MTSC algorithms to the direct dimension independent transformations of univariate classifiers. After that, we apply timeXplain framework [8] to interpret the decision making process of these transformations. We carefully choose the interpretability parameters and explain each interpretation depending on the nature of the specific dataset. We discuss the results and draw further conclusions for the future work.

2   Related Work

One of the notable applications of TSC is the usage of distance measures. In this respect, [1] thoroughly reviewed different measures. Through kNN or SVM, these distance measures can be utilized for TSC. The sktime library provides many distance measures and TSC algorithms in this respect. The papers [3,9] present research for MTSC using the data from UEA archive.

The closest to our paper is [9] which uses the approach to build an ensemble of univariate classifiers over dimensions for MTSC. They compare the predictive performance of the ensemble classifier to deep learning models, traditional algorithms using DTW [5] and combined methods on UEA datasets [3].

In contrast, our paper is focused on medical datasets to look for better interpretability and datasets with small number of dimensions. Instead of building an ensemble of univariate classifiers, we directly apply univariate classifiers over dimensions, compare them to bespoke MTSC algorithms and investigate the decision making process of these classifiers on classification results.

To the best of our knowledge, our paper is the pioneering one which is focused on specific datasets from UEA archive using the transformations of univariate classifiers for MTSC and attempting to interpret these classifiers.

3   Empirical Evaluation

We selected multivariate time series datasets from UEA archive [3] which are formatted to be of equal length, include no missing values and have predefined train/test splits. The descriptions of them are as follows:

AtrialFibrillation - This is a Physionet dataset of two-channel ECG recordings. The class labels are n, s and t. The class n is described as a non-termination atrial fibrillation The class s is described as an atrial fibrillation that self terminates at least one minute after the recording process. The class t is described as immediate termination (within a second of recording ending) [3].

ERing - This data is generated with a prototype finger ring, called eRing, that can be used to detect hand and finger gestures. There are six classes for six postures. The classes are *hand open*, *fist*, *two*, *pointing*, *ring* and *grasp* [3].

SelfRegulationSCP1 - It is obtained from the dataset Ia of BCI II competition. The data were taken from a healthy subject. The subject was asked to move a cursor up and down on a computer screen, while his cortical potentials were taken. There are two classes: (cortical) positivity, (cortical) negativity [3]. **SelfRegulationSCP2** - It is obtained from the dataset Ib of BCI II competition. The data were taken from an artificially ventilated ALS patient [3].

We performed binary/multiclass classification on these datasets. Classes were fairly balanced, thus, we did not apply any imbalanced learning strategy.

In our work, we use *column ensembling* strategy from [2]. It enables univariate classifiers to be applied on multivariate time series data. Such univariate classifiers are called *dimension independent* classifiers or *adapted* classifiers for MTSC. We use the same classifiers from [2]. All classifiers are run with user-provided and/or default parameters and random seed of 1 for reproducibility.

3.1 Classification Results

We utilize the same classification metrics provided in [2]. Results are reported in Table 1: among the classifiers defined in [2], TSF refers to **TSF**, kNN to **kNN**, R-F to **RISF**, c-S to **cBOSS**, W-L to **WEASEL**, M-L to **MrSEQL** and M-E to **MUSE** where the last two classifiers are bespoke MTSC algorithms.

Table 1. Performance of classifiers in AtrialFibrillation, ERing, SelfRegulationSCP1 and SelfRegulationSCP2 datasets.

Clas.	AtrialFibrillation				ERing				SelfRegulationSCP1				SelfRegulationSCP2			
	Acc	F1	AUC	Rec	Acc	F1	AUC	Rec	Acc	F1	AUC	Rec	Acc	F1	AUC	Rec
TSF	0.26	0.22	0.41	0.26	0.91	0.91	0.98	0.91	0.86	0.85	0.95	0.86	0.5	0.49	0.49	0.5
R-F	0.26	0.26	0.34	0.26	0.87	0.86	0.97	0.87	0.72	0.71	0.82	0.72	0.46	0.46	0.47	0.46
kNN	0.26	0.22	0.42	0.26	0.82	0.82	0.95	0.82	0.79	0.79	0.85	0.79	0.48	0.48	0.47	0.48
c-S	0.26	0.28	0.39	0.26	0.86	0.86	0.98	0.86	0.78	0.78	0.95	0.78	0.45	0.44	0.47	0.45
W-L	0.4	0.41	0.52	0.39	0.95	0.95	0.99	0.95	0.77	0.76	0.91	0.77	0.53	0.53	0.54	0.53
M-L	0.26	0.30	0.42	0.26	0.87	0.87	0.99	0.87	0.87	0.87	0.99	0.87	0.50	0.49	0.46	0.50
M-E	0.26	0.27	0.38	0.26	0.97	0.97	0.99	0.97	0.97	0.97	0.99	0.97	0.51	0.42	0.53	0.51

In AtrialFibrillation, WEASEL outperforms all other classifiers. The classifier that comes close to WEASEL is MrSEQL. The other adapted classifiers perform similarly to MrSEQL and MUSE. One reason for the general low performance is that a train set size is much smaller than a length of the series (15 vs. 640).

In ERing dataset, MUSE outperforms the rest of the classifiers. The best adapted classifier performing closely to MUSE is WEASEL.

The last two datasets have higher time series length (896 and 1152) than the other datasets. In SelfRegulationSCP1, MUSE outperforms all other classifiers. In SelfRegulationSCP2, WEASEL outperforms all others, while MUSE becomes the closest to WEASEL. SelfRegulationSCP2 has the highest length, thus, the performance of WEASEL as a dimension independent classifier is significant here. The reason for lower performance of the other classifiers in SelfRegulation-SCP2 is that the data is taken from an artificially respirated ALS patient and as compared to the healthy subject of SelfRegulationSCP1, that patient may not have smooth time series which can be fitted effectively by the chosen classifiers.

We observe that even if MUSE is winner on SelfRegulationSCP1 and ERing, dimension independent approaches still show decent or comparable performance on these datasets. They are even winners on AtrialFibrillation and SelfRegulationSCP2. For instance, WEASEL is winner on AtrialFibrillation and SelfRegulationSCP2, runner-up on ERing and average on SelfRegulationSCP1. TSF holds the third place on ERing, SelfRegulationSCP1 and SelfRegulationSCP2.

4   Interpretation of the Classifier Models

The paper [4] provides an overview of ML model interpretability approaches such as intrinsic/post-hoc and global/local. The *timeXplain* [8] is a post-hoc local explanation framework for univariate time series classifiers. Post-hoc local approaches use perturbation such that they study a ML model in the vicinity of the input to obtain an explanation. This input is called the *specimen*. By slightly changing features of the specimen, an insight is gained as to which of these features are important for the model, and an explanation is obtained. To realize this, mapping functions need to be defined that describe how the specimen is to be perturbed [8]. The *timeXplain* defines mappings that build on the time domain, frequency domain, and statistics of time series. In our work, we use time domain or time-slice mappings. The idea behind time-slice mapping is to partition the time series x into d' contiguous subseries, called *slices*, whose lengths differ by at most 1 [8]. Then the function $\tau : \{1,\ldots,d\} \rightarrow \{1,\ldots,d'\}$ assigns each time t a slice number $\tau(t)$. To disable a slice of x, which is actually a fragment of the specimen, one cannot just replace it with missing values as most models cannot deal with. Instead, a slice is replaced with the respective slice of a so-called *replacement time series* $r \in I$, where I is the input space [7]. The replacements are defined according to an (unbiased) background set S of sample time series, e.g., test data. We use the default replacement strategy [8].

To find out impacts of different fragments, we use *multi-run averaging and background set splitting* [8]. It finds more fragments that are influential in the prediction of x and handles inhomogeneities in S by asking the model to discern x from n different samples from S and then average the impacts [8]. For each dataset, we utilized 500 samples for each of 5 runs in our experiments. For all datasets, we found d' as 10 after tests between 10 and 100 in steps of 10.

Since, we study the time domain, we utilize TSF on dimensions of the specimen for explainability. According to [8], TSF is better interpreted by time slices. In comparison, WEASEL should be interpreted in the frequency domain [8].

In Fig. 1a, time-slice explanations for the specimen from AtrialFibrillation dataset is shown. The true class of the specimen is n - non-termination. However, the predictions for both dimensions are t - immediate termination. In Fig. 1a, red color shows slices which conduce towards the true class and blue color symbolizes

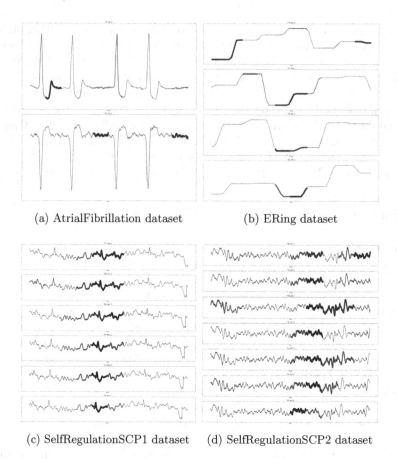

(a) AtrialFibrillation dataset   (b) ERing dataset

(c) SelfRegulationSCP1 dataset   (d) SelfRegulationSCP2 dataset

Fig. 1. Time slices contribute towards (red) or against (blue) the true class of the specimen. (a) Predictions: [*t, t*], True class: *n*; (b) Predictions: [*pointing, fist, fist, fist*], True class: *fist*; (c) Predictions: [*positivity, positivity, positivity, positivity, positivity, positivity*], True class: *positivity*; (d) Predictions: [*positivity, negativity, positivity, negativity, positivity, positivity, negativity*], True class: *positivity*. (Color figure online)

slices which conduce against the true class. For the 2nd dimension (from the top), two (almost smooth) slices conduce against the true class resulting in wrong prediction. In the 1st dimension, however, there are slices which conduce both against and towards the true class, especially the two peaks.

Figure 1b shows time-slice explanations for the specimen from ERing dataset. The true class is *fist*. Predictions are [*pointing, fist, fist, fist*]. During the fist position, a hollow is formed in the middle of a hand. Thus, the 2nd, 3rd and 4th dimensions (from the top) have red colors on the bottom of the hollow which indicates contribution towards the true class. However, the 1st dimension has peaks instead, thus, that dimension is wrongly classified as *pointing*.

Explanations for the specimen from SelfRegulationSCP1 is shown in Fig. 1c. The start and end of the series do not have much impact on the classification. Specific time slices in the middle conduce towards the true class for all dimensions. Thus, the prediction for each dimension is same as the true class (*positivity*). Only the beginning of the 2nd dimension contributes against the true class which is negligible if we consider the prediction for this dimension.

In contrast to SelfRegulationSCP1, the specimen of SelfRegulationSCP2 in Fig. 1d is obtained from artificially ventilated ASL patient. The undulant behavior in dimensions, especially blue colored slices in the middle of series conduce against the true class. In the 1st, 2nd, 3rd, 5th, 6th and 7th dimensions, there are red colored slices which follow the blue ones. However, for some dimensions their impact is not as big as the blue colored slices resulting in the contribution against the true class (*positivity*), i.e., the 2nd, 4th and 7th dimensions.

5 Conclusion

We utilized multi-run averaging and background set splitting strategy to increase the impact from different slices. This can be tuned to certain level with more runs and more samples if we need to achieve higher impacts from slices. However, the important point here is a choice of the right number of fragments d'. We obtained the value for d' after straightforward testing. According to [8], taking d' too low or too high may reduce quality of the explanation. A thorough investigation of d' can be an interesting step for the future work.

For the sake of reproducing the results obtained in this work, our source code is published in *ipynb* files in a public repository[1].

Acknowledgement. This work was supported by the Fraunhofer Internal Programs under Grant No. Attract 042-601000. We thank to the providers of timeXplain library (Felix Mujkanovic and others) for their source code and data set donors of the UEA archive (Anthony Bagnall and others) for their valuable datasets.

References

1. Abanda, A., Mori, U., Lozano, J.A.: A review on distance based time series classification. Data Min. Knowl. Disc. **33**(2), 378–412 (2019)
2. Babayev, R., Wiese, L.: Benchmarking classifiers on medical datasets of UEA archive. In: Proceedings of AI Health WWW 2021: International Workshop on AI in Health (2021)
3. Bagnall, A., et al.: The UEA multivariate time series classification archive (2018). arXiv preprint arXiv:1811.00075 (2018)
4. Du, M., Liu, N., Hu, X.: Techniques for interpretable machine learning. Commun. ACM **63**(1), 68–77 (2019)
5. Giorgino, T.: Computing and visualizing dynamic time warping alignments in R: the dtw package. J. Stat. Softw. **31**(7), 1–24 (2009). https://doi.org/10.18637/jss.v031.i07

---

[1] https://github.com/CavaJ/MTSInterpretability.

6. Löning, M., Bagnall, A., Ganesh, S., Kazakov, V., Lines, J., Király, F.J.: sktime: a unified interface for machine learning with time series. In: Workshop on Systems for ML at NeurIPS 2019 (2019)
7. Lundberg, S.M., Lee, S.I.: A unified approach to interpreting model predictions. In: Advances in Neural Information Processing Systems, pp. 4765–4774 (2017)
8. Mujkanovic, F., Doskoč, V., Schirneck, M., Schäfer, P., Friedrich, T.: timeXplain - a framework for explaining the predictions of time series classifiers. arXiv preprint arXiv:2007.07606 (2020)
9. Ruiz, A.P., Flynn, M., Bagnall, A.: Benchmarking multivariate time series classification algorithms. arXiv preprint arXiv:2007.13156 (2020)

ADBIS 2021 Workshop: Data-Driven Process Discovery and Analysis – SIMPDA

SIMPDA: Data-Driven Process Discovery and Analysis

Paolo Ceravolo[1], Maurice van Keulen[2],
and Maria Teresa Gomez Lopez[3]

[1] Università degli Studi di Milano, Italy
[2] University of Twente, The Netherlands
[3] University of Seville, Spain

Description. With the increasing automation of business processes, growing amounts of process data are becoming available. This opens new research opportunities for business process data analysis, mining, and modeling. The aim of the Eleventh IFIP WG 2.6 International Symposium on Data-Driven Process Discovery and Analysis (SIMPDA 2021) is to offer a forum where researchers from different communities and the industry can share their insights into this hot new field. In business process management (BPM), many monitoring and controlling activities rely on data. The prevalent trend has been collecting these data manually to precisely shape them on the analytics to be applied. Today, data collection is often automated thanks to process-aware information systems or event logs processing. While early data-driven approaches focused on process discovery or reengineering, interest has risen in using data-driven approaches in monitoring to prevent dysfunctional behavior or optimize performances. Thus, there is a shift from design time-oriented phases (e.g., discovery, analysis, and improvement), where data is exploited in offline mode, to runtime-oriented phases (e.g., monitoring), where data is used in real-time to forecast process behavior, performance, and outcomes. These topics were central to SIMPDA 2021. In accordance with our historical tradition of proposing SIMPDA as a symposium, the workshop featured a number of keynotes illustrating new approaches and shorter presentations on recent research. Our thanks go to the authors, to the Program Committee, and to those who participated in the organization or the promotion of this event. We are very grateful to the Università degli Studi di Milano, the University of Seville, the University of Twente, and the IFIP, for supporting this event.

Keynote Presentations. "Process innovations for breast cancer care" by Maurice Van Keulen highlighted how the health sector is continuously innovating. Powering many of such innovations is the high quality digital registration in an Electronic Patient Dossier (EPD) of symptoms, diagnoses, treatments, test results, images, interpretations, and outcomes, to which the full breadth of techniques from AI and analytics can be applied. This talk provided an overview of several research projects focusing on breast cancer diagnostic processes carried out by the University of Twente and the Hospital Group Twente in the Netherlands. Diagnostic processes may deviate from the standard processes due to unforeseeable events and the complexity of the organization, or are intended to deviate, for example, by giving priority to severe cases. Process discovery from event data in electronic health records cannot only shed light on the patient flows but with a new approach can also discover subpopulations with a significantly different flow. Another aspect of the quality of diagnostic processes are the protocols that hospitals often set. Clinicians involved in breast cancer care primarily communicate with each other in natural language through letters and reports for which the international BIRADS protocol exists. NLP technology has the capability of checking the adherence to protocols for quality assurance as well as for extraction of structured data ranging from specific variables to a patient's medication history and comorbidities. The administrative pressure on medical professionals for keeping the data in EPDs accessible, and of high quality, often comes at the expense of attention for the actual diagnosis, inter-clinician communication, and 'hands on the bed'. The research strives to improve the quality of the diagnostic processes both directly and by relieving this pressure. "Data Warehouse and Data Lake Technologies: Selected Challenges still to be Researched" by Prof. Robert Wrembel presented a subjective view of unsolved research problems in the field of data warehouses and their successors, i.e., data lakes. In recent years, we have observed a widespread of new data sources, especially all types of social media and IoT devices, which produce huge volumes of structured data. All these kinds of data are commonly referred to as big data. The outlined problems concern handling the evolution of data sources at the ETL layer, optimizing executions of ETL processes, metadata management, and data quality. Research on these problems has been conducted for decades resulting in multiple extraordinary solutions; however, the most challenging problems still remain to be solved and big data make these problems more challenging.

Selected Papers. The first paper is a full research paper titled "Process Mining Encoding via Meta-Learning for an Enhanced Anomaly Detection". It proposes an approach using meta-learning for enhanced anomaly detection. The methodology is based on several steps to establish meta-learning. Thereby, the authors use meta-feature extraction and meta-target definition to create a meta-feature database. Furthermore, the meta-model is established using a meta-learner approach. Based on this, an encoding technique is recommended for ML anomaly detection. To validate the findings, several event logs are synthetically generated. The meta-learning approach is applied to the event logs. The results show that the meta-learning approach outperforms the other methods in most cases for the detection of anomalies. The second paper is a short paper titled "OCEL: A Standard for Object-Centric Event Logs". It presents a new formalism for the definition of Object Centric Logs for an enrichment of the discovery process. The format is implemented in either JSON or XML. An object is mapped to an entity of the log. So, each event log contains a list of objects, and the properties of the object are written only once in the log (and not replicated for each event). Some challenges remain open for the OCEL format. The format does not provide extensions or consistency checks based on advanced constraints (such as the number of objects per event or mandatory attributes at the event/object level). The main challenge lies in widening the adoption of object-centric process mining techniques such as process discovery. The few proposed techniques such as MVP models have shown promising results. Object-centric process mining techniques lay their foundation on top of object-centric event logs. Hence, the proposal of a reasonable format can lead to extraction tools for object-centric event logs and increase the adoption of object-centric process mining.

Organization

Chairs

Paolo Ceravolo	Università degli Studi di Milano, Italy
Maurice van Keulen	University of Twente, The Netherlands
Maria Teresa Gomez Lopez	University of Seville, Spain

Program Committee

Han van der Aa	University of Mannheim, Germany
Wil Van der Aalst	RWTH Aachen University, Germany
Chintan Amrit	University of Amsterdam, The Netherlands
Helen Balinsky	Hewlett Packard Laboratories, UK
Karima Boudaoud	University of Nice Sophia Antipolis, France
Faiza Allah Bukhsh	University of Twente, the Netherlands
Angelo Corallo	Università del Salento, Italy
Benoît Depaire	Hasselt University, Belgium
Chiara Di Francescomarino	Fondazione Bruno Kessler, Italy
Carlos Fernandez-Llatas	Universitat Politècnica de València, Spain
Maria Leitner	University of Vienna and AIT Austrian Institute of Technology, Austria
Fabrizio Maria Maggi	Free University of Bozen-Bolzano, Italy
Haralambos Mouratidis	University of Brighton, UK
Luisa Parody	Universidad Loyola Andalucía, Spain
Mirjana Pejic-Bach	University of Zagreb, Croatia
Tamara Quaranta	Amazon Dublin, Ireland
Robert Singer	FH JOANNEUM, Austria
Pnina Soffer	University of Haifa, Israel
Angel Jesus Varela Vaca	University of Seville, Spain
Edgar Weippl	University of Vienna, Austria

Process Mining Encoding via Meta-learning for an Enhanced Anomaly Detection

Gabriel Marques Tavares[1]([⊠])[ID] and Sylvio Barbon Junior[2][ID]

[1] Università degli Studi di Milano (UNIMI), Milan, Italy
gabriel.tavares@unimi.it
[2] Londrina State University (UEL), Londrina, Brazil
barbon@uel.br

Abstract. Anomalous traces diminish the event log's quality due to bad execution or security issues, for instance. Focusing on mitigating this phenomenon, organizations spend efforts to detect anomalous traces in their business processes to save resources and improve process execution. Conformance checking techniques are usually employed in these situations. These methods rely on the comparison of the event log obtained and the designed process model. However, in many real-world environments, the log is noisy and the model unavailable, requiring more robust techniques and expert assistance to perform conformance checking. The considerable number of techniques and reduced availability of experts pose an additional challenge to detecting anomalous traces for particular event log scenarios. In this work, we combine the representational power of encoding with a Meta-learning strategy to enhance the detection of anomalous traces in event logs towards fitting the best discriminative capability between common and irregular traces. Our method extracts meta-features from an event log and recommends the most suitable encoding technique to increase the anomaly detection performance. We used three encoding techniques from different families, 80 log descriptors, 168 event logs, and six anomaly types for experiments. Results indicate that event log characteristics influence the representational capability of encodings differently. Our proposed Meta-learning method outperforms the baseline reaching an F-score of 0.73. This performance demonstrates that traditional process mining analysis can be leveraged when matched with intelligent decision support approaches.

Keywords: Anomaly detection · Meta-learning · Encoding · Process mining · Recommendation

The authors would like to thank CNPq (National Council for the Scientific and Technological Development) for their financial support under Grant of Project 420562/2018-4 and 309863/2020-1 and the program "Piano di sostegno alla ricerca 2020" funded by Università degli Studi di Milano.

L. Bellatreche et al. (Eds.): ADBIS 2021 Short Papers, Workshops and Doctoral Consortium, CCIS 1450, pp. 157–168, 2021.
https://doi.org/10.1007/978-3-030-85082-1_15

1   Introduction

Organizations rely on the correct execution of business processes to achieve their goals. However, anomalous instances in event logs are harmful to process quality. This way, stakeholders are interested in detecting and mitigating anomalies so that business processes correspond to their expected behavior. Detecting anomalies is not only beneficial for resource-saving but also to avoid security issues [23]. Process Mining (PM) is an area devoted to extracting valuable information from organizational business processes. Within PM, conformance checking methods are dedicated to finding anomalies. Conformance techniques compare process models and event logs, quantifying deviations and, consequently, identifying anomalies [21]. Traces not complying with the model are interpreted as anomalous, either from a control-flow or data-flow perspective. Although conformance checking supports the recognition of anomalous traces, the methods are model-dependent, hindering their applicability since the model might not be available in many scenarios.

Several approaches have been proposed for anomaly detection in business processes. In [6], the authors use likelihood graphs to model process behavior and support anomaly detection. The method is applicable to control- and data-flow perspectives, although the quality of discovered models limits its performance. Recently, many methods are relying on Machine Learning (ML). In [17], the authors use an autoencoder to model process behavior and detect irregular cases. The authors from [18] use a deep neural network trained to predict the next event. An activity with a low probability score (extracted from the network) is recognized as an anomaly. In [22], the authors use language-inspired trace representations to model process behavior. Cases isolated in the feature space are identified as abnormal.

As pointed by the authors in [4], a suitable encoding technique is crucial for the quality of posterior methods applied to the event log. This way, by finding the appropriate encoding, one can improve the identification of anomalous instances. In other words, a suitable encoding technique allows the best representation of typical behavior and adjusted discriminant capacity of anomalous traces. In this work, we propose a Meta-learning (MtL) strategy to recommend the best encoding for a given event log, maximizing the number of identified anomalies by fitting each particular event log pattern based on its features. MtL has been applied as a recommender system, succeeding in emulating expert decisions [13] for a wide range of applications. Taking advantage of structural and statistical light-weight meta-features from event logs, we propose an MtL approach to suggest the superior encoding technique from a group. Our MtL approach was built using 80 meta-features, trained over 168 event logs (meta-instances) for guiding the superior one of three promising encoding methods (meta-target). Moreover, the proposed approach is easily scalable, allowing the inclusion of additional encoding techniques. Results show the MtL approach outperforms current baselines and can improve the detection of anomalous traces independently of the applied learning algorithm.

The remainder of this paper is organized as follows. Section 2 presents the MtL-based methodology proposed in this paper. Moreover, the section explores the event logs, the extracted features, and the encoding techniques used along with the MtL approach. Section 3 compares our method's performance with two baselines and discusses the implications of the different anomalies. Lastly, Sect. 4 concludes the paper and highlights our contributions.

2 Methodology

This section presents the proposed methodology to enhance anomaly detection in event logs. The method is based on the combination of encoding representational power with MtL as the learning paradigm. We also present the design details and the materials used for experiments.

2.1 Meta-learning for Anomaly Detection in Process Mining

In this work, we investigate encoding methods that boost the performance of traditional algorithms for the anomaly detection task in business processes. For that, our proposed approach relies on MtL. The primary assumption is that the event log characteristics, i.e., descriptors, support the choice of the best encoding method. The best encoding is the technique that produces the highest anomaly detection rates when combined with a given ML-induced model. The boosted capability provided by a particular encoding method relies on the correctly and effectively discriminative capacity to represent traces [4]. However, identifying the best encoding is very tricky, depending on the expert's experience in the particular domain. Here, we follow the assumption that this experience could be emulated using MtL.

Figure 1 presents an overview of our approach. Starting from the event logs, the first step is the *Meta-feature extraction*, which mines descriptors that characterize the event logs. Next, we submit the event logs to encoding techniques. The encoding methods work at the trace level, and the encoded traces serve as input for an ML algorithm aiming to detect anomalies. Hence, we assess the performance metric (namely, F-score) that ranks the encoding algorithms for each event log. This step is called *Meta-target definition*, where each event log is submitted to all encoding methods, and the best encoding (meta-target) is identified. Then, the *Meta-database creation* joins the two previous steps. Here, a database is created using meta-features and the meta-targets extracted from the logs. Consequently, each meta-instance is a set of log descriptors associated with an encoding technique that leverages anomaly detection performance for that event log. Once the meta-database is created, we induce a *Meta-model* in the *Meta-learner* step. The meta-model is the final product of our workflow. Given a new meta-instance (log descriptors), the meta-model indicates the best encoding technique for that event log.

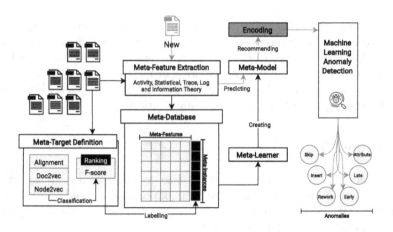

Fig. 1. Overview of the proposed approach.

2.2   Event Logs – Meta-instances

The more instances available, the more representative is the meta-database as it contains more examples of business process behaviors. Experiments on anomaly detection benefit from labeled datasets since one can compute traditional performance metrics to evaluate if anomalous cases are indeed captured. Considering these constraints, we built our meta-database from two groups of synthetic event logs composed of a wide range of behaviors originating from 12 different process models and affected by six types of anomalies.

The first group of event logs was initially presented by Nolle et al. [18] and replicated in [2] and [22]. Six models were generated using the PLG2 tool [8]. PLG2 randomly generates process models representing several complementary business patterns such as sequential, parallel, and iterative control-flows. Moreover, PLG2 allows the configuration of the number of activities, breadth and width, hence, providing a complex set of models that capture diverse behavior. One additional process model, P2P, extracted from [19] was added to the pool. Then, the authors adopted the concept of likelihood graphs [6] to introduce long-term control-flow dependencies. The likelihood graphs are able to mimic complex relations between event to event transitions and attributes attached to these events. This way, the control-flow perspective is constrained by probability distributions that coordinate the model simulation. For instance, an activity may follow another given a probability. The combination of stochastic distributions with a set of process models leverages the similarity between produced event logs and real-world logs. Four event logs were simulated in each process model, generating a total of 28 logs. The final step added anomalies to the traces within the synthetic event logs, which is a traditional practice in related work [5,6]. We applied six anomaly types for all event logs with a 30% incidence: i) Skip: a sequence of 3 or less necessary events is skipped; ii) Insert: 3 or less random activities are inserted in the case; iii) Rework: a sequence of 3 or less necessary events is executed twice; iv) Early: a sequence of 2 or fewer events executed too early, which is then skipped later in the case; v) Late: a sequence of 2 or fewer events executed too late; vi) Attribute: an incorrect attribute value is set in 3 or fewer events.

The second group of synthetic event logs was proposed by Barbon et al. [4]. The authors also used the PLG2 tool to create five process models representing scenarios of increasing complexity, i.e., a higher number of activities and gateways. Then, the process models were simulated using the *Perform a simple simulation of a (stochastic) Petri net* ProM plug-in[1], producing 1000 cases for each log. As a post-processing step, the same anomalies used for the previous set of logs were applied in this set but with different configurations. The authors implemented four anomaly incidences (5%, 10%, 15% and 20%). Moreover, the dataset contains binary and multi-class event logs, meaning that some logs incorporate normal behavior and only one anomaly type (binary), and some logs contain both normal behavior and all anomalies at the same time (multi-class). The latter configuration is especially challenging given the higher complexity as more behaviors are present in the same log. In total, this set contains 140 event logs.

For all event logs, anomalies sit on the event level, but they can be easily converted to the case level. That is, cases containing events affected by any anomaly are considered anomalous cases. Table 1 shows the event log statistics for all event logs used in this work. As demonstrated, the set of logs presents a significant variation in the number of cases, events, activities, trace lengths, and variants. These characteristics support the creation of a heterogeneous metadatabase, increasing business process representability.

Table 1. Event log statistics: each log contains different levels of complexity

Name	#Logs	#Cases	#Events	#Activities	Trace length	#Variants
P2P	4	5k	38k–43k	25	5–14	513–655
Small	4	5k	43k–46k	39	5–13	532–702
Medium	4	5k	28k–31k	63	1–11	617–726
Large	4	5k	51k–57k	83	8–15	863–1143
Huge	4	5k	36k–43k	107	3–14	754–894
Gigantic	4	5k	28k–32k	150–155	1–14	693–908
Wide	4	5k	29k–31k	56–67	3–10	538–674
Scenario1	28	1k	10k–11k	22–380	6–16	426–596
Scenario2	28	1k	26k	41–333	23–30	1k
Scenario3	28	1k	43k–44k	64–348	39–50	1k
Scenario4	28	1k	11k–13k	83–377	1–30	383–536
Scenario5	28	1k	18k–19k	103–406	1–37	637–737

---

[1] http://www.promtools.org/doku.php.

2.3 Log Descriptors – Meta-feature Extraction

Extracting high-quality descriptors is fundamental for the performance of our meta-model. Moreover, meta-feature extraction should have a low computational cost, otherwise, the MtL pipeline is unjustified. This way, we selected a group of lightweight features that contains reliable representational capacities. To retrieve a multi-perspective view of event logs, we extract features from several process layers: activities, traces, and logs. These features were first proposed in [3], which combines business process features from different sources.

Three subgroups capture activity-level descriptors: all activities, start activities, and end activities. Each subgroup contains 12 features: number of activities, minimum, maximum, mean, median, standard deviation, variance, the 25th and 75th percentile of data, interquartile range, skewness, and kurtosis coefficients. For trace-level descriptors, we extracted features related to trace lengths and trace variants. 29 features compose the trace length group: minimum, maximum, mean, median, mode, standard deviation, variance, the 25th and 75th percentile of data, interquartile range, geometric mean and standard variation, harmonic mean, coefficient of variation, entropy, and a histogram of 10 bins along with its skewness and kurtosis coefficients. Regarding the trace variants, we obtained 11 features: mean, standard variation, skewness coefficient, kurtosis coefficient, the ratio of the most common variant to the number of traces, and ratios of the top 1%, 5%, 10%, 20%, 50% and 75% to the total number of traces. Finally, for log-level descriptors, we extracted the number of traces, unique traces, and their ratio, along with the number of events.

Overall, 80 features were extracted from the event logs. They capture complementary elements of business processes and encode information such as statistical dispersion, probability distribution shape and tendency, and log complexity.

2.4 Encodings – Meta-target Definition

In this work, the application of encoding for anomaly detection in PM is a fundamental step towards building the meta-database. Ultimately, the encodings are the meta-targets associated with log features. Given a log and its meta-features, we associate it with an encoding that maximizes the anomaly detection performance. Therefore, encoding techniques play a major role as they can excel in detecting anomalous instances for certain types of log behaviors. The application of encoding in PM has already been explored by several researches [4,10,15,20]. Barbon et al. [4] extensively evaluated trace encoding methods using feature quality metrics to assess encoding capacity. Moreover, the authors submit the encoding methods to a classification task for anomaly detection. The work proposes the application of three encoding families to event logs: PM-based encoding, word embedding, and graph embedding. The PM-based encodings are conformance checking techniques that compare an event log to a process model, measuring deviance and producing fitness results [21]. Word embeddings can naturally be applied in event logs when considering activities and traces as words and sentences [2, 10]. These techniques rely on context information captured by neural networks' weights when trained for context prediction. Lastly, graph embeddings are techniques that encode graph information, such as nodes, vertices, and their attributes. Graph embeddings are particularly interesting in the PM domain as they can represent process models (with limitations such as not capturing concurrency) and traces, modeling entity links and long-term relations.

Considering the three encoding families presented in [4], we selected one encoding method from each family. This way, we aim to reduce the representative bias and evaluate if there is a relation between encoding families and log behavior. For PM-based encodings, we used alignments as it has been considered the state-of-the-art conformance checking method [9]. Alignments compare the event log and process model and measure the deviations between the two. For that, it relates traces to valid execution sequences allowed by the model. This evaluation unfolds into three types. Synchronous moves are observed when both the trace and model can originate a move. Model-dependent moves are originated only from the model, and log-dependent moves are derived from traces but are not allowed by the model. Synchronous moves represent the expected behavior when model and log executions agree. The alignment technique searches for an optimal alignment, i.e., when the fewest number of the model- and log-moves are necessary. This process, measured by a cost function, produces a fitness value and other statistics regarding the states consumed by the model.

Word embedding techniques in PM have mostly relied upon *word2vec* and *doc2vec* to encode traces. In [2], the authors propose the word2vec encoding in conjunction with One-class Classification to detect anomalies in business processes. The authors in [10] use both word2vec and doc2vec to encode activity and trace information, respectively. In this work, we adopt the doc2vec encoding technique as it has been used to encode traces in similar works. Moreover, doc2vec is an extension of word2vec adapted to documents, independently of their length. Word2vec creates numerical representations for words and, for that, a neural network is trained to reconstruct the linguistic context of words in a corpus [16]. The word embeddings come from the weights of the induced neural network. The main advantage is that words appearing in similar contexts produce similar encoding vectors. However, this method is limited to unique word representations. Doc2vec extends word2vec by adding a paragraph vector in the encoding process [14]. This way, the document context is captured by the encoding.

For the graph embedding family, we employed *node2vec*, another encoding technique built on top of word2vec. Node2vec's primary goal is to encode graph data while maintaining graph structure. Given a graph, node2vec performs random walks starting from different nodes [12]. This process creates a corpus, which is used as input for word2vec. The second-order random walks balance a trade-off between breadth and width, capturing neighbor and neighborhood information. Hence, the method can represent complex neighborhoods given its node exploration approach.

Using the three defined encodings (alignment, doc2vec and node2vec), we performed the *Meta-target definition* step shown in Fig. 1. The goal is to identify which encoding enhances the detection of anomalous traces in event logs. For that, we applied a traditional ML pipeline where each log has its traces divided into an 80%/20% holdout strategy, 80% of traces are used for model inferring while 20% of traces are used for testing, i.e., testing if the trained model can correctly label traces. This process is repeated 30 times. We employed the Random Forest algorithm [7] due to its robustness and wide use in similar works by the ML community. The average F-score is obtained after 30 iterations, and the encoding techniques are ranked according to their average F-score. Thus, the best encoding technique for a given log is the one that produces the highest F-score value in the anomaly detection task. We chose F-score as the ranking metric as it successfully balances different performance perspectives.

2.5   Meta-database and Meta-model

We create the meta-database with the meta-features extracted from the logs and the defined meta-targets. Once the meta-database is built, the meta-learner step occurs. The meta-learner embeds a traditional ML pipeline, that is, the meta-database is submitted to an ML algorithm that infers a meta-model. In this scenario, we also employed the Random Forest algorithm with the same parameters (holdout and iterations). The algorithm produces a meta-model used to recommend the suitable encoding for an event log considering its meta-features.

3   Results and Discussion

In this section, we present the performance of our approach and compare it with a baseline performance. Moreover, we develop a discussion regarding the impact of the anomalies in the encoding.

3.1   Meta-learning Performance

First, we report the results of the meta-target definition step, i.e., ranking the encodings for each meta-instance. Since we are following a data-driven strategy, it is worth observing the balance regarding data quality. For that, Fig. 2a shows the frequency of the encoding techniques in each position. The ranking is built using the F-scores obtained by each encoding algorithm. This analysis brings insights about the balanced scenario when selecting an encoding technique, illustrating the "no free lunch theorem" [1]. The alignment method was the best encoding for 54 event logs, while doc2vec was optimal for 56 logs and node2vec for 57 logs. These results highlight a balanced distribution between the best-ranked encodings. Regarding individual encoding, we note that node2vec has similar frequencies in all positions. On the other hand, doc2vec frequency in the third position is high, that is, this encoding was the worst for 87 event logs. Overall, alignments were the most stable method since it appears more frequently in the second position than the others.

Figure 2b reports the performance of our method for the task of recommending the encoding technique that leverages anomaly detection in event logs. Considering the lack of literature in the area, we compare the MtL performance with two baselines: majority and random selection. Majority regards the encoding method with the highest frequency in the first position, hence always recommending the node2vec encoding. Random selection works by arbitrarily selecting one of the possible targets for each event log. From both accuracy and F-score perspectives, our approach outperforms the others with a large advantage. The meta-model obtains an average accuracy of 74% and an average F-score of 0.73. The violin visualization also demonstrates the MtL robustness since the density curve is compressed, i.e., most recordings are near the average mark. The majority approach produced accuracy and F-score averages of 34% and 0.17, respectively. The random method achieves 33% accuracy and 0.32 F-score. The density curves are more stretched, implying less robustness, which is expected given its random nature. It is worth mentioning that these results report the performance for selecting the best encoding method.

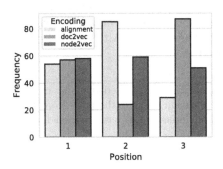
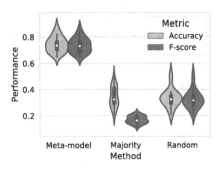

(a) Encodings position frequency. (b) Recommendation performance.

Fig. 2. Encoding ranking extracted from the Meta-target definition step. The ranking is ruled by the F-score obtained by recommending a suitable encoding technique. Given a new event log, the Meta-model recommends the best encoding considering the Meta-features derived from the log. Figure 2b demonstrates the Meta-model performance in comparison with baseline approaches.

3.2 Anomaly Analysis

As introduced in Sect. 2.2, the event logs contain six different anomaly types. Moreover, a subset of the logs is struck by all six anomalies at the same time. Considering the anomaly perspective, Table 2 reports the F-score performance of all three encodings and compares with the MtL approach. Naturally, detecting anomalous instances in logs affected by *all* anomalies is the most difficult task.

Hence, the F-score values are the worst in this scenario. Nonetheless, we observe that MtL reports the highest mean F-score, reaching 0.489. It is followed by alignments (0.468), doc2vec (0.429) and node2vec (0.427). The performance rises considerably in the other anomaly types as the problem is binary in these cases. *Insert, rework* and *skip* are the most detectable anomalies because they deeply affect the control-flow perspective of traces, therefore, this behavior change is easily captured by the encodings. For these anomalies, MtL reaches the highest performance values, producing F-score values close to 1. For *rework* and *skip* anomalies, node2vec ties with the MtL approach. Alignment follows the previous approaches, while doc2vec is the worst encoding for these anomalies. The encoding order changes when observing *early* and *late* anomalies. In these scenarios, MtL remains as the best technique, reaching 0.944 F-score for *early* and 0.94 F-score for *late*, but now is followed more closely by doc2vec, which reaches 0.942 and 0.939 for *early* and *late*, respectively. Finally, MtL and doc2vec tie as the best approaches for the *attribute* anomaly, followed by node2vec and alignment. Interpreting performance from the anomaly perspective reinforces the hypothesis that encodings perform differently in different scenarios, that is, log behavior is determinant when choosing the appropriate encoding. Hence, the MtL efficiency in this experiment exposes the influence of event log behavior on the encoding representational power. The results indicate that anomaly detection is enhanced when the relationship between event log descriptors and encodings is appropriately mapped. This mapping is mastered by our proposed MtL method, which outperforms the use of fixed encodings for all event logs.

Table 2. Comparison of anomaly detection performance using fixed encoding methods and MtL recommendation. Mean and standard deviation (in parenthesis) F-score values are reported for each anomaly type. Bold values indicate the best method for each anomaly.

Encoding	All	Attribute	Early	Insert	Late	Rework	Skip
Alignment	0.468 (0.15)	0.919 (0.04)	0.931 (0.04)	0.975 (0.02)	0.933 (0.03)	0.97 (0.02)	0.98 (0.02)
doc2vec	0.429 (0.21)	**0.931** (0.03)	0.942 (0.03)	0.932 (0.03)	0.939 (0.03)	0.943 (0.03)	0.945 (0.03)
node2vec	0.427 (0.11)	0.926 (0.04)	0.925 (0.04)	0.985 (0.01)	0.928 (0.04)	**0.99** (0.01)	**0.988** (0.01)
MtL	**0.489** (0.15)	**0.931** (0.03)	**0.944** (0.03)	**0.989** (0.01)	**0.94** (0.03)	**0.99** (0.01)	**0.988** (0.01)

We compared the F-score obtained by classifying all event logs using statistical analysis grounded on the non-parametric Friedman test to determine any significant differences between the usage of a unique encoding technique and meta-recommended ones. We used the post-hoc Nemenyi test to infer which differences are statistically significant [11]. As Fig. 3 shows, differences between populations are significant. Furthermore, there are no significant differences within two groups: doc2vec and node2vec, and node2vec and alignment. All other differences are significant. Thus, MtL for recommending individual encoding methods to maximize the predictive performance achieved superior results statistically different from the usage of only one encoding. In other words, the performance obtained using MtL was statistically superior to a single encoding technique.

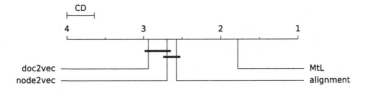

Fig. 3. Nemenyi post-hoc test (significance of $\alpha = 0.05$ and critical distance of 0.361) considering the F-score obtained from all event log classifications.

4   Conclusion

Organizations are interested in detecting anomalous instances in their business processes as a method to leverage process quality, avoid resource waste, and mitigate security issues. In this work, we proposed to combine encoding techniques with MtL to enhance the detection of anomalous traces in event logs. Our strategy relies on a powerful set of meta-features extracted from the event logs. We showed its viability by recommending the best encoding technique with an F-score of 0.73. MtL boosted the anomaly detection by fitting the optimal encoding technique for each event log, statistically outperforming the usage of a single encoding technique. Moreover, our method is highly scalable, which can lead to incremental advancements in the area. For future works, we plan to include more encoding techniques and propose additional features for the meta-feature extraction step.

References

1. Adam, S.P., Alexandropoulos, S.-A.N., Pardalos, P.M., Vrahatis, M.N.: No free lunch theorem: a Review. In: Demetriou, I.C., Pardalos, P.M. (eds.) Approximation and Optimization. SOIA, vol. 145, pp. 57–82. Springer, Cham (2019). https://doi.org/10.1007/978-3-030-12767-1_5
2. Barbon, S., Jr., Ceravolo, P., Damiani, E., Omori, N.J., Tavares, G.M.: Anomaly detection on event logs with a scarcity of labels. In: 2020 2nd International Conference on Process Mining (ICPM), pp. 161–168 (2020)
3. Barbon, S., Jr., Ceravolo, P., Damiani, E., Tavares, G.M.: Using meta-learning to recommend process discovery methods (2021). https://arxiv.org/abs/2103.12874
4. Barbon Junior, S., Ceravolo, P., Damiani, E., Marques Tavares, G.: Evaluating trace encoding methods in process mining. In: Bowles, J., Broccia, G., Nanni, M. (eds.) DataMod 2020. LNCS, vol. 12611, pp. 174–189. Springer, Cham (2021). https://doi.org/10.1007/978-3-030-70650-0_11
5. Bezerra, F., Wainer, J.: Algorithms for anomaly detection of traces in logs of process aware information systems. Inf. Syst. **38**(1), 33–44 (2013)
6. Böhmer, K., Rinderle-Ma, S.: Multi-perspective anomaly detection in business process execution events. In: Debruyne, C., et al. (eds.) OTM 2016. LNCS, vol. 10033, pp. 80–98. Springer, Cham (2016). https://doi.org/10.1007/978-3-319-48472-3_5
7. Breiman, L.: Random forests. Mach. Learn. **45**(1), 5–32 (2001)
8. Burattin, A.: PLG2: multiperspective processes randomization and simulation for online and offline settings (2015)

9. Carmona, J., van Dongen, B.F., Solti, A., Weidlich, M.: Conformance Checking. Relating Processes and Models, Springer, Cham (2018)
10. De Koninck, P., vanden Broucke, S., De Weerdt, J.: act2vec, trace2vec, log2vec, and model2vec: representation learning for business processes. In: Weske, M., Montali, M., Weber, I., vom Brocke, J. (eds.) BPM 2018. LNCS, vol. 11080, pp. 305–321. Springer, Cham (2018). https://doi.org/10.1007/978-3-319-98648-7_18
11. Demšar, J.: Statistical comparisons of classifiers over multiple data sets. J. Mach. Learn. Res. **7**, 1–30 (2006). http://dl.acm.org/citation.cfm?id=1248547.1248548
12. Grover, A., Leskovec, J.: Node2vec: scalable feature learning for networks. In: Proceedings of the 22nd ACM SIGKDD International Conference on Knowledge Discovery and Data Mining, KDD 2016, pp. 855–864. ACM, New York (2016)
13. He, X., Zhao, K., Chu, X.: AutoML: a survey of the state-of-the-art. Knowl. Based Syst. **212**, 106622 (2021)
14. Le, Q., Mikolov, T.: Distributed representations of sentences and documents. In: Xing, E.P., Jebara, T. (eds.) Proceedings of the 31st International Conference on Machine Learning. Proceedings of Machine Learning Research, Beijing, China, 22–24 June 2014, vol. 32, pp. 1188–1196. PMLR (2014)
15. Leontjeva, A., Conforti, R., Di Francescomarino, C., Dumas, M., Maggi, F.M.: Complex symbolic sequence encodings for predictive monitoring of business processes. In: Motahari-Nezhad, H.R., Recker, J., Weidlich, M. (eds.) BPM 2015. LNCS, vol. 9253, pp. 297–313. Springer, Cham (2015). https://doi.org/10.1007/978-3-319-23063-4_21
16. Mikolov, T., Chen, K., Corrado, G.S., Dean, J.: Efficient estimation of word representations in vector space. CoRR abs/1301.3781 (2013)
17. Nolle, T., Luettgen, S., Seeliger, A., Mühlhäuser, M.: Analyzing business process anomalies using autoencoders. Mach. Learn. **107**(11), 1875–1893 (2018)
18. Nolle, T., Luettgen, S., Seeliger, A., Mühlhäuser, M.: BINet: multi-perspective business process anomaly classification. Inf. Syst. 101458 (2019)
19. Nolle, T., Seeliger, A., Mühlhäuser, M.: BINet: multivariate business process anomaly detection using deep learning. In: Weske, M., Montali, M., Weber, I., vom Brocke, J. (eds.) BPM 2018. LNCS, vol. 11080, pp. 271–287. Springer, Cham (2018). https://doi.org/10.1007/978-3-319-98648-7_16
20. Polato, M., Sperduti, A., Burattin, A., Leoni, M.d.: Time and activity sequence prediction of business process instances. Computing **100**(9), 1005–1031 (2018)
21. Rozinat, A., van der Aalst, W.: Conformance checking of processes based on monitoring real behavior. Inf. Syst. **33**(1), 64–95 (2008)
22. Tavares, G.M., Barbon, S.: Analysis of language inspired trace representation for anomaly detection. In: Bellatreche, L., et al. (eds.) TPDL/ADBIS -2020. CCIS, vol. 1260, pp. 296–308. Springer, Cham (2020). https://doi.org/10.1007/978-3-030-55814-7_25
23. van der Aalst, W., de Medeiros, A.: Process mining and security: detecting anomalous process executions and checking process conformance. Electron. Notes Theor. Comput. Sci. **121**, 3–21 (2005). Proceedings of the 2nd International Workshop on Security Issues with Petri Nets and Other Computational Models (WISP 2004)

OCEL: A Standard for Object-Centric Event Logs

Anahita Farhang Ghahfarokhi[1]([✉]), Gyunam Park[1], Alessandro Berti[1,2],
and Wil M. P. van der Aalst[1,2]

[1] Process and Data Science Chair, RWTH Aachen University, Aachen, Germany
{farhang,gnpark,a.berti,wdaalst}@pads.rwth-aachen.de
[2] Fraunhofer Institute for Applied Information Technology,
Sankt Augustin, Germany

Abstract. The application of process mining techniques to real-life
information systems is often challenging. Considering a Purchase to Pay
(P2P) process, several case notions such as order and item are involved,
interacting with each other. Therefore, creating an event log where events
need to relate to a single case (i.e., process instance) leads to convergence (i.e., the duplication of an event related to different cases) and
divergence (i.e., the inability to separate events within the same case)
problems. To avoid such problems, object-centric event logs have been
proposed, where each event can be related to different objects. These
can be exploited by a new set of process mining techniques. This paper
describes OCEL (Object-Centric Event Log), a generic and scalable format for the storage of object-centric event logs. The implementation of
the format can use either JSON or XML, and tool support is provided.

Keywords: Object-centric event logs · Object-centric process mining

1 Introduction

Process mining is a field of data science bridging the gap between model-based
analysis and data-oriented analysis. Process mining techniques include process
discovery techniques to discover process models describing the event log, conformance checking algorithms that compare process models with event logs, and
model enhancement methods that enrich the process model with some information inferred from the event log [12].

Event logs are the starting points to apply process mining techniques. Several
approaches have been proposed towards having a standard for event logs that are
summarized in Fig. 1. The most successful one is XES, which has been accepted
as the IEEE standard in 2014 for the storage of (traditional) event logs [13].
Numerous event logs have been recorded using the XES, and several approaches
have focused on extraction of complex data to obtain XES event logs [2,3,11].
However, although the XES standard can capture the behavior of the events with
a single case notion, in real processes such as O2C (Order to Cash) supported by
ERP systems, multiple case notions are involved, and XES cannot record that.

© Springer Nature Switzerland AG 2021
L. Bellatreche et al. (Eds.): ADBIS 2021 Short Papers, Workshops and Doctoral Consortium,
CCIS 1450, pp. 169–175, 2021.
https://doi.org/10.1007/978-3-030-85082-1_16

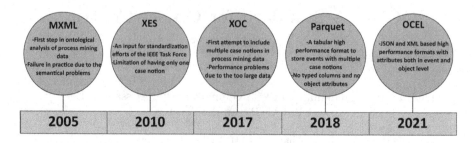

Fig. 1. A timeline demonstrating the proposed standards to store event data.

In extracting an event log from some information system, challenges such as convergence (an event is related to multiple cases) and divergence (repeated execution of the activities with the same case notion) may occur. These problems will affect the results of process mining techniques such as process discovery and lead to not real results. It is not possible to avoid completely from these problems; however, it is worthwhile to consider them in process mining analysis. To describe these challenges, consider the example shown in Fig. 2 where we have two object types, i.e., *order* and *item*, and three activities, i.e., *place order*, *check item*, and *pack item*. To apply process mining techniques, we have two possible case notions (i.e., *order* and *item*). For example, here, there are 100 *orders* and 1000 *items*. Each *item* refers to one *order* and *place order* is executed per *order*. So it is executed 100 times.

- *Convergence*: Suppose that we want to use *item* as the case notion and *place order* as the activity. The number of *items* is 1000; therefore, in applying traditional process mining techniques, we need 1000 times *place order* instead of 100. This is related to convergence, where an event includes multiple cases.
- *Divergence*: If we use *order* as the case notion and consider *pack item* as an activity, for the same case notion we will have many *pack item* events. Each *check item* should be followed by the *pack item*. However, we cannot distinguish between the different *items* within an *order*, and random arrangement may exist between these two activities. This is called divergence challenge, where the order between two activities is lost.

Object-centric event logs have been proposed to address the above challenges in the extraction of process-related information from information systems. Most

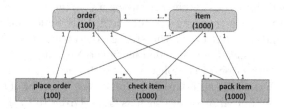

Fig. 2. A fragment of the relationships between case notions (i.e., *order* and *item*) and activities (i.e., *place order*, *check item*), and *pack item*.

Table 1. Informal representation of the events of an OCEL.

Id	Activity	Timestamp	Order	Item	Package	Customer	Resource	Price
e_1	place order	2020-07-09 08:20:01.527+01:00	$\{o_1\}$	$\{i_1, i_2, i_3\}$	∅	$\{c_1\}$	Alessandro	200.0
e_2	confirm order	2020-07-10 09:23:01.527+01:00	$\{o_1\}$	∅	∅	∅	Anahita	302.0
e_3	check availability	2020-07-10 17:10:08.527+01:00	$\{o_1\}$	$\{i_1\}$	∅	∅	Gyunam	125.0
...	...	...	...	...	...	...	...	...

Table 2. Informal representation of the objects of an OCEL.

Id	Type	Product	Color	Age	Job
i_1	item	iPod	silver		
c_1	customer			young	teacher
...	...	...	...	...	...

of the studies in object-centric process mining is focused on process discovery such as *artifact-centric modeling* [4,9]. Some studies have been done on the extraction of object-centric event logs from information systems [1,5,6,10]. This also includes contributions related to the storage format. For example, in [6] authors propose a meta-model, called OpenSLEX, that integrates process and data perspectives. It can generate different views from the database flexibly. New storage formats such as XOC have been proposed as for object-centric event logs [7,8]. An XOC log contains a list of events. Each event is characterized by some attributes, by a list of related objects, and by the state of the object model. However, this format suffers from performance issues since the size of the event log scales badly with the number of events and objects related to the process. Moreover, the attributes of an object are replicated with all the events that are related to such an object.

In this paper, we provide a new format (OCEL) to store object-centric event logs to address these limitations, focusing on its specification, serialization in JSON and XML formats, and tool supports. In http://ocel-standard.org/, the formal definition of OCEL standard is available along with the detailed conceptualization and specification.

The rest of the paper is organized as follows. Section 2 specifies the OCEL format. Section 3 proposes an implementation, tool support, and a set of event logs for the OCEL format. Finally, Sect. 4 concludes the paper.

2   Specification

This section introduces the specification of OCEL based on its formal definition introduced in http://ocel-standard.org/. Based on the definition of OCEL, an event log consists of information about events and objects involved in the events. An event contains an identifier, an activity, a timestamp, some additional attributes, and is related to some objects. The events are ordered based on the timestamp. The objects can have properties themselves. A sample of the informal representation of OCEL is shown in Tables 1 and 2. Here, we introduce the meta-model for the specification of OCEL.

The meta-model for the specification of OCEL is shown in Fig. 3 as a UML class diagram. The log, event, and object classes define the high-level structure of logs. The description for each class is as follows:

- **Log:** The log class contains sets of events and objects. A log contains global elements such as global log, global event, and global object elements. First, a global log element contains the version, attribute names, and object types that compose the log. Second, a global event element specifies some default values for the elements of events. Finally, a global object element specifies some default values for the elements of objects.
- **Event:** An event represents an execution record of an underlying business process. It associates multiple elements (e.g., an identifier, an activity, a timestamp, and relevant objects) and possibly optional features (e.g., event attributes). For example, each row in Table 1 shows an event.
- **Object:** An object indicates the information of an object instance in the business process. It contains required (e.g., type) and optional (e.g., *color* and *size*) elements. For example, each row in Table 2 shows an object.

The element class specifies the elements of the high-level classes. It is composed of a key and value(s). The key is string-based, whereas the value may be of type *string*, *timestamp*, *integer*, *float*, and *boolean*.

An element can be nested, i.e., a parent element can contain child elements. Among nested elements we have *lists* and *maps*.

We can state some advantages in comparison to the existing formats for the storage of object-centric event logs:

- In comparison to tabular formats (such as CSVs), the information is strongly typed.
- Support to lists and maps elements, in comparison with existing formats (XOC, tabular formats, OpenSLEX) that do not properly support them.
- Decoupling of the attributes at the object level from the event level. This helps to avoid replication of the same information, in comparison to the XOC format.

In the next section, we will propose the serialization of OCEL based on the definition, which is fully described in the website of OCEL and the tool support.

3   Serialization and Tool Support

The specification has been serialized in both the JSON and XML formats. An example of JSON format is shown in Listing 1.1. This example shows a part of the data, describing event *place order* and object i_1. "_INVALID_" denotes the default values to be used when activity and type information is missing in event and object, respectively. Furthermore, we provide tool support for the OCEL,

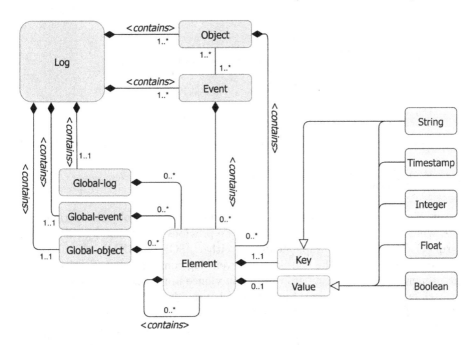

Fig. 3. The UML class diagram for the complete meta-model for the OCEL format.

for both serializations, for the Python[1] and Java programming language[2]. In our tool, the user can import and export OCELs in both serializations and validate the JSON/XML log files according to the serialization schema. Some other features, such as flattening OCELs into traditional event logs, are also possible [1]. To show more examples of OCEL, we have provided some non-trivial OCELs related to the SAP IDES system in both formats.

The current implementations load the entire object-centric event log in memory. This could represent a problem when managing big logs. However, the JSON implementation of the standard could be transferred to a document database such as MongoDB or ElasticSearch in a straightforward way.

Listing 1.1. JSON-OCEL example.

```
{" ocel : global−log " :  {
    " ocel : version " :  " 1.0 " ,
    " ocel : attribute −names " :  [
        " resource " ,  " price " ,  " product " ,  " color " ,  " age " ," job " ] ,
    " ocel : object −types " :  [
        " customer " ,  " item " ,  " order " ,  " package " ]
    } ,
    " ocel : global−event " :  {
        " ocel : activity " :  " __INVALID__ "
    } ,
    " ocel : global−object " :  {
        " ocel : type " :  " __INVALID__ "
    } ,
    " ocel : events " :  {
```

---

[1] Standalone library; https://github.com/OCEL-standard/ocel-support.
[2] ProM 6.10 nightly build; package: *OCELStandard*.

```
"e1": {
  "ocel:activity": "place_order",
  "ocel:timestamp": "2020-07-09 08:20:01.527+01:00",
  "ocel:omap": ["i1", "o1", "i2", "i3", "c1"],
  "ocel:vmap": {"resource": "Alessandro", "price": 200.0}
} },
"ocel:objects": {
  "i1": {
    "ocel:type": "item",
    "ocel:ovmap": {"color": silver, "product": iPad}
  }, } }
```

4   Conclusion

In this paper, we presented OCEL as a format for storing object-centric event logs, which overcomes the limitations of the previous proposals (e.g., XOC and Parquet). The format is implemented in either JSON and XML. An object is mapped to an entity of the log. So, each event log contains a list of objects, and the properties of the objects are written only once in the log (and not replicated for each event). Furthermore, tool support is provided (Python and Java).

Some challenges remain open. The format does not provide consistency checks based on advanced constraints (such as the number of objects per event). The main challenge lies in the adoption of object-centric process mining techniques such as object-centric process discovery. Moreover, the support for streams in the context of object-centric event logs is missing. Furthermore, case studies are needed to illustrate the usage of OCEL in real-world processes.

Acknowledgments. We thank the Alexander von Humboldt (AvH) Stiftung for supporting our research. Funded by the Deutsche Forschungsgemeinschaft (DFG, German Research Foundation) under Germany's Excellence Strategy–EXC-2023 Internet of Production – 390621612.

References

1. Berti, A., van der Aalst, W.M.P.: Extracting multiple viewpoint models from relational databases. In: Data-Driven Process Discovery and Analysis - 8th IFIP WG 2.6 International Symposium, vol. 379 (2019)
2. Calvanese, D., Kalayci, T.E., Montali, M., Santoso, A.: Obda for log extraction in process mining. In: Reasoning Web Summer School (2017)
3. Calvanese, D., Montali, M., Syamsiyah, A., van der Aalst, W.M.P.: Ontology-driven extraction of event logs from relational databases. In: Reichert, M., Reijers, H.A. (eds.) BPM 2015. LNBIP, vol. 256, pp. 140–153. Springer, Cham (2016). https://doi.org/10.1007/978-3-319-42887-1_12
4. Cohn, D., Hull, R.: Business artifacts: a data-centric approach to modeling business operations and processes. IEEE Data Eng. Bull. **32**, 3–9 (2009)
5. González López de Murillas, E., Hoogendoorn, G.E., Reijers, H.A.: Redo log process mining in real life: data challenges & opportunities. In: Teniente, E., Weidlich, M. (eds.) BPM 2017. LNBIP, vol. 308, pp. 573–587. Springer, Cham (2018). https://doi.org/10.1007/978-3-319-74030-0_45

6. de Murillas, E.G.L.: Process mining on databases: extracting event data from real-life data sources (2019)
7. Li, G., de Carvalho, R.M., van der Aalst, W.M.P.: Automatic discovery of object-centric behavioral constraint models. In: Abramowicz, W. (ed.) BIS 2017. LNBIP, vol. 288, pp. 43–58. Springer, Cham (2017). https://doi.org/10.1007/978-3-319-59336-4_4
8. Li, G., de Murillas, E.G.L., de Carvalho, R.M., van der Aalst, W.M.P.: Extracting object-centric event logs to support process mining on databases. In: Mendling, J., Mouratidis, H. (eds.) CAiSE 2018. LNBIP, vol. 317, pp. 182–199. Springer, Cham (2018). https://doi.org/10.1007/978-3-319-92901-9_16
9. Nooijen, E.H.J., van Dongen, B.F., Fahland, D.: Automatic discovery of data-centric and artifact-centric processes. In: La Rosa, M., Soffer, P. (eds.) BPM 2012. LNBIP, vol. 132, pp. 316–327. Springer, Heidelberg (2013). https://doi.org/10.1007/978-3-642-36285-9_36
10. Simović, A.P., Babarogić, S., Pantelić, O.: A domain-specific language for supporting event log extraction from ERP systems. In: International Conference on Computers Communications and Control. IEEE (2018)
11. Valencia-Parra, Á., Ramos-Gutiérrez, B., Varela-Vaca, A.J., López, M.T.G., Bernal, A.G.: Enabling process mining in aircraft manufactures: extracting event logs and discovering processes from complex data. In: International Conference on Business Process Management (2019)
12. van der Aalst, W.M.P.: Process Mining. Springer, Heidelberg (2016). https://doi.org/10.1007/978-3-662-49851-4
13. Verbeek, H.M.W., Buijs, J.C.A.M., van Dongen, B.F., van der Aalst, W.M.P.: XES, XESame, and ProM 6. In: Soffer, P., Proper, E. (eds.) CAiSE Forum 2010. LNBIP, vol. 72, pp. 60–75. Springer, Heidelberg (2011). https://doi.org/10.1007/978-3-642-17722-4_5

ADBIS 2021 Workshop: Modern Approaches in Data Engineering and Information System Design – MADEISD

MADEISD: Modern Approaches in Data Engineering and Information System Design

Ivan Luković[1], Slavica Kordić[2], and Sonja Ristić[2]

[1] University of Belgrade, Serbia
[2] University of Novi Sad, Serbia

Description. For decades, there has been the open issue of how to support information management processes so as to produce useful knowledge and tangible business values from data being collected. One of the hot issues in practice is still how to effectively transform large amounts of daily collected operational data into useful knowledge, from the perspective of declared company goals, and how to set up the information design process aimed at production of effective software services. Nowadays, we have great theoretical potentials for application of new and more effective approaches in data engineering and information system design. However, it is more likely that real deployment of such approaches in industry practice is far behind their theoretical potentials.

The main goal of the 3rd Workshop on Modern Approaches in Data Engineering and Information System Design (MADEISD 2021) is to address open questions and real opportunities for various applications of modern approaches and technologies in data engineering and information system design so as to develop and implement effective software services in support of information management in various organization systems. The intention is to address the interdisciplinary character of a set of theories, methodologies, processes, architectures, and technologies in disciplines such as data engineering, information system design, big data, NoSQL systems, and model-driven approaches in the development of effective software services. MADEISD 2021 received nine submissions and after a rigorous selection process, we accepted four papers for presentation at the workshop and publication in these proceedings.

Selected Papers. The authors of the paper "A RESTful Privacy-aware and Mutable Decentralized Ledger" advocate that blockchain technology has gained massive attention due to its decentralized, transparent, and verifiable features. However, data stored on the blockchain is publicly available, immutable, and may link to the data owner, thus creating privacy management and data modification major challenges. The authors propose in the paper a RESTful decentralized storage framework that provides data privacy and mutability. They have designed a protocol that exploits metadata and pointers stored on the blockchain, while corresponding encrypted data are stored off-chain, so that data owners are able to control their data.

In the paper "Process of medical dataset construction for machine learning - multifield study and guidelines", the authors focus on investigating various practical aspects of medical data acquisition and annotation, as well as various methods of collaboration between IT and medical teams to build datasets that fulfill the desired quality, quantity, and time requirements. Based on their five projects in diverse medical fields, in which the dataset construction procedure was iteratively optimized, the authors give a set of guidelines and good practices to be followed when building new medical datasets.

In the next paper by the same authors "Segmentation quality refinement in large-scale medical image dataset with crowd-sourced annotations", the focus is on achieving high classification accuracy with low variance in a medical image classification system. To achieve this, there is a need for a large training data set with a suitable quality score. The authors present a study on the use of various consistency checking methods to refine the quality of annotations, where it is assumed that tagging was done by volunteers in the crowd-sourcing model. The aim was to evaluate the fitness of the approach in the medical field and the usefulness of the innovative web tool proposed by the authors, called MedTagger (designed to facilitate large-scale annotation of magnetic resonance images), as well as the accuracy of crowd-source assessment using the tool, in comparison to expert classification.

The authors of the paper "Natural Semantics for Domain-Specific Language" present an existing simple domain-specific language for representing the motion of a robot in an orthogonal two-dimensional system and define its natural semantics by their own method proposed in the paper. They propose a method for extending the semantic definition for a given language. Thus, they define natural semantics for both basic and extended (modified) versions of a language. The achieved results will be included as a new topic in the course Formal Semantics, as a support in the education of young IT experts.

Organization

Chairs

Ivan Luković   University of Belgrade, Serbia
Slavica Kordić   University of Novi Sad, Serbia
Sonja Ristić   University of Novi Sad, Serbia

Program Committee

Paulo Alves   Instituto Politècnica de Bragança, Portugal
Moharram Challenger   University of Antwerp, Belgium
Boris Delibašić   University of Belgrade, Serbia
João Miguel Lobo Fernandes   University of Minho, Portugal
Krešimir Fertalj   University of Zagreb, Croatia
Krzysztof Goczyła   Gdańsk University of Technology, Poland
Ralf-Christian Härting   Aalen University, Germany
Dušan Jakovetić   University of Novi Sad, Serbia
Miklós Krész   InnoRenew CoE and University of Primorska, Slovenia
Dragan Maćoš   Beuth University of Applied Sciences Berlin, Germany
Zoran Marjanović   University of Belgrade, Serbia
Sanda Martinčić-Ipšić   University of Rijeka, Croatia
Cristian Mihaescu   University of Craiova, Romania
Nikola Obrenović   University of Novi Sad, Serbia
Maxim Panov   Skolkovo Institute of Science and Technology, Russia
Rui Humberto Pereira   Polytechnic Institute of Porto, Portugal
Aleksandar Popović   University of Montenegro, Montenegro
Patrizia Poščić   University of Rijeka, Croatia
Adam Przybyłek   Gdansk University of Technology, Poland
Kornelije Rabuzin   University of Zagreb, Croatia
Igor Rožanc   University of Ljubljana, Slovenia
Nikolay Skvortsov   Russian Academy of Sciences, Russia
William Steingartner   Technical University of Košice, Slovakia
Vjeran Strahonja   University of Zagreb, Croatia
Slavko Žitnik   University of Ljubljana, Slovenia

Natural Semantics for Domain-Specific Language

William Steingartner$^{(\boxtimes)}$ and Valerie Novitzká

Faculty of Electrical Engineering and Informatics, Technical University of Košice,
Košice, Slovakia
{william.steingartner,valerie.novitzka}@tuke.sk

Abstract. In this paper, we present an existing simple domain-specific language for representing the motion of a robot in an orthogonal two-dimensional system and we define its natural semantics. The language contains basic statements for moving the robot in specified directions, actually without flow control constructs such as a loop or conditional. In our approach, we show how to extend the semantic definition for a given language. We define natural semantics for both basic and extended (modified) versions of a language. Our new method of natural semantics constructed in this way becomes a natural part of the spectrum of semantic methods for a given type of languages. The achieved results will be included as a new and current topic in the course Formal Semantics as a support in the education of young IT experts. A correct method with proven and verified properties is also a great assumption for the future construction of visual semantic technologies to support university education.

Keywords: Domain-specific language · Formal methods · Modeling language · Natural semantics · Semantic function · University didactics

1 Introduction

Teaching in the field of formal semantics requires permanent actualization of topics in connection with modern software approaches. This means taking into account new techniques in software development and related formal methods. One such approach is the use of domain-specific languages that are domain-dependent. Domain-specific languages (DSL) form a class of programming or specification languages to describe and solve tasks of a limited, specific problem domain. A more complex problem area can be solved by creating multiple DSL languages that form one family of DSL languages [9]. They offer substantial gains in expressiveness and ease of use compared with general-purpose programming languages in their domain of application [17].

This work was supported by the Project KEGA 011TUKE4/2020: "A development of the new semantic technologies in educating of young IT experts".

L. Bellatreche et al. (Eds.): ADBIS 2021 Short Papers, Workshops and Doctoral Consortium,
CCIS 1450, pp. 181–192, 2021.
https://doi.org/10.1007/978-3-030-85082-1_17

A domain-specific language is simply a language that is optimized for a given class of problems, called a domain. It is based on abstractions that are closely aligned with the domain for which the language is built [12,29]. A DSL uses the concepts and rules from the field or domain. DSLs can be dedicated to a particular problem domain, a particular problem representation technique or a particular solution technique. The basic idea behind DSLs is to offer means which would allow expressing solutions in the idioms and at the abstraction level of the problem domain [6,24].

One of the ways to use DSL is also in teaching programming, where we just need to define the basic language with only the necessary commands. In practice, approaches using the Karel the Robot method [7,22] or the LOGO language [3] are successfully used to teach programming. For simplicity, it is enough to assume that the basic commands for movement in the specified direction and related elements are sufficient to control the robot. For such a simplified language, denotational and structural operational semantics were formulated in the teaching module here [11].

In our paper, we present how to formulate and define natural semantics for the mentioned language for the description of robot's movement. Because the topic of DSL is still actual, we consider also the complete definition of semantic methods for a given language as important and interesting topic in educating of young IT experts.

The structure of this paper is the following: we present in Sect. 2 some notes about the background and some related works, Sect. 3 contains basic concepts about the natural semantics, Sect. 4 contains the basic definition of the language for the description of robot's movement. In Sect. 5, we formulate and define the natural semantic for the mentioned language, in Sect. 6 we extend the definition of the basic language and we define a natural semantics for this extended language. Finally, the Sect. 7 concludes our paper.

2   Teaching Background and Some Related Works

Teaching the basics of algorithms and programming thanks to supporting tools such as Karel has a long tradition and brings positive results. The basic programming course at the Faculty of Electrical Engineering and Informatics at the Technical University in Košice is available on the website [1]. But teaching programming alone cannot work without formal foundations. Therefore, we consider it necessary to follow up on this course with the formal foundations of the semantics of programming languages. Although Karel is essentially understood as an imperative language, the extension of this domain to domain-specific languages opens up greater opportunities for students to understand the properties of languages.

For a better understanding of formal methods, we are currently focusing our research on the visualization of these methods. This research is covered by the KEGA project, which we refer to on the first page of this article. As part of the project, we have already implemented some software tools that are used to visualize formal semantics for imperative languages. Some results are presented in the works [26,27]. Because we want to continue this trend and prepare the basis for the visualization of semantic methods for domain-specific languages, we consider it necessary to expand the set of semantic methods in this field.

Our formulation of natural semantics for a given domain-specific language is the original result. In this approach, we present how natural semantics is formulated for a language and what its properties are.

Our next goal is the design and development of a software module that will allow the input program to be transformed into byte-code, with which it will be possible to visualize the individual steps of the robot. At its core, this module will contain a compiler from the robot language to the appropriate semantic methods (natural, operational or denotational), including an interactive option to set input values, coordinates, etc. The resulting visualization will be realized at once or in individual steps with the possibility of storing the results (static image, structured sequence of steps, LaTeX source code, etc.).

Now, when the contact of educators and students is limited mostly to the online environment, the interactive teaching software tools play a crucial rôle in the teaching process. The fact that theoretically oriented courses, such as Semantics of Programming Languages (and other courses focused on formal methods and theoretical foundations of computer science [18]) require special support for students, has forced a reassessment of the course teaching method and the design and development of new interactive teaching software tools. Similar software tools (even for other courses) were created to simplify the understanding process of the basics, for example, the course on Formal Languages [15,19], Dynamic Geometry [10], Formal Logic [25], Data Structures and Algorithms [20], Object-oriented Programming [28], Operating Systems [8] or general formal methods in software development [16].

3   Basic Concepts about Natural Semantics

The formal representation of the semantics helps to provide an unambiguous definition and precise meaning of a program [4]. In an operational semantics, we are concerned with how to execute programs and not merely what the results of execution are [21]. Operational semantics provides an observation of how the states are modified during the execution of the statements. There are well-known two approaches to operational semantics:

- natural semantics, which describes how the overall results of execution are obtained;
- structural operational semantics, which describes how the individual steps of the computation arise.

Structural operational semantics, sometimes called a small-step (operational) semantics was formulated and defined by Gordon Plotkin in [23]. An operational (and denotational) approach to the robot movement language was introduced in [11].

For both kinds of operational semantics, the meaning of statements is specified by a transition system and particular steps are expressed as transition relations which describe how the execution takes place. For natural semantics, the following transition relation is defined:

$$\langle S, s \rangle \rightarrow s', \tag{1}$$

where S is a statement and s, s' are general notations (metavariables) for an initial and final state of execution.

Natural semantics, also known as big-step semantics, was defined by Gilles Kahn in [13]. This method is mostly used for imperative languages but it has nice application also in area of domain specific languages, e.g. [2]. Our aim is to define an approach of natural semantics for the kind of domain-specific language presented in this paper.

In the process of defining a semantic method, we will proceed in the usual way. For a given syntactic domain of a language, we formulate semantic equations in the form of rules and define the appropriate semantic function that assigns a state change to the syntactic element (statement). To write the rules of natural semantics, we will use the usual notation according to [21], i.e.

$$\frac{\langle S_1, s_1 \rangle \rightarrow s_1', \cdots, \langle S_n, s_n \rangle \rightarrow s_n'}{\langle S, s \rangle \rightarrow s'} \ \text{(name)} \tag{2}$$

where $S_1, \ldots S_n$ are immediate constituents of S or are statements constructed from the immediate constituents of S. This scheme of rule has a number of premises written above the solid line and one conclusion below the solid line. Rules with an empty set of premises are called axioms.

4   The Definition of Language

In this section, we introduce the language that we focus on. A robot coordination language (language without a concrete name) is a domain-specific language used for describing the movement of a robot in a defined orthogonal two-dimensional system. The main idea comes from standard languages used in tuition of programming: mostly LOGO, Karel (or Karel the Robot) and similar languages.

The syntactic notation used is based on BNF. First, we list the syntactic domains (syntactic categories):

$n \in$ **Num**—numerals
$S \in$ **Statm**—statements

We assume that the structure of numerals and variables is given elsewhere; for example, numerals might be strings of digits. As usual, natural numbers have no internal structure from the semantic point of view, but syntactically they can be represented with a regular expression $[0, \ldots, 9]^+$. Then, the only structured elements in the language are statements. The structure of statements is given as follows:

$$S ::= \textbf{left} \mid \textbf{right} \mid \textbf{up} \mid \textbf{down} \mid$$
$$\textbf{left } n \mid \textbf{right } n \mid \textbf{up } n \mid \textbf{down } n \mid \qquad (3)$$
$$\textbf{reset} \mid \textbf{skip} \mid S; S$$

The first four statements express the movement of the robot in a given direction by one position. The other four directional statements, with the argument n, express the movement in a given direction by a specific number of steps. The statement **reset** represents an immediate return to the starting (default or home) position, which is defined by user at the beginning. The **skip** statement is an empty statement (the statement does not perform any action). The last pattern is the composition (sequence) of statements.

5   Semantics of the Language

As semantic domains, we consider here standard numerical sets for naturals with zero ($\mathbb{N}_0$) and integers ($\mathbb{Z}$). More complex semantic domains we construct from those basic ones by using standard set operations.

Numerals expressing the natural numbers are used in statements for providing the number of steps to be done. For providing the semantics of numerals (to determine the number represented by a numeral), a simple semantic function is used:

$$[\![\;]\!] : \textbf{Num} \to \mathbb{N}_0. \qquad (4)$$

Any numeral (numeric string) is converted to the natural number (or zero value) using standard semantic rules, e.g. [21]. We follow also the convention, that e.g. 1 is an element in syntactic domain **Num** and its semantic value is given by applying the semantic function, i.e. $[\![1]\!] = 1$.

As a state space for movements of a robot, we consider a two-dimensional orthogonal system where a robot can move in four directions stepping on particular coordinates. One position of the robot on the concrete coordinate is considered as an actual state. Then a state is simply a point p in the orthogonal system:

$$p \in \mathbb{Z} \times \mathbb{Z}.$$

For our purposes, we denote the semantic domain as **Point** $= \mathbb{Z} \times \mathbb{Z}$.

A change of robot's position is considered as a change of state. This change is expressed by new coordinates in a point. For example, if the robot steps one step left, then the horizontal coordinate is decremented. For an access to coordinates, projections functions are used. In the mentioned example, if the robot steps from a position p one step left, we calculate new coordinates as follows:

$$p' = p\left[\pi_1(p) \ominus 1, \pi_2(p)\right],$$

where

- π_1 and π_2 are well-known projections defined on a Cartesian product:

$$\pi_1 : A \times B \to A,$$
$$\pi_2 : A \times B \to B,$$

- p is a starting position of robot,
- p' is a new position where robot moved,
- notation $p = [x_0, y_0]$ stands for concrete position given by its coordinates,
- notations $p' = p[f(x_0), y_0]$ or (alternatively for another coordinate) $p' = p[x_0, f(y_0)]$ represent an actualization of a given position where some operation f is applied to the coordinates from a position p,
- symbols $+$, $-$ represent syntactic operators, and symbols $\oplus$, $\ominus$ (eventually $\otimes$) stand for real mathematical operations.

The meaning of statements is summarized as a function from **Point** to **Point**. We define:

$$\mathscr{S}_{ns} : \textbf{Statm} \to (\textbf{Point} \to \textbf{Point}). \tag{5}$$

Then function for every statement is given by

$$\mathscr{S}_{ns}[\![\,S\,]\!]\, p = \begin{cases} p', \text{ if } \langle S, p \rangle \to p', \\ \bot, \quad \text{otherwise.} \end{cases} \tag{6}$$

Here, the tuple $\langle S, p \rangle$ is a configuration in natural semantics and $\langle S, p \rangle \to p'$ is a transition relation. For the initial position, we use a notation p^*.

We note that the function (6) is defined as total if we consider a possible infinite state space for a robot. However, in real applications, we usually work with a limited orthogonal system (a kind of its subsystem), then the mentioned function is defined as a partial one:

$$\textbf{Point} \rightharpoonup \textbf{Point}.$$

For n-step statements we follow the idea that the first one step is realized and then the remaining number of steps is repeated until we reach an empty step. We show this approach on the statement **left** n. For the other commands, this approach is analogous.

The statement **left** n is semantically equivalent to the statements

$$\textbf{left}; \textbf{left } m, \tag{7}$$

for $[\![\,n\,]\!] = [\![\,m\,]\!] \oplus \mathbf{1}$, and $[\![\,n\,]\!], [\![\,m\,]\!] \in \mathbb{N}_0$.

Proof. This equivalence can be proved by mathematical induction on the number of steps. We denote the property (7) as $P(n)$.

1. We consider that this property holds for the number of steps equal to one, so we prove $P(1)$:

$$\textbf{left } 1 \equiv \textbf{left}; \textbf{left } 0 \tag{8}$$

Since the statement **left** 0 does not do anything (the robot does not move in any direction, even not to the left at all), is this statement equivalent to **skip** trivially. So the first step of induction holds.

2. Now we prove the induction step $P(k) \Rightarrow P(k+1)$ for $k \in \mathbb{N}$.

- For any k we have:

$$\mathbf{left}\ k \equiv \mathbf{left}; \mathbf{left}\ k' \tag{9}$$

  where $k = k' + 1$.
- Then for $k + 1$ we get:

$$\mathbf{left}\ k + 1 \equiv \mathbf{left}; \mathbf{left}\ k, \tag{10}$$

  hence

$$\mathbf{left}\ k + 1 \equiv \mathbf{left}; \mathbf{left}; \mathbf{left}\ k', \tag{11}$$

  where $k + 1 = (k' + 1) + 1 = k' + 2$ and this induction step holds. *Qed*

We start with formulating the rules for one-step operations:

$$\frac{}{\langle \mathbf{left}, p \rangle \to p\,[\pi_1(p) \ominus \mathbf{1}, \pi_2(p)]}\ (\mathrm{left}_{ns})$$

$$\frac{}{\langle \mathbf{right}, p \rangle \to p\,[\pi_1(p) \oplus \mathbf{1}, \pi_2(p)]}\ (\mathrm{right}_{ns})$$

$$\frac{}{\langle \mathbf{up}, p \rangle \to p\,[\pi_1(p), \pi_2(p) \oplus \mathbf{1}]}\ (\mathrm{up}_{ns})$$

$$\frac{}{\langle \mathbf{down}, p \rangle \to p\,[\pi_1(p), \pi_2(p) \ominus \mathbf{1}]}\ (\mathrm{down}_{ns})$$

Following are the rules for n steps. Here we consider that $[\![\, n \,]\!] = [\![\, m \,]\!] \oplus \mathbf{1}$ and $[\![\, n \,]\!] \geq \mathbf{1}$. As we showed, statements where argument is 1 are semantically equivalent to those without argument.

$$\frac{\langle \mathbf{left}, p \rangle \to p' \quad \langle \mathbf{left}\ m, p' \rangle \to p'\,[\pi_1(p) \ominus [\![\, m \,]\!], \pi_2(p)]}{\langle \mathbf{left}\ n, p \rangle \to p\,[\pi_1(p) \ominus [\![\, n \,]\!], \pi_2(p)]}\ (\mathrm{left\text{-}n}_{ns})$$

for $p' = p\,[\pi_1(p) \ominus \mathbf{1}, \pi_2(p)]$,

$$\frac{\langle \mathbf{right}, p \rangle \to p' \quad \langle \mathbf{right}\ m, p' \rangle \to p'\,[\pi_1(p) \oplus [\![\, m \,]\!], \pi_2(p)]}{\langle \mathbf{right}\ n, p \rangle \to p\,[\pi_1(p) \oplus [\![\, n \,]\!], \pi_2(p)]}\ (\mathrm{right\text{-}n}_{ns})$$

for $p' = p\,[\pi_1(p) \oplus \mathbf{1}, \pi_2(p)]$,

$$\frac{\langle \mathbf{up}, p \rangle \to p' \quad \langle \mathbf{up}\ m, p' \rangle \to p'\,[\pi_1(p), \pi_2(p) \oplus [\![\, m \,]\!]]}{\langle \mathbf{up}\ n, p \rangle \to p\,[\pi_1(p), \pi_2(p) \oplus [\![\, n \,]\!]]}\ (\mathrm{up\text{-}n}_{ns})$$

for $p' = p\,[\pi_1(p), \pi_2(p) \oplus \mathbf{1}]$,

$$\frac{\langle \mathbf{down}, p \rangle \to p' \quad \langle \mathbf{down}\ m, p' \rangle \to p'\,[\pi_1(p), \pi_2(p) \ominus \llbracket\, m \,\rrbracket]}{\langle \mathbf{down}\ n, p \rangle \to p\,[\pi_1(p), \pi_2(p) \ominus \llbracket\, n \,\rrbracket]} \quad (\text{down-n}_{ns})$$

for $p' = p\,[\pi_1(p), \pi_2(p) \ominus \mathbf{1}]$.

A special statement for resetting the initial position of robot (can be understood as a "teleportation") to an initial position:

$$\frac{}{\langle \mathbf{reset}, p \rangle \to p^*} \quad (\text{reset}_{ns})$$

for given starting position p^*. This coordinate (state) shall be defined by user.

The rule for the sequence of statements that are executed sequentially:

$$\frac{\langle S_1, s \rangle \to s'' \quad \langle S_2, s'' \rangle \to s'}{\langle S_1; S_2, s \rangle \to s'} \quad (\text{comp}_{ns})$$

The last one is the rule for an empty statement. This statement does not do anything and its role is primarily in proofs and in description of empty steps:

$$\frac{}{\langle \mathbf{skip}, p \rangle \to p} \quad (\text{skip}_{ns})$$

6   Extended Language

The language introduced in Sect. 4 we extend (and slightly modify) with the concepts known from LOGO or Karel the Robot to make this language closer to real languages. This means that we consider in this model a movement direction given by an angle. In this sense, we add into the language the statements for moving forward (one and more steps) and turning left or right. However, in this approach, the statements for moving in concrete directions need not be present in the language. Then syntax is as follows:

$$S ::= \mathbf{turn\ left} \mid \mathbf{turn\ right} \mid \mathbf{forward} \mid \mathbf{forward}\ n \mid \\ \mathbf{reset} \mid \mathbf{skip} \mid S; S \tag{12}$$

We note, that for now we do not consider some standard control-flow operations as loops and conditionals. These elements will be discussed in further research.

Changes in the syntax often require modification of semantic rules (similar in [5]). Here, the configuration in natural semantics shall be extended. Configuration must express the current position with the direction. Direction is given by an angle of movement in the orthogonal system. However, the angles can be assigned (defined) by convention, we follow the idea in [11] and we consider the following directions: **0** represents the up, **90** represents the right, **180** represents the down and **270** the left movement direction. In this model, the robot can turn only **90** degrees and the values of angles we consider as elements of the set **Angle** $= \mathbb{Z}$.

A configuration is an element of the semantic domain defined as follows:

$$\textbf{Config} = \textbf{Point} \times \textbf{Angle}.$$

Then elements of this semantic domain are tuples consisting of a point and actual angle: $\langle p, \alpha \rangle$.

The semantic function is then defined as follows

$$\mathscr{S}'_{ns} : \textbf{Statm} \to (\textbf{Config} \rightharpoonup \textbf{Config}) \tag{13}$$

and the function for every statement is given by

$$\mathscr{S}'_{ns}[\![S]\!] \langle p, \alpha \rangle = \begin{cases} \langle p', \alpha' \rangle, \text{ if } \langle S, p, \alpha \rangle \to \langle p', \alpha' \rangle, \\ \bot, \qquad \text{otherwise.} \end{cases} \tag{14}$$

Here, the tuple $\langle S, p, \alpha \rangle$ is a new configuration in natural semantics for the extended language and $\langle S, p, \alpha \rangle \to \langle p', \alpha' \rangle$ is a new transition relation.

The semantic rules for the language defined in (12) are the following:

$$\frac{}{\langle \textbf{turn left}, p, \alpha \rangle \to \langle p, (\alpha \oplus \textbf{270}) \bmod \textbf{360} \rangle} \ (\text{turn-l}_{ns})$$

$$\frac{}{\langle \textbf{turn right}, p, \alpha \rangle \to \langle p, (\alpha \oplus \textbf{90}) \bmod \textbf{360} \rangle} \ (\text{turn-r}_{ns})$$

$$\frac{}{\langle \textbf{forward}, p, \alpha \rangle \to \langle p [\pi_1(p) \oplus \sin \alpha, \pi_2(p) \oplus \cos \alpha], \alpha \rangle} \ (\text{forward}_{ns})$$

$$\frac{\langle \textbf{forward}, p, \alpha \rangle \to \langle p'', \alpha \rangle \quad \langle \textbf{forward } m, p'', \alpha \rangle \to \langle p', \alpha \rangle}{\langle \textbf{forward } n, p, \alpha \rangle \to \langle p [\pi_1(p) \oplus [\![n]\!] \sin \alpha, \pi_2(p) \oplus [\![n]\!] \cos \alpha], \alpha \rangle} \ (\text{forward-n}_{ns})$$

In the rule (forward-n$_{ns}$), we consider the following states:

- $p'' = p [\pi_1(p) \oplus \sin \alpha, \pi_2(p) \oplus \cos \alpha]$,
- $p' = p'' [\pi_1(p) \oplus [\![m]\!] \sin \alpha, \pi_2(p) \oplus [\![m]\!] \cos \alpha]$ for $[\![n]\!] = [\![m]\!] \oplus \textbf{1}$.

We note that the values of $\sin \alpha$ and $\cos \alpha$ are either $-\textbf{1}$, **0** or **1**, as all angles α in our case are divisible by **90**.

The rules for remaining statements are as follows:

$$\frac{}{\langle \mathbf{reset}, p, \alpha \rangle \to \langle p^*, \mathbf{0} \rangle} \; (\text{e-reset}_{ns})$$

$$\frac{}{\langle \mathbf{skip}, p, \alpha \rangle \to \langle p, \alpha \rangle} \; (\text{e-skip}_{ns})$$

$$\frac{\langle S_1, p, \alpha \rangle \to \langle p'', \alpha'' \rangle \quad \langle S_2, p'', \alpha'' \rangle \to \langle p', \alpha' \rangle}{\langle S_1; S_2, p, \alpha \rangle \to \langle p', \alpha' \rangle} \; (\text{e-comp}_{ns})$$

We note here, that for the statement **reset**, an initial configuration shall be defined. We consider that a state p^* is given by the user and an initial angle is always zero, hence an initial configuration is then $\langle p^*, \mathbf{0} \rangle$.

When we compare both versions of the language and their semantic representations, we can observe the following equivalences. The **forward** statement is equivalent to individual direction statements if the condition is met that in a given configuration, the direction angle has a value corresponding to the direction for a particular movement statement (if they would exist in this configuration):

$$\alpha = 0 \qquad \langle \mathbf{forward}, p, \alpha \rangle \equiv \langle \mathbf{up}, p, \alpha \rangle$$
$$\alpha = 90 \qquad \langle \mathbf{forward}, p, \alpha \rangle \equiv \langle \mathbf{right}, p, \alpha \rangle$$
$$\alpha = 180 \qquad \langle \mathbf{forward}, p, \alpha \rangle \equiv \langle \mathbf{down}, p, \alpha \rangle$$
$$\alpha = 270 \qquad \langle \mathbf{forward}, p, \alpha \rangle \equiv \langle \mathbf{left}, p, \alpha \rangle$$

7   Conclusion

In this paper, we have shown our approach to semantic modeling of a selected domain-specific language. For a given existing language expressing the robot's motion in the orthogonal coordinate system, we have defined and formulated an approach in natural semantics that is equivalent to the existing denotational and operational approaches. Our method of semantic modeling required some modifications and extensions over the existing approaches, which we explained and documented.

We want to continue our further research in this area by constructing a proof of equivalence between particular semantic methods. Also, a very interesting step seems to be to extend the language with some standard flow-control constructs and to explore more possible directions for movement of robot (based on values of the angle). For such an extended language, it will also be necessary to construct the proof of equivalence and verify the properties of the language.

Both the current results and the results of future research will become part of the course Formal Semantics, which will extend the course with other interesting and current topics in the field of semantic modeling. In this course, we plan to prepare additional semantic technologies (inspired by e.g. [10, 14, 28]) for the visualization of given semantic methods as a support for blended learning and distance education.

Acknowledgment. The authors express their gratitude to prof. Marjan Mernik for the original idea of how to formulate a domain-specific language for a given application domain. The authors also thank Dániel Horpácsi and Judit Kőszegi for their approach to DSL semantics for the robot that motivated us in our research.

References

1. Základy algoritmizácie a programovania (2021). (in Slovak). https://kurzy.kpi.fei. tuke.sk/zap/index.html. Accessed 18 June 2021
2. Benčík, M., Dedera, L.: Natural semantics of battle management languages. In: 2019 Communication and Information Technologies (KIT), pp. 1–4 (2019). https:// doi.org/10.23919/KIT.2019.8883485
3. Blaho, A., Kalaš, I.: Imagine Logo. Computer Press (2006)
4. Challenger, M., Mernik, M., Kardas, G., Kosar, T.: Declarative specifications for the development of multi-agent systems. Comput. Stand. Interfaces **43**, 91–115 (2016). https://doi.org/10.1016/j.csi.2015.08.012
5. Chodarev, S., Porubän, J.: Development of custom notation for XML-based language: a model-driven approach. Comput. Sci. Inf. Syst. **14**(3) (2017). https://doi. org/10.2298/CSIS170116036C
6. Dedera, L.: Domain-specific languages for command and control systems. Sci. Milit. J. **5**(1), 40–46 (2010)
7. Freiberger, U.: Karel. Eine Übersicht über verschiedene Entwicklungen, die auf der Idee von "Karel, the Robot" basieren (2002). https://docplayer.org/3612024-Karel-eine-uebersicht-ueber-verschiedene-entwicklungen-die-auf-der-idee-von-karel-the-robot-basieren.html. Accessed 16 Mar 2021
8. Genči, J., Bilanová, Z., Deák, A., Vrábel, M.: Project and team based teaching of system programming in the course of operating systems. In: 2017 15th International Conference on Emerging eLearning Technologies and Applications (ICETA), pp. 1–6 (2017)
9. Gronbeck, R.: Eclipse Modeling Project: A Domain-Specific Language (DSL) Toolkit. Addison Wesley Professional (2009)
10. Herceg, D., Radaković, D., Ivanović, M., Herceg, D.: Possible improvements of modern dynamic geometry software. Comput. Tools Educ. (2), 72–86 (2019). https://doi.org/10.32603/2071-2340-2019-2-72-86. http://cte.eltech.ru/ojs/index. php/kio/article/view/1600
11. Horpácsi, D., Kőszegi, J.: Formal semantics (2014). https://regi.tankonyvtar.hu/ en/tartalom/tamop412A/2011-0052_05_formal_semantics/index.html. Accessed 14 Dec 2020
12. Höver, K.M., Borgert, S., Mühlhäuser, M.: A domain specific language for describing S-BPM processes. In: Fischer, H., Schneeberger, J. (eds.) S-BPM ONE 2013. CCIS, vol. 360, pp. 72–90. Springer, Heidelberg (2013). https://doi.org/10.1007/ 978-3-642-36754-0_5
13. Kahn, G.: Natural semantics. In: Brandenburg, F.J., Vidal-Naquet, G., Wirsing, M. (eds.) STACS 1987. LNCS, vol. 247, pp. 22–39. Springer, Heidelberg (1987). https://doi.org/10.1007/BFb0039592
14. Kloboves, K., Mihelič, J., Bulić, P., Dobravec, T.: FPGA-based SIC/XE processor and supporting toolchain. Int. J. Eng. Educ. (6A) (2018)
15. Kollár, J.: Computron VM: identification of expert knowledge in virtual computer architecture development. In: CSE 2012: International Scientific Conference on Computer Science and Engineering, pp. 87–94 (2012)

16. Korečko, Š, Sorád, J., Dudláková, Z., Sobota, B.: A toolset for support of teaching formal software development. In: Giannakopoulou, D., Salaün, G. (eds.) SEFM 2014. LNCS, vol. 8702, pp. 278–283. Springer, Cham (2014). https://doi.org/10.1007/978-3-319-10431-7_21

17. Mernik, M., Heering, J., Sloane, A.: When and how to develop domain-specific languages. ACM Comput. Surv. **37**(4), 316–344 (2005). https://doi.org/10.1145/1118890.1118892

18. Mihályi, D., Peniašková, M., Perháč, J., Mihelič, J.: Web-based questionnaires for type theory course. Acta Electrotech. Inf. **17**(4), 35–42 (2017)

19. Mihelič, J., Dobravec, T.: SicSim: a simulator of the educational SIC/XE computer for a system-software course. Comput. Appl. Eng. Educ. **23**(1), 137–146 (2015)

20. Mocinecová, K., Steingartner, W.: Software support for visualizing of the graph algorithms in a novel approach in educating of young IT experts. Trans. Internet Res. **16**(2), 14–23 (2020), http://ipsitransactions.org/journals/papers/tir/2020jul/p3.pdf

21. Nielson, H.R., Nielson, F.: Semantics with Applications: An Appetizer (Undergraduate Topics in Computer Science). Springer, Berlin, Heidelberg (2007). https://doi.org/10.1007/978-1-84628-692-6

22. Pattis, R.: Karel The Robot: A Gentle Introduction to the Art of Programming. Wiley, Hoboken (1981)

23. Plotkin, G.: The origins of structural operational semantics. J. Log. Algebraic Programm. **60–61**, 3–15 (2004). https://doi.org/10.1016/j.jlap.2004.03.009

24. Raja, A., Lakshmanan, D.: Domain specific languages. Int. J. Comput. Appl. **1** (2010). https://doi.org/10.5120/37-640

25. Schreiner, W., Steingartner, W., Novitzká, V.: A novel categorical approach to the semantics of relational first-order logic. Symmetry **12**(10) (2020). https://doi.org/10.3390/sym12101584

26. Steingartner, W.: Support for online teaching of the Semantics of Programming Languages course using interactive software tools. In: 18th International Conference on Emerging eLearning Technologies and Applications (ICETA) 2020, pp. 665–671 (2020). https://doi.org/10.1109/ICETA51985.2020.9379225

27. Steingartner, W.: On some innovations in teaching the formal semantics using software tools. Open Comput. Sci. **11**(1), 2–11 (2021). https://doi.org/10.1515/comp-2020-0130

28. Vaclavkova, M., Kvet, M., Sedlacek, P.: Graphical development environment for object programming teaching support. In: INFORMATICS 2019 - IEEE 15th International Scientific Conference on Informatics, Proceedings, pp. 77–82. IEEE (2019). https://doi.org/10.1109/Informatics47936.2019.9119284

29. Voelter, M., et al.: DSL engineering - designing, implementing and using domain-specific languages (2013). http://dslbook.org

A RESTful Privacy-Aware and Mutable Decentralized Ledger

Sidra Aslam[1,2(✉)] and Michael Mrissa[1,2]

[1] Faculty of Mathematics, Natural Sciences and Information Technology,
University of Primorska, Glagoljaška ulica 8, 6000 Koper, Slovenia
[2] InnoRenew CoE, Livade 6, 6310 Izola, Slovenia
{sidra.aslam,michael.mrissa}@innorenew.eu

Abstract. During the last decade, blockchain technology has gained massive attention due to its decentralized, transparent, and verifiable features. However, data stored on the blockchain is publicly available, immutable, and may link to the data owner, thus making privacy management and data modification major challenges. In this paper, we present a RESTful decentralized storage framework that provides data privacy and mutability. To do so, it combines blockchain with distributed hash table, role-based access control, ring signature, and multiple encryption mechanisms. We designed a protocol that exploits metadata and pointers stored on the blockchain, while corresponding encrypted data are stored off-chain, so that data owners are able to control their data. Each peer in our framework offers RESTful APIs to operate, thus ensuring interoperability over the Web. In this paper, we present the operation of our framework and its components that enable data protection at runtime. We also evaluate its performance with time measurements from our proof-of-concept implementation.

Keywords: Blockchain · Security · Privacy

1 Introduction

For several decades, people have been depending on centralized solutions that act as Trusted Third Parties (TTPs) to exchange information and to transfer assets through the Internet. These TTPs are responsible for securing data exchanges and they collect massive amounts of privacy-sensitive information from their users. However, a TTP becomes a single point of failure (SPOF) and is more vulnerable to security breaches and attacks [13]. As a solution to overcome this issue, blockchain [9] has gained massive attention due to its decentralized, transparent and immutable features. Indeed, blockchain allows participants to exchange information and store transactions without any TTP. Concretely, a blockchain is a chain of blocks that contain transactions, and each block is linked to the previous one with a cryptographic signature generated

© Springer Nature Switzerland AG 2021
L. Bellatreche et al. (Eds.): ADBIS 2021 Short Papers, Workshops and Doctoral Consortium,
CCIS 1450, pp. 193–204, 2021.
https://doi.org/10.1007/978-3-030-85082-1_18

using a hash function. Adding a block to the chain relies on a consensus algorithm [3], which ensures that the same copy of the transactions in the block are validated by enough (in general, the majority) participants. For the validation to happen, different consensus algorithms (e.g. proof of work, proof of stake, etc.) are available nowadays, with different characteristics (computational cost, complexity, etc.). However, the availability of the recorded data to everyone in the blockchain network raises issues when it comes to privacy-sensitive data [8]. Besides this, the immutability property of blockchain guarantees that the data records stored in transactions are tamper-proof, i.e. they can neither be deleted nor be mutated, which can be seen as a limiting factor.

In this paper, we aim at addressing these challenges with a single framework that integrates the following contributions:

- a solution for decentralized data storage that combines blockchain and Distributed Hash Table (DHT) to allow for data updates,
- a Role-Based Access Control (RBAC) solution to manage access to privacy-sensitive data,
- a flexible encryption design that allows to choose between multiple types of encryption while storing and querying data on the blockchain,
- a proof-of-concept implementation with performance evaluation that demonstrate the feasibility of our solution.

The rest of the paper is organized as follows. Section 2 presents the motivating scenario that highlights the research problem. In Sect. 3, we discuss existing work and their limitations before highlighting the originality of our contribution. Section 4 presents our framework and its components and explains how its provides privacy-preserving, secure, and decentralized data management. Section 5 describes the experimental results that confirm the feasibility of our proposed solution. Finally, Sect. 6 concludes this paper and lays ground for future work.

2 Motivating Scenario and Research Problem

In this section, we present a wooden furniture supply chain scenario that motivates the need for data updates and highlights our research problems. In the following, we identified 6 actors that involve in the wood supply chain: 1) **Wood cutting company** identifies specific trees and cuts them into logs. 2) **Transport company** transports wood logs to warehouse. 3) **Storage warehouse company** stores logs temporarily. 4) **Furniture assembly company** assemble the furniture. 5) **Furniture shop company** displays the assembled furniture 6) **Customer** purchases wooden furniture and verifies product origin.

Our wood supply chain scenario highlights the need for trust and traceability in the supply chain process. Existing solutions rely on Radio Frequency Identification (RFID) technology to enable electronic traceability of wood in the supply chain. Generally, this traceability framework needs a third-party centralized database framework to collect and store RFID data, which leads to a Single Point Of Failure (SPOF). Decentralized storage solutions can solve the problem of a single point of failure. In particular, blockchain is a decentralized and distributed ledger technology that stores records of users in such a way that makes them accessible to all participants without the risk of SPOF. A blockchain consists of a linear sequence of blocks. The contents of each block[1] contains a hash of the contents of the previous one to prevent the modification of stored transactions [11]. If the previous block is modified, the new hash one could generate from the content upon verification would not match the one stored in the next block. This design provides the blockchain with its immutability feature [6]: once data has been stored, no one can modify it.

However, our wood supply chain scenario highlights that actors need to insert, retrieve and delete data about their business activities, and at the same time, they need to be able to modify data, while keeping the proof that data was inserted. There is a need to develop a solution that overcomes the immutability characteristic of blockchain to allow update and delete operations on stored data. At the same time, the developed solution must fine-grained access control, as data access permissions vary depending on the data requester identity (data owner, business partners, client). In the following, we identify the research problems that raise from the scenario discussed above:

– **Data modification management:** In our case study, actors may need to modify data in blockchain (e.g. number of logs and product type). However, blockchain does not allow data modification, once it has been added to the chain due to its immutability nature. The challenge consists in overcoming this limitation while keeping the properties that make the blockchain interesting.
– **Data security and access control:** Blockchain stores data publicly and allows anyone to access it. In our context, a decentralized solution should ensure data privacy and protect privacy-sensitive data from unauthorized access.

In the following, we discuss the limitations of decentralized solutions supporting privacy-aware data access and data update.

3   Related Work

In this section, we present the related work and its limitations to store and update data on the blockchain. In [7], the authors present a blockchain-based framework to share data between stakeholders. Data hash sum is stored on blockchain while original data is stored on a MySQL database. However, MySQL databases are centralized and they are not as scalable as DHTs to store large amounts of data [5]. The authors in [14], propose a blockchain-based supply chain framework to maintain food traceability using a smart contract. However, product data is accessible publicly and immutable, which leads to privacy and data updates issues. In [12], the authors combine blockchain with DHT for secure IoT data

---

[1] Except the first block called genesis block.

sharing. Blockchain is used to store access control permissions, which is publicly visible and raise privacy issues. However, access control permissions are unable to modify due to blockchain immutability feature. The authors in [1], present a blockchain-based data storage for PingER (Ping End-to-End Reporting). The proposed solution use permissioned blockchain to store metadata of PingER files whereas corresponding data are stored on DHT without any encryption mechanisms. Additionally, this framework stores monitoring agent name and upload locations of the file on the blockchain, which raises data security and privacy issues. Inspired by the PingER metadata structure, our framework extends metadata structure and enables privacy and security management that ensures authorized access control and privacy protection.

As a summary, we have identified the most relevant work related to blockchain and DHT data storage. To the best of our knowledge, this is the first paper that provides decentralized data storage, data mutability, manages access to privacy-sensitive data, multiple types of encryption, and message sender anonymity at the same time in a single solution. In the following, we discuss the steps of our proposed framework in detail.

4   Proposed Framework

In this paper, we propose a secure privacy-aware decentralized framework that supports role-based access control, data mutability and actor's anonymity. Each actor, as a peer of the framework, runs the same code that is structured into a set of components. The following subsections describe each of these components in detail as depicted in Fig. 1.

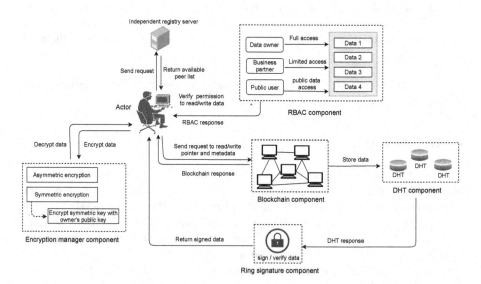

Fig. 1. Overview of a peer architecture.

4.1   General Overview

Our proposed framework allows wood supply chain actors to store data, read data upon request, and communicate with other actors through HTTP calls. Figure 1 shows the layout of our framework and its components organized around a main program. In our framework different actors are running the main program and they connect to each other using their APIs, after an initial call to the registry_server to get the list of available peers.

Let us consider the following example: an actor such as a wood cutter logs in our framework to store the number of logs cut on this day. When the program starts, the wood cutter will send a 'POST' request to the /peers resource of the registry_server to add its public key and URL (Uniform Resource Locator) to the list of connected peers[2]. Then, it will call the /peers resource with the 'GET' method and retrieve the list of connected peers. It will then connect with other available peers to download a copy of the blockchain (/chain resource, method 'GET', available on each peer). Upon request, the main component will call the rbac_manager component to verify the current actor's permission such as wood cutter to perform read and write operations. Indeed, each actor's roles, resources, and permission are defined in this rbac_manager component. Our framework allows the authorized actor to choose different types of encryption methods while storing data and generates a public key, private key, or symmetric key accordingly. Before storing the data, the main component will call the encrypt_manager component to encrypt the entered data with the current actor's public key or symmetric key depends on the selected encryption method. For each actor, the encrypt_manager component is responsible for generating public and private keys. This encrypted data sent to off-chain (key-value) storage called DHT_manager component, while corresponding pointer and metadata are stored on a blockchain_manager component (pointer is the hash of the data).

An authorized actor allows to create, update, delete, and read data using the pointer stored on the public ledger. A request (/chain/<block_no>, method 'GET') to the main component might call the ring_signature component to sign data anonymously, only in the case where the data is privacy-sensitive and the role of the requester requires anonymization. Accordingly, a request to (/chain/<block_no>, method 'POST'), will create a block, or update it if it already exists, and process the contents sent in the request message.

The following subsections describe each of these components in detail.

4.2   Framework Components

In this section, we describe in detail the components of the proposed framework including decentralized data storage, authorized data access, ensure data traceability, and maintains the actor's anonymity.

---

[2] Please note that the registry server can easily be replaced by a decentralized discovery protocol like Chord4S [4].

RBAC Manager Component: we use a Role-Based Access Control (RBAC) model to manage access to privacy-sensitive data. The RBAC model is based on the following four parameters: user, role, resource, and permission. In RBAC, users are actors related to the application. Roles are the application's functions that allow to access resources based on the given permissions. A permission is an authorization to access one or more resources within the application [2].

In our work, we define the following users, roles, resources, and permissions that assigns permissions to the user based on their role in our wood supply chain scenario.

- *Users:* In our framework, we need to define RBAC users according to the actors of the wood supply chain. Therefore, we define the following users: wood cutter, transporter, warehouse manager, furniture assembler, furniture seller, and customer.
- *Roles:* According to the different actions our supply chain users can perform on the architecture, we define the following roles: 1) **Data owner** any user[3] can be data owner. Data owners can add, read, modify and delete data about their products. For example, a wood cutter would act as "data owner" and insert information such as (trees-cut:20, type:oak,). 2) **Business partner** the business partner role allows specific users (chosen by the data owner) to access data that is not available to anyone. For example, a furniture assembler would act as "business partner" and might be allowed to read from the previous example: (trees-cut:20, type:oak). 3) **Public reader** the public reader role gives access to all public data. For example, a customer would act as "public reader" and might be allowed to read (type:oak).
- *Resources:* In our framework, user can access resources according to defined roles and permissions. In our framework, we define the following resources: 1) **DHT:** user can access DHT resource to add data about their business activities. For example, a wood cutter has a role "data owner" and store information such as (trees-cut:20, type:oak). 2) **Blockchain:** User allows to access blockchain resource to read data. For example, a customer has a role "public reader" and might be allowed to read information such as (type:oak).
- *Permissions:* We define permissions to restrict user's actions to access resources. For example, from previous example. a wood cutter has a role "data owner" and has a "permission" to write, read, update, and delete data such as (trees-cut:30, type:maple), whereas transport company would act as a "business partner" and has only "permission" to read information such as (trees-cut:30).
- *Rules and policies:* Our framework defines rules and policies that controls access to the data such as private data, privacy-sensitive data, and public data. Our `rbac_manager` component is responsible to authenticate role of current login actor. It also ensures if current role has permission to access resource or not as denoted by *verify_permission (role, operation, resource)*. For example, wood cutter has a role 'business partner' logs into the framework to store data on blockchain. The **main** component calls the method

---

[3] Except the end client that has read-only access.

`authenticate(actor, role)` to authenticate that if a 'business partner' role exists in our `rbac_manager` component or not. After role authentication, the `rbac_manager` component verifies the permissions of actions for current login actor's role such as if (`actor_role == 'owner'`), then "owner" has permission to perform read, write, update, and delete all types of data on the blockchain. In case, if (`actor_role == 'business_partner'`), then our framework allows just to read some data such as privacy-sensitive and public data. If (`actor_role == 'public_user'`), then our framework provides access to just read public data.

Our framework provides filter access based on role such as wood cutter as a 'business partner' has not permission to write, update, and delete data. We maintain data security by limiting unnecessary access to sensitive data based on each actor's role. Please note that although this simple RBAC model answers the requirements of our scenario, more elaborate models could be plugged in without changing anything in the framework design.

Blockchain Component: We use `blockchain_manager` component to manage metadata and pointer of encrypted data. Our proposed metadata structure consists of the data entry date, data entry time, and data pointer. The main components of the blockchain include block transaction, consensus algorithms, and metadata extension. Each component is explained as follows.

- *Block transaction:* Each block contains the block header, consensus signature, hash of the previous block, timestamps, verified metadata, and pointer of the actual data. In our framework, actors will connect to the framework and call `initialize(chain)` method to copy the blockchain if there will be any other available actors on the network, otherwise genesis block will be created and added to the blockchain. A blockchain is composed of a chain of the blocks where each block is comprised of many transactions [10]. Each transaction is broadcast on the network for verification and miners verifies the transaction through signature. Then, the verified transactions are added to the block of the blockchain. After storing verified metadata and pointer on the blockchain, our framework returns the block number to the data owner. The proposed framework allows data owner to access specific block to perform data update and delete operations.
- *Consensus mechanism:* It is used in our `blockchain_manager` component to establish the agreement on one state of the data in a distributed network. It ensures that the same copy of the data is replicated to all nodes in the blockchain network. Further, it verifies the transactions from this block and prevents the attacker to change the state of the data. Our framework uses a proof of work consensus mechanism to add each block to the blockchain. To do so, miners solve the complex puzzle and receive a reward such as a new coin to validate the block. Miners validate the transactions in a block and add this block to the blockchain. Proof of work consensus mechanism prevents a malicious actor to compromise more than half of the hashing power on the

blockchain. The process to verify the proof and its correctness is easy and fast. In the following we define the proposed metadata structure.

- *Metadata extension:* In [1], the authors allow storing metadata in the blockchain. We follow a similar approach and store the metadata information for each piece of data to maintain product traceability and actors' trust. In our framework (see Fig. 1), we have an RBAC_manager component to restrict user's actions on the data and we use a blockchain component to store metadata and pointer of actual data that are stored on the DHT component. We use REST APIs (/chain) that allow actors to copy blockchain and to store and read data on the distributed framework. We propose a metadata extension that relies on paper [1], to handle privacy constraints on data. To do so, we propose to encrypt user's sensitive information (e.g. location) with encryption mechanisms, and we store this encrypted data on offline storage (DHT). In our sample scenario, actual data on DHT consists of an actor's name, product identity, product location, quantity, and wood type.

DHT Component: In the proposed framework the encrypted data of each actor is stored on off-blockchain (key, value) storage called DHT. We implement a DHT component of our framework by using the Kademlia library. DHT is comprised of network of nodes that enable actors to write/read data associated with a given key. Actor's data are randomized across the nodes of the network and replicated to eliminate the chance of data loss. Our proposed framework records the date and time of each new data entered by the actor. This enables a network to keep track of the product and maintains the order of product entries.

Encryption Manager Component: In our framework, the encrypt_manager component is responsible for data encryption and decryption according to the selected encryption method. Our framework allows actors to choose encryption methods for each data write operation. If the data owner chooses the asymmetric encryption method then data will be encrypted with the owner's public key and stored encrypted data on DHT. A public key is accessible publicly while the private key is kept private by the key's owner to decrypt the data. If the data owner chooses the symmetric encryption method then data will be encrypted with a symmetric key and this symmetric key again will be encrypted with the owner's public key to ensure that only the data owner can access it later. Both encrypted symmetric key and encrypted data will be stored on DHT.

Ring Signature Component: It is an option here to actor's ensure anonymity within a group. A signature is created by any member from a set of public keys called a ring. Therefore, the identity of the signer remains hidden and no one can identify that who is the actual signer of the data. In our framework, the data owner can allow other actors to read their data upon request by using (/chain/<block_no>, method 'GET'). To read data, we rely on encryption according to data reading requirements: 1) **Private data** will not be shared with anyone. Therefore, it will be encrypted with the owner's public key, so only the owner can decrypt data using their private key. 2) **Privacy-sensitive data** is shared with only a specific number of users. It will be encrypted using the receiver's public key, so later data can be decrypted only with the corresponding private key. The data owner will also sign data by using ring signature to remain anonymous within a group, An authorized requester can read data and verify the signature. 3) **Public data** is available to anyone. It will optionally be signed by ring signature or encrypted with the data owner's public key to guarantee data ownership.

5   Implementation and Evaluation

This section discusses the implementation and evaluation of our proposed work. We implemented the key components of our framework by using an open source blockchain library[4] and the Kademlia DHT library[5]. The blockchain library is used to achieve consensus on a distributed network and creation of blocks. While, we used the DHT to store and retrieve data link with a key in a network of peer nodes. We performed all the experimental process using Python 3. The experiments are performed on the data (privacy-sensitive, private, and public) entered by the actors into the framework.

We evaluated the key components of our proposed framework on 64-bit Microsoft Windows Operating System with 16 GB of memory. In the following, we discuss the qualitative security and privacy analysis as well as quantitative performance evaluation. 1) **Security analysis:** according to the design of proposed framework, only authorized actors are allowed to access the system to perform write, read, update, and delete operations. A malicious user cannot modify existing data unless he/she controls more computation power than all other miners. Our framework ensures following security properties: we achieve `confidentiality` using asymmetric and symmetric encryption. We encrypt data with the owner's public key and store the corresponding pointer on the blockchain to achieve `integrity`. Our framework archives `availability` through the access control model. We ensures `non-repudiation` by adding metadata to the chain. 2) **Linking attack:** our framework uses a unique public key for each transaction. It prevents an attacker to link multiple data and transactions with the same ID. 3) **Modification attack:** in our solution, data owner has ability to encrypt data with their public key and store hash of the encrypted data on the blockchain. It also records evidence of data entry date and data entry time to trace last modification of data. An attacker can not modify owner's data. 4) **Privacy:** our proposed solution ensures that the owner owns and control their private data. Actor's private data will not be shared with other actors on the network. We encrypted privacy-sensitive and public data using requester public key to protect the data from malicious actor. In our proposed solution, we

---

[4] https://github.com/satwikkansal/python_blockchain_app/tree/ibm_blockchain_post.

[5] https://github.com/bmuller/kademlia.

achieve anonymity using ring signature. 4) **Scalability** currently, we tested our prototype with six actors and achieve reasonable performance. Our framework is flexible and scalable to work with a large number of actors.

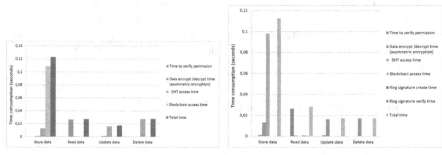

(a) Time overhead of asymmetric encryption without ring signature

(b) Time overhead of asymmetric encryption with ring signature

Fig. 2. Overall time overhead for asymmetric encryption

Performance Evaluation: We evaluate the time overhead to verify permission, data encryption/decryption using a symmetric or asymmetric method, DHT access, blockchain access, and overall total time while data store, read, update and delete operations. Figure 2a and Fig. 2b outline the time processing for both asymmetric encryption without ring signature and asymmetric encryption with ring signature. The results demonstrate that the total time of asymmetric encryption without ring signature is larger than the total time of asymmetric encryption with ring signature while store, update and delete data. We calculated the overall time for symmetric encryption as depicted in Fig. 3a and Fig. 3b. We compare results symmetric encryption without ring signature with symmetric encryption using ring signature. It is seen from the results that the total time of storing and deleting data for symmetric encryption without ring signature is larger than the symmetric encryption with ring signature. The total time to read data for symmetric encryption without ring signature is less than the symmetric encryption with ring signature. Total time to update data for both Fig. 3a and Fig. 3b are not much affected by the ring signature and symmetric encryption.

We also calculated average, standard deviation, min, and max value for both asymmetric and symmetric encryption while store, read, update, and delete data. We ran our prototype 50 times and experimental results show that asymmetric encryption gives a standard deviation of 0,022 s and symmetric encryption has a standard deviation of 0,023 s during data storing operation. To read data, asymmetric encryption has a minimum value of 0,124 s and symmetric encryption gives 0,142 s. For data update operation, asymmetric encryption has maximum value of 0,068 s and symmetric encryption gives 0,052 s maximum value. Experimental results clearly show that our proposed framework achieves a low overhead that is acceptable for the actor.

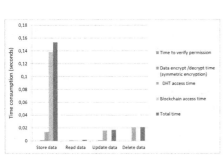

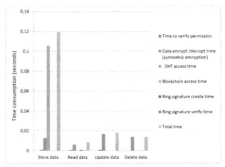

(a) Time overhead of symmetric encryption without ring signature

(b) Time overhead of symmetric encryption with ring signature

Fig. 3. Overall time overhead for symmetric encryption

6   Conclusion

In this paper, we illustrate the need for privacy-aware decentralized data storage, access control, data mutability, and actor anonymity in the wood supply chain scenario. Our framework enables this by combining the blockchain with DHT, role-based access control, and multiple encryption mechanisms that allow only authorized actors to access and modify their data without disclosing their identity on a distributed ledger. Thanks to its RESTful (between peers) and component-based (inside a peer) design, our framework is fully reusable across the wide diversity of possible application domains and use cases. We also presented a performance evaluation regarding its operation. Our simulation results demonstrate that our framework shows promising results and achieves an acceptable overhead. To the best of our knowledge, this research is the first work that integrates this combination of technologies in a single framework. In future work, we plan to compare our solution to similar blockchain implementations. Furthermore, we will study how the behaviour of our prototype evolves over larger number of peers, and devise optimizations to improve its performance over large scale networks, in real or simulated environments.

Acknowledgment. The authors gratefully acknowledge the European Commission for funding the InnoRenew project (Grant Agreement #739574) under the Horizon2020 Widespread-Teaming program and the Republic of Slovenia (Investment funding of the Republic of Slovenia and the European Regional Development Fund). They also acknowledge the Slovenian Research Agency ARRS for funding the project J2-2504.

References

1. Ali, S., Wang, G., White, B., Cottrell, R.L.: A blockchain-based decentralized data storage and access framework for pinger. In 2018 17th IEEE International Conference on Trust, Security and Privacy in Computing and Communications/12th IEEE International Conference on Big Data Science and Engineering (TrustCom/BigDataSE), pp. 1303–1308. IEEE (2018)
2. Bertino, E.: RBAC models - concepts and trends. Comput. Secur. **22**(6), 511–514 (2003)
3. Dinh, T.T.A., Liu, R., Zhang, M., Chen, G., Ooi, B.C., Wang, J.: Untangling blockchain: a data processing view of blockchain systems. IEEE Trans. Knowl. Data Eng. **30**(7), 1366–1385 (2018)
4. He, Q., Yan, J., Yang, Y., Kowalczyk, R., Jin, H.: A decentralized service discovery approach on peer-to-peer networks. IEEE Trans. Serv. Comput. **6**(1), 64–75 (2011)
5. Khamphakdee, N., Benjamas, N., Saiyod, S.: Performance evaluation of big data technology on designing big network traffic data analysis system. In: 2016 Joint 8th International Conference on soft computing and Intelligent Systems (SCIS) and 17th International Symposium on Advanced Intelligent Systems (ISIS), pp. 454–459. IEEE (2016)
6. Vinod Kumar, M., Iyengar, N.C.S.: A framework for blockchain technology in rice supply chain management. Adv. Sci. Technol. Lett. **146**, 125–130 (2017)
7. Longo, F., Nicoletti, L., Padovano, A., d'Atri, G., Forte, M.: Blockchain-enabled supply chain: an experimental study. Comput. Ind. Eng. **136**, 57–69 (2019)
8. Moser, M.: Anonymity of bitcoin transactions. In: Münster Bitcoin Conference (MBC), Münster, Germany, July 2013
9. Nakamoto, S.: Bitcoin: A peer-to-peer electronic cash system (2008). https://bitcoin.org/bitcoin.pdf
10. Nofer, M., Gomber, P., Hinz, O., Schiereck, D.: Blockchain. Bus. Inf. Syst. Eng. **59**(3), 183–187 (2017)
11. Pazaitis, A., De Filippi, P., Kostakis, V.: Blockchain and value systems in the sharing economy: the illustrative case of backfeed. Technol. Forecast. Soc. Chang. **125**, 105–115 (2017)
12. Shafagh, H., Burkhalter, L., Hithnawi, A., Duquennoy, S.: Towards blockchain-based auditable storage and sharing of IoT data. In: Proceedings of the 2017 on Cloud Computing Security Workshop, pp. 45–50 (2017)
13. Wang, S., Zhang, Y., Zhang, Y.: A blockchain-based framework for data sharing with fine-grained access control in decentralized storage systems. IEEE Access **6**, 38437–38450 (2018)
14. Westerkamp, M., Victor, F., Küpper, A.: Blockchain-based supply chain traceability: token recipes model manufacturing processes. In: 2018 IEEE International Conference on Internet of Things (iThings) and IEEE Green Computing and Communications (GreenCom) and IEEE Cyber, Physical and Social Computing (CPSCom) and IEEE Smart Data (SmartData), pp. 1595–1602. IEEE (2018)

Segmentation Quality Refinement in Large-Scale Medical Image Dataset with Crowd-Sourced Annotations

Jan Cychnerski◉ and Tomasz Dziubich^(✉)◉

Computer Vision and Artificial Intelligence Laboratory,
Department of Computer Architecture,
Faculty of Electronics, Telecommunications and Informatics,
Gdańsk University of Technology, Gdańsk, Poland
{jan.cychnerski,tomasz.dziubich}@eti.pg.edu.pl

Abstract. Deployment of different techniques of deep learning including Convolutional Neural Networks (CNN) in image classification systems has accomplished outstanding results. However, the advantages and potential impact of such a system can be completely negated if it does not reach a target accuracy. To achieve high classification accuracy with low variance in medical image classification system, there is needed the large size of the training data set with suitable quality score. This paper presents a study on the use of various consistency checking methods to refine the quality of annotations. It is assumed that tagging was done by volunteers (crowd-sourcing model). The aim of this work was to evaluate the fitness of this approach in the medical field and the usefulness of our innovative web tool, called MedTagger, designed to facilitate large-scale annotation of magnetic resonance (MR) images, as well as the accuracy of crowd-source assessment using this tool, comparing to expert classification. We present the methodology followed to annotate the collection of kidney MR scans. All of the 156 images were acquired from the Medical University of Gdansk. Two groups of students (with and without medical educational background) and three nephrologists were engaged. This research supports the thesis that some types of MR image annotations provided by naive individuals are comparable to expert annotation, but this process could be shortened in time. Furthermore, it is more cost-effective in the simultaneous preservation of image analysis accuracy. With pixel-wise majority voting, it was possible to create crowd-sourced organ segmentations that match the quality of those created by individual medical experts (mAP up to 94% ±3.9%).

Keywords: Annotation medical data · Crowd-sourcing · Deep learning · Medical imaging · Cooperative work

---

https://cvlab.eti.pg.gda.pl/.

© Springer Nature Switzerland AG 2021
L. Bellatreche et al. (Eds.): ADBIS 2021 Short Papers, Workshops and Doctoral Consortium, CCIS 1450, pp. 205–216, 2021.
https://doi.org/10.1007/978-3-030-85082-1_19

1   Introduction

Machine learning techniques are increasingly used in supporting medical diagnostics. For many of them, one of the essential steps is collecting data to aim to create learning, testing, and validation sets. The building process of large learning sets of dependable data has the crucial importance in the further construction of classifiers for medical applications. The medical data collection splits up into substages: acquiring, anonymization, curation, exploration, labeling, and quality control [10].

Well-balanced datasets with trustworthy annotations are necessary to train generalizable models. Annotations are considered as a special class of metadata that applies to the diagnosis and anatomical or pathological regions contained within a singular image. The number and size of public medical shared databases are not as large as they could be. One of the problems is the low and limited availability of medical data due to legal aspects related to the protection of patients' privacy (including anonymization process). However, new medical datasets are still appearing, which to a certain extent is due to the creators of online platforms for organizing competitions in the field of machine learning (e.g., Kaggle[1], Miccai workshops[2], Medical decathlon[3]).

Another burden is the inconsistency of either image or annotation formats. Most of the medical data are saved in DICOM format, but the internal data representation depends on the device manufacturer. In practice, some data fields could be filled wrongly or left blank, some with sensitive data and others with non inter-operable one [9]. Using of external format converters can feed distortions in annotations [14].

The next issue is the varying degrees of difficulty and labor intensitivity of assessing the gathered results, especially in a semantic segmentation. This kind of segmentation of well-known objects is quite easy through untrained people in opposite to the recognition of human organs on 3D-MR images. It is clear that the annotation process can be provided by medical specialists, but the number of them is limited and the cost of involvement is extremely high. On the other side, the accuracy and precision of the annotations are crucial. Therefore, there is a need to create new tools and methods to assist specialists and volunteers in the tagging process.

The objective of our work is to propose an effective process that can reduce the time necessary for acquisition of the annotations, to check of the usefulness and applicability of four agreement methods based on crowdsourcing in medical application, and finally, to evaluate the suitability of disparate groups of volunteers regarding the used selection tool. We propose to resolve the major problem outlined above by designing and implementing a new distributed software, called MedTagger, which an aim is to speed up the process of medical annotation gathering. More specifically, our major contributions include the following:

---

[1] https://www.kaggle.com/competitions.

[2] https://www.miccai2021.org/en/miccai2021-challenges.html.

[3] http://medicaldecathlon.com/.

- Improved method for creating high-quality, large-scale datasets based on crowdsourcing methodology,
- Evaluation of various algorithms (i.e., majority voting, k-Means, DBSCAN and Gaussian Mixture Models) to analyze collected data in the scope of their consistency,
- Comparision of two groups of volunteers (with and without medical background) to generated datasets in terms of quality and annotation time.

The paper is organized as follows. Related work is characterized in Sect. 2. Evaluation of quality control methods and description of our experimental setup are described in Sect. 3. In Sect. 4 results and discussion about the accuracy of selected methods and the usefulness of crowdsourcing in medical image segmentation. Summary of the paper and our propositions of future work directions in Sect. 5.

2 Related Works

One of the basic techniques to build annotated large datasets is simply downloading through the Internet (by using keywords and crawlers) and combining after the preprocessing phase. Unfortunately, this method based on web crawling is unsuitable and inefficient in medical data set construction. The second approach involves an augmentation technique to extend an existing small dataset and to build a synthetic large one. The input dataset is called empirical, output - synthetic. Many researchers point out that it is possible to create a synthetic image dataset similar to empirical images, which holds perceptually and qualitatively, although quantitatively there are differences in class, shape and color distributions [1,2].

One of the modern augmentation methods is generative adversarial network (GANs [6,7]). In the medical field, using this approach is acceptable using but the best effect can be achieved when we operate on empirical datasets. This is due to the fact that the medical image dataset for an machine learning application, besides adequate volume, has annotation, truth, and reusability. Annotation is connected to some specific object and contains data elements, metadata, and an identifier (so called an imaging examination). The classification labels or ground truth of each imaging examination, should be as accurate and reproducible as possible. It is highly desirable that newly created datasets meet the set of rules, referred to as the FAIR Data Principles (Findable, Accessible, Interoperable, and Reusable) [4].

Researchers can be satisfied with small, strictly targeted data sets to effectively respond to specific questions that result in papers and further funding. In contrast, the medical industry desires data of sufficient volume, diversity and quality to build classifiers to operate in different online production environments (due to potential approval by the relevant organizations). Medical researchers and radiologists are using various applications to make annotations on three- dimensional images, e.g., Geos[4], 3D Slicer[5], 3D Suite[6]. One of the most interesting software for described purpose was Visual Concept Extraction Challenge in Radiology (VISCERAL). It combines all desired features for cooperative work non-medical users. Unfortunately, the site was shut down in 2017 due to lack of funds. The comprehensive comparison can be find in [13].

This situation leads towards the third approach: creating or using shared platforms for work that requires human intelligence. These are emerging important tools in the machine learning community thanks to which can enter the selection of the original images and mark them using methods based on volunteering and crowdsourcing. An example is the Amazon Mechanical Turk service[7]. Sadly, this well-known platform requires a lot of manual labour to adopt and use medical data. License and some legal notices can also be troublesome. In analyzed field Supervisely platform[8] can be a better solution. Its key feature is to perform initial selection, i.e. the user roughly marks the object, the system makes a more precise selection, after which the user can make the rectification. Both of them are commercial. The second one enables an academic use, but we have to share data with the owner.

The question of dependability of the selections remains as well. For this purpose, the system (owner) rejects the selection from users who make errors on individual listings. The decision on exclusion is made on the basis of random checks. However, despite their high usability, these platforms do not address the consistency of data formats from medical devices. Each of the mentioned above tools and platforms has its own advantages. Although most of them have a common disadvantage - a very high entry threshold due to the unintuitive and complicated user interface, particular for potential volunteers.

We propose the use of a dedicated platform for the acquisition and annotation of medical imaging (including three-dimensional) that enables team cooperation of both medically educated people and those without a special education. NVIDIA Clara[9] is a full stack CUDA-based and scalable platform for the development and deploying of healthcare and medical imaging applications and workflows. It is an open-source environment used for research activities and projects in the field of medical imaging. It provides a federated learning concept and the ability to facilitate the collaborative training of local learning models without compromising on the data security. NVidia Clara provides efficient tools to speed up the annotation process, but does not include the concept of crowdsourcing. Cabezas et al. [3] had been proposed an interactive object segmentation approach to determine the level of expertise. This approach is rather simple and relies on pointing out a background and foreground on the image. Based on this, the weights of individual workers have been established. Our aim was a

---

[4] http://research.microsoft.com/en-us/projects/geos.

[5] https://www.slicer.org.

[6] https://imagej.net/Fiji.

[7] https://www.mturk.com/.

[8] https://supervise.ly/.

[9] https://developer.nvidia.com/clara.

choice and assessment of a more cost-efficient technique to assure high-quality annotations. The main purpose of the crowdsourcing data collection process is to prepare a set of comparable quality to the expert group.

3   Evaluation of Quality Refinement Methods in Crowdsourcing Segmentation

To assure high-quality annotations, the advanced validation mechanism is demanded. In theory, the medical doctor's task is to check each of the markings in terms of their correctness. In practice and on a real scale, such a situation could be very difficult to implement due to the very large number of annotations. Therefore, validation should occur in critical situations and the system itself should select scans that need to be controlled due to, for example, inconsistencies of the introduced signs by different volunteers. For this, we need method(s) that allow us to evaluate the consistency of the annotations. We analysed various methods for dependability improvement and selected: majority voting, Gaussian Mixture Models, K-means [11], DBSCAN [5] methods and as the most suitable for our goals. The GMM (soft/fuzzy clustering), K-means (hard clustering) and DBScan (density-based clustering) methods are related to clustering. In the number of centroids is K=1, the algorithm comes down to calculating the mean value of all values from the data set, which is equivalent to the majority voting algorithm for the regression problem.

3.1   Proposed Method

For showing that in some cases annotations from a naive user could be comparable to well-educated people, we propose the following method:

- data collection - MRI scans/images/data preparation by medical doctors (experts), anonymization, expert labeling;
- image annotation - providing of annotations by users - cohorting, test-bed preparation, labeling;
- automatic quality monitoring - using a selected technique to increase of data dependability (e.g. majority voting, ISO 2859, ANSI/ASQC Z1.4);

In a real world workflow, there is one more stage: manual quality refinement, in which reassessment of single images should be performed by the medical expert in the case of large variance of volunteers' annotations. In this paper, we are focusing on the assessment of the annotation conformity and extraction of an optimal method. As part of the study, participants were asked to perform determinations on MR abdominal scans that had been previously prepared by independent radiologists. MR scans were presented to users in the coronal view. Each of the scans contained from 22 to 50 slices, and one of the exemplary scans was of significantly lower quality than the others and was made with a lower dose of radiation. A total of 156 slices were available for labeling. The task to perform was to segment the left and right kidneys on each of the available scans. The left human kidney was to be marked with a polygonal tool and the right kidney with a rectangular tool (bounding box). During the research, the method of using the tool was not imposed, nor were there any requirements as to the quality of the introduced labels. Each person had to make markings on all available sections in their own way and with their own precision.

3.2   Data Set

Two radiologists selected and annotated kidney MRI scans, which consisted of a series of ca. 30 slices - T2-weighted turbo spin echo sequence from the Gdansk University of Medicine PACS (Picture Archiving and Communication System). The chosen examination had a mixture of kidney pathology comprising megaureters and retention. They selected images with predetermined criteria: the normal slice (N = 87) had no pathology or did not contain kidney's area. Severely abnormal images (N = 69) were determined as having grossly and mildly abnormal findings. All images were anonymized and uploaded into storage through MedTagger frontend site (described later) for the study duration to allow remote access. "Healthy" images are those classified as normal, "nonhealthy" images are those classified as mildly or severely abnormal. The MR scans were derived from adults (N = 2) and children (N = 3).

3.3   Group Selection

For assessment correlation between the available methods and different groups of people, two teams of volunteers were asked to cooperate, reflecting the practical use of crowdsourcing. The first group (A) was representing people unrelated to the problem domain (non-medical educational background, A = 35 persons). The second group (B) that was used in the research is a group of people indirectly related to the field of the problem, although not characterized by high experience. This group consisted of B = 13 students of the medical faculty at the Medical University of Gdansk. The vast majority of these people have had prior contact with medical imaging and have knowledge of the anatomy of the human body in the domain.

3.4   MedTagger Platform Architecture

In order to carry out the experiment, we designed a distributed environment, called MedTagger[10], to meet the performance and scalability requirements. In the system, the following users' roles are available: a volunteer and a doctor. Most of the functionalities are common to both groups, however, two of them are limited to the doctor's role: scan uploading and validation. The labeling is the basic and most important functionality of the system. Each user can use a short visual tutorial before going to the labeling screen. Details of annotated

---

[10] https://kask.eti.pg.gda.pl/medtagger.

objects (ROI) were not explained. MedTagger supports the annotation of three-dimensional imaging, but in this experiment there was used only the ability to display a series of subsequent slices from a single scan (context preservation). The volunteer has the option of scrolling between cross-sections in search of anomalies and previewing of former annotations.

MedTagger is based on client - server architecture and consists of following modules: Application server, Communication Server (a load balancer), Processing Nodes (asynchronously performing costly computation tasks) and SQL and NoSQL databases.

3.5 Used Metrics

There are several metrics for evaluating the segmentation of the medical image. Each of these metrics are based on the statistical interpretation of the prediction regarding the ground truth. In the paper [12] there are presented 20 metrics from different categories with guidelines for their application. Accordingly, IoU, precision and mAP were chosen.

Intersection over Union (IoU), also known as the Jaccard Index, is a statistic used for comparing the similarity and diversity of sample sets. In semantics segmentation, it is the ratio of the intersection of the pixel-wise classification results with the ground truth to their union. IoU metric is defined in Eq. 1.

$$IoU = \frac{TP}{TP + FP + FN} \tag{1}$$

where TP - True Positive, TN - True Negatives, FP - False Positive and FN - False Negative.

Precision is defined in Eq. 2:

$$precision = \frac{TP}{TP + FP} \tag{2}$$

To calculate the average precision (AP), using the precision for uniformly sampled recall values (e.g., 0.0, 0.1, 0.2, ..., 1.0), precision values are recorded. The average of these precision values are referred to as the AP. Similarly, mean average precision (mAP) is the mean of the AP values, calculated on a per-class basis. We compute the mean Average Precision (mAP) of the predicted bounding boxes and polygons at different IoU ratios with ground truth boxes. mAP gives the mean average precision of predicted object locations across all object predictions, matched with ground truth object prediction, and gives each object equally importance. The impact of limiting the set of volunteers on the quality of labeling (regardless of educational background) was carried out. For comparisons, subsets of 100%, 50% and 25% of people with the best labeling quality characteristics, measured by compliance, were used successively. The compliance was determined on the basis of the estimated set of ground truth of all persons and comparing it with the labels introduced by the given person. Each participant was subjected to an analysis of the values of specificity and sensitivity, which allowed to determine the quality of his/her work. This approach complies with the method of the limit of acceptable quality and allows for a insight at the quality of the gathered data by the crowdsourcing method. Based on specificity and sensitivity, it is possible to evaluate a quality measure of each person participating in the study. This measure can be determined by the square distance from the ideal labeling person, characterized by a sensitivity of 1 and a specificity of 0 (Eq. 3):

$$quality = (specificity + sensitivity - 1)^2 \qquad (3)$$

where

$$specificity = \frac{TN}{TN + FP} \qquad (4)$$

$$sensitivity = \frac{TP}{TP + FN} \qquad (5)$$

4   Results and Discussion

4.1   Segmentation and Detection Accuracy

The four prior selected algorithms were compared in terms of the IoU metrics, both with labels indicated with a polygonal tool and bounding box.

Table 1 shows the mean coefficient for all participants, the half and the first quartile with the lowest variance. In the table, the mean average precision mAP (at the threshold 0.75) was included (best results in bold).

Table 1. Comparison of selected method to refine annotation quality

Selection tool	Group	Metrics	Agreement method			
			Maj. Voting	GMM	k-Means	DBSCAN
Bounding-box	100%	IoU	0.667 ± 0.118	0.649 ± 0.122	0.664 ± 0.120	0.648 ± 0.121
		mAP	0.839 ± 0.100	0.821 ± 0.097	0.835 ± 0.102	0.817 ± 0.100
	50%	IoU	0.720 ± 0.041	0.701 ± 0.048	0.707 ± 0.041	0.701 ± 0.048
		mAP	0.887 ± 0.046	0.880 ± 0.040	0.861 ± 0.052	0.880 ± 0.040
	25%	IoU	**0.743 ± 0.043**	0.719 ± 0.044	0.728 ± 0.055	0.717 ± 0.042
		mAP	0.896 ± 0.450	0.889 ± 0.410	0.885 ± 0.610	**0.889 ± 0.410**
Polygon tool	100%	IoU	**0.829 ± 0.071**	0.599 ± 0.122	0.609 ± 0.129	0.563 ± 0.087
		mAP	0.814 ± 0.075	0.765 ± 0.052	0.802 ± 0.066	0.756 ± 0.068
	50%	IoU	0.689 ± 0.075	0.651 ± 0.076	0.667 ± 0.071	0.590 ± 0.067
		mAP	0.864 ± 0.044	0.831 ± 0.075	0.853 ± 0.049	0.794 ± 0.070
	25%	IoU	0.692 ± 0.078	0.664 ± 0.077	0.671 ± 0.075	0.624 ± 0.086
		mAP	**0.871 ± 0.037**	0.856 ± 0.021	0.860 ± 0.038	0.811 ± 0.086

For a bounding box tool and both metrics, the results for all methods are very similar. The Majority Voting method achieved a slight advantage (IoU less than 3% and mAP@0.75 less than 2%), however, the difference is relative small. Additionally, it is worth noting that the data quality is significantly better when only the 50% and the 25% most cohesive people are considered. Moreover, the limitation of the set of people resulted in a reduction of the variance (from 12% to 4–5%), although none of the analyzed methods obtained a significant qualitative advantage. In the case of usage of a polygon tool shows a significant advantage of the Majority Voting method using the entire group of volunteers. The difference between the full group and the most cohesive is 4.2% only. The results also show the weakness of the DBSCAN method for analyzing more complex data. This algorithm has the worst performance, with a score nearly 25% worse than MV.

The polygon annotation for the left kidney labels shows the weakness of the DBSCAN method. The result is lower by ca. 8% (mAP) compared to the best method (MV). It is also worthy of note the very low metrics generated by this method compared to competing algorithms. The K-Means method obtained a very good result, although it was Gaussian Mixture Models that obtained the lowest variance of precision (2.1%) for the most cohesive group of annotators.

4.2 Impact of Domain-Specific Knowledge

In the second experiment, an analysis of the impact of expertise on the quality of annotations was evaluated. The following section presents the quality analysis of the label dataset from the annotating process of the right kidney with a bounding box and the left one with a polygonal tool. The results for both groups (A - non-medical background, B - medical background) were presented in Table 2. The used metrics are the same as in the previous experiment.

Table 2. Comparison of domain knowledge impact in the annotation accuracy: group A - non-domain, group B - domain

Selection tool	Metrics	Group	Agreement method			
			Maj. Voting	GMM	k-Means	DBSCAN
Bounding-box	IoU	A	0.676 ± 0.107	0.670 ± 0.111	0.672 ± 0.104	0.668 ± 0.110
		B	**0.803 ± 0.052**	0.774 ± 0.061	0.791 ± 0.058	0.773 ± 0.062
	mAP	A	0.832 ± 0.100	0.829 ± 0.090	0.841 ± 0.079	0.825 ± 0.096
		B	**0.940 ± 0.055**	0.897 ± 0.078	0.921 ± 0.062	0.893 ± 0.085
Polygon tool	IoU	A	0.819 ± 0.057	0.584 ± 0.061	0.584 ± 0.089	0.577 ± 0.079
		B	**0.905 ± 0.011**	0.718 ± 0.046	0.737 ± 0.051	0.740 ± 0.032
	mAP	A	0.792 ± 0.081	0.751 ± 0.100	0.788 ± 0.088	0.770 ± 0.079
		B	**0.905 ± 0.039**	0.870 ± 0.050	**0.905 ±0.039**	0.867 ± 0.050

The outcomes show a clear advantage in the quality of labeling of the group (B) of people with expertise in the medical field, regardless of the tool, metrics and methods used (average difference ca. 10%). This group also generates a smaller standard deviation for both the IoU and the mean precision. Regarding individual algorithms, the Majority Voting method is of the highest quality again. In the case of the usage of a more difficult and laborious polygonal tool, a significant difference can be seen between the Majority Voting method and other algorithms (e.g., 15% IoU: MV - 0.905 vs. DBScan 0.74). For B group

and the poloygon tool, the MV algorithm represents both a high mean value
and a very low variance. To check the annotation quality difference between the
groups (A non-medical and B - medical background), the quality expressed by
formula (3) was computed and presented on Fig. 1 (best results in bold). The
MV method was used in each case: the left one (a) for right kidney labels marked
with a bounding box and the right one for left kidney labels marked with polygonal tool. A single measurement (represented by a blue or orange dot) means
the average time spent by the volunteer on a single image annotation and the
quality of his/her labeling.

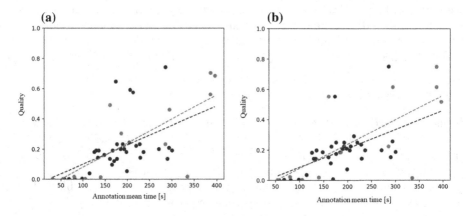

Fig. 1. Annotation quality vs mean time of labeling of per user. (a) bounding box (b)
polygon tool: blue - group A, orange - group B (Color figure online)

The quality of the annotations clearly correlates with the time needed to
prepare them. As we can see, a group of people in the medical field requires a
long period to be able to prepare the labels. It shows that time can also be a
determinant of annotators.

5   Summary

Crowdsourcing challenges include inaccuracy with anatomic variations and
pathologies, quality control, and ethical issues such as sharing medical images
with the crowd. Heim [8] compared the segmentation of the liver performed by
non-experts, medical students, technicians with domain knowledge, and radiologists by using polygon tool and MTurk platform. Accuracy was similar between
these groups, however, the crowd needed more time. The refined segmentation outlines from multiple volunteers were merged to the final segmentation
by applying the pixel-wise Majority Voting approach.

   Our research confirms the potential and strength of the Majority Voting
method, which, despite very simple assumptions, presented very efficient outcomes in almost all studies. K-means method can also be assessed as quite good,

as in many studies it has become only slightly worse than the MV. DBSCAN algorithm is distinguished by a relatively very low efficiency compared to other solutions. This feature is particularly conspicuous when a more difficult polygonal tool is engaged.

An important aspect of the data collection process based on the crowdsourcing method, is the educational background of people who will take part in the image annotation. People with the primary expertise in the field can provide about 10% better results in the context of the IoU score as well as the mean precision (mAP@0.75). When the bounding box tool is used, it is possible to utilize the annotations provided by nonmedical persons. The most coherent dataset does not depend on the applied agreement method and assures at the low variance level. This conclusion is in line with Heim's research [8], who claims that the experts had different annotation quality depending on the available tool.

In conclusion, the proposed crowd-algorithm frame- work demonstrates that crowdsourcing can be used to create accurate segmentations with a similar quality to those created by single medical expert annotators on the condition pointing out large anatomical structures and using an easy selection tool. Further work will be focused on smaller organ annotation and greater group diversity.

Acknowledgements. This work has been partially supported by Statutory Funds of Electronics, Telecommunications and Informatics Faculty, Gdansk University of Technology, and grants from National Centre for Research and Development (Internet platform for data integration and collaboration of medical research teams for stroke treatment centers, PBS2/A3/17/2013).

References

1. Barth, R., IJsselmuiden, J., Hemming, J., Van Henten, E.: Synthetic bootstrapping of convolutional neural networks for semantic plant part segmentation. Comput. Electron. Agric. **161**, 291–304 (2019). https://doi.org/10.1016/j.compag.2017.11.040. https://www.sciencedirect.com/science/article/pii/S0168169917307664. Big-Data and DSS in Agriculture
2. Brzeski, A., Grinholc, K., Nowodworski, K., Przybyłek, A.: Evaluating performance and accuracy improvements for attention-OCR. In: Saeed, K., Chaki, R., Janev, V. (eds.) CISIM 2019. LNCS, vol. 11703, pp. 3–11. Springer, Cham (2019). https://doi.org/10.1007/978-3-030-28957-7_1
3. Cabezas, F., Carlier, A., Charvillat, V., Salvador, A., Giro-I-Nieto, X.: Quality control in crowdsourced object segmentation. In: Proceedings - International Conference on Image Processing, ICIP 2015-December(May), pp. 4243–4247 (2015). https://doi.org/10.1109/ICIP.2015.7351606
4. Cocos, A., Masino, A., Qian, T., Pavlick, E., Callison-Burch, C.: Effectively crowdsourcing radiology report annotations. In: Proceedings of the Sixth International Workshop on Health Text Mining and Information Analysis, pp. 109–114. Association for Computational Linguistics, Lisbon, September 2015. https://doi.org/10.18653/v1/W15-2614

5. Ester, M., Kriegel, H.P., Sander, J., Xu, X.: A density-based algorithm for discovering clusters in large spatial databases with noise. In: Proceedings of the Second International Conference on Knowledge Discovery and Data Mining, KDD 1996, pp. 226–231. AAAI Press (1996)
6. Frid-Adar, M., Diamant, I., Klang, E., Amitai, M., Goldberger, J., Greenspan, H.: GAN-based synthetic medical image augmentation for increased CNN performance in liver lesion classification. Neurocomputing **321**, 321–331 (2018). https://doi.org/10.1016/j.neucom.2018.09.013. https://www.sciencedirect.com/science/article/pii/S0925231218310749
7. Goodfellow, I., et al.: Generative adversarial networks. Commun. ACM **63**(11), 139–144 (2020). https://doi.org/10.1145/3422622
8. Heim, E., et al.: Large-scale medical image annotation with crowd-powered algorithms. J. Med. Imaging **5**(03), 1 (2018). https://doi.org/10.1117/1.jmi.5.3.034002. https://www.ncbi.nlm.nih.gov/pmc/articles/PMC6129178/
9. Kohli, M.D., Summers, R.M., Geis, J.R.: Medical image data and datasets in the era of machine learning-whitepaper from the 2016 C-MIMI meeting dataset session. J. Digit. Imaging **30**(4), 392–399 (2017). https://doi.org/10.1007/s10278-017-9976-3
10. Montagnon, E., et al.: Deep learning workflow in radiology: a primer (2020). https://doi.org/10.1186/s13244-019-0832-5
11. Press, W.H., Teukolsky, S.A., Vettering, W.T., Flannery, B.P.: Numerical Recipes the Art of Scientific Computing, 3rd edn. Cambridge University Press (2007). https://doi.org/10.1017/CBO9781107415324.004
12. Taha, A.A., Hanbury, A.: Metrics for evaluating 3D medical image segmentation: analysis, selection, and tool. BMC Med. Imaging **15**(1), 29 (2015). https://doi.org/10.1186/s12880-015-0068-x
13. Jimenez-del Toro, O., et al.: Cloud-based evaluation of anatomical structure segmentation and landmark detection algorithms: Visceral anatomy benchmarks. IEEE Trans. Med. Imaging **35**(11), 2459–2475 (2016). https://doi.org/10.1109/TMI.2016.2578680
14. Wong, S.C., Gatt, A., Stamatescu, V., McDonnell, M.D.: Understanding data augmentation for classification: when to warp? In: 2016 International Conference on Digital Image Computing: Techniques and Applications, DICTA 2016 (2016). https://doi.org/10.1109/DICTA.2016.7797091

Process of Medical Dataset Construction for Machine Learning - Multifield Study and Guidelines

Jan Cychnerski[ID] and Tomasz Dziubich[✉][ID]

Computer Vision and Artificial Intelligence Laboratory,
Department of Computer Architecture,
Faculty of Electronics, Telecommunications and Informatics,
Gdańsk University of Technology, Gdańsk, Poland
{jan.cychnerski,tomasz.dziubich}@eti.pg.edu.pl

Abstract. The acquisition of high-quality data and annotations is essential for the training of efficient machine learning algorithms, while being an expensive and time-consuming process. Although the process of data processing and training and testing of machine learning models is well studied and considered in the literature, the actual procedures of obtaining data and their annotations in collaboration with physicians are in most cases based on the personal intuition and suppositions of the researchers.

This article focuses on investigating various practical aspects of medical data acquisition and annotation, as well as various methods of collaboration between IT and medical teams to build datasets that fulfill the desired quality, quantity, and time requirements. Based on five projects undertaken by the authors in diverse medical fields, in which the dataset construction procedure was iteratively optimized, a set of guidelines and good practices to be followed when building new medical datasets was developed as described.

Keywords: Medical data acquisition · Dataset construction · Data annotation · Machine learning workflow · Guidelines

1 Introduction

Building a biomedical application is a complex issue in which eight basic steps can be distinguished. The first two involve gathering a skilled project team and obtaining institutional review board approval. The team is usually composed of a project manager (e.g., physician, data scientist), clinical expertise (e.g., surgeon or hepatologist), imaging expertise (e.g., radiologist) and technical expertise (e.g., data scientist, software engineer). Next, it is expected to carry out a data

---

https://cvlab.eti.pg.gda.pl/.

© Springer Nature Switzerland AG 2021
L. Bellatreche et al. (Eds.): ADBIS 2021 Short Papers, Workshops and Doctoral Consortium, CCIS 1450, pp. 217–229, 2021.
https://doi.org/10.1007/978-3-030-85082-1_20

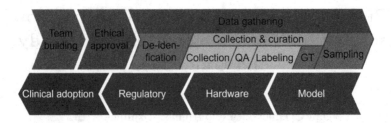

Fig. 1. Typical medical machine learning system creation workflow. Orange components are the subject of this paper. (Color figure online)

selection and anonymization process, followed by its collection and curation. The remaining four steps concern creating training/testing sets, designing a classifier, determining the best software and hardware environment, receiving clinical validation and certification, and deploying the product in medical care facilities [15].

The aim of this paper is to introduce practical guidelines which could be found helpful in the mentioned *collection and curation* of data within the process of deploying a biomedical application, as presented in Fig. 1). Based on our five finished projects in the scope of computer-aided diagnosis algorithm development, we performed a multi-field study over the problem of construction of medical datasets that fulfill their desired quality, quantity of raw data and annotations, with respect to the desired project time span. Existing scientific papers are mostly based on individual experiments (i.e., prototype projects) and analysis of available scientific papers. In this article, we present first-hand knowledge derived from our biomedical projects carried out in the period from 2008 to 2020.

2   Related Works

In [13] the entire process of **data collection** is reported. As medical information is strictly sensitive, it is crucial to ensure that no private details are exposed to third-party software, using de-identification tools (anonymization, pseudonymization).

As the data is effectively anonymized, the further action is to aggregate it, which is called **data collection** and subsequently review it by supervising medical expert in order to find out whether all the data is suitable to be used as proper datasets for artificial intelligence models afterwards. This is what is called the **data curation**. Data **collection and curation** is the fundamental step of a building machine learning solutions. In respect to that, the project team ought to endeavour to put emphasis on this stage. Verified data is then inspected in the data **quality assurance** and exploration step, where a look into the raw dataset is taken. This might yield specific features or global trends of the dataset, which could be advantageous in planning further work.

With all data successfully prepared, it is then passed on to the medical experts responsible for **data labeling**. The specialists examine several case looking for desired lesion and create either annotations, which are the graphic symbols of depiction, or tags, which refer to descriptive form of indication. It is also salient to establish a reference standard (**ground truth**) so that future classifiers' outputs can be properly evaluated. The last stages consist of data sampling, model training, testing, and deployment [6,10].

In this paper, we want to focus on the **data collection and curation** step. The workflow presented above lacks an aspect concerning the establishment of a clinical study protocol and data profile, in which it is explicitly expressed what actions are to be taken in case of any ambiguity. It is essential to ensure that the whole project team works according to the fundamental rules implicit in the protocol (e.g., strict definition of the boundaries of an examined organ).

There are several common problems concerning the data collection and curation process. **Intra-observer variability** is a factor which is characterized by the degree to which measurements taken by the same specialist are consistent. Inter-observer variability, on the contrary, specifies the variation between results obtained from two or more specialists inspecting the same case.

In [8], 199 X-ray angiography samples were used in the study. A partial lack of **reproducibility** of the results was noted. It was stated that in an evaluation of changes in a vascular tree the inter-observer variability between observers showed poor agreement for lesion classification (expressed as Fleiss' kappa, $\kappa = 0.33$), length of lesion ($\kappa = 0.35$) and branch point involvement in stenosis ($\kappa = 0.39$). For classification of vessel calcification and thrombus the variability was equal to $\kappa = 0.53$ and $\kappa = 0.60$ respectively. Similar results were presented by e.g., [7,9,14] regarding other organs, modalities and specialisations.

These findings can be attributed to a **quality of data acquisition**, experience of angiography experts – mainly - lack of strict definitions. Further investigation, preferably by panel assessment, might be performed only after an agreement upon complete and detailed definitions for each angiographic variable. Besides, if an estimation of procedural success rates and risk for procedural complications is wanted, it is necessary to debate operator experience and their clinical variables (i.e. their devotion and involvement in the project).

Another problem in the data collection and curation step is the **time pressure** and process of labeling being laborious and time-consuming. The pressure is associated with the fact, that in most cases taking part in a project is a secondary occupation of specialists. This problem could be dealt with by increasing the number of labeling individuals, although such move would cause an increase of an inter-observer error. There have also been attempts to support medical experts by volunteers in a form of crowdsourcing (e.g. Hasty, Supervisely, Playment, SageMaker, COCO Annotator, Mechanical Turk).

With a view to problems presented above, various solutions can be found in literature. In order to enhance the work of specialists, a **Human-in-the-Loop approach** is used (e.g., Nvidia Clara[1], AggNet [2]). It works by parallelizing labeling of data and classifier trainings. In the initial phase, created classifiers help with rough annotation, which is then corrected by an expert. Such process

---

[1] https://developer.nvidia.com/blog/federated-learning-clara/.

is repeated in each subsequent, improved iteration of an AI model, ending up with a significant increase in time of tagging. In [12], a four to tenfold increases in average glomerular annotation speed between the initial and end iterations were achieved.

The tools presented above accelerate both integration of data and models and process of tagging per user. However, they do not decrease the intra- and inter-observer errors, neither are they capable of quality monitoring. In our work we present a multi-field study and review of finished projects concerning the step of data collection and curation, results regarding fulfilment of project indicators and encountered errors in each project. The aim of this article is to present project guidelines that will allow to achieve planned amount of valuable medical data in specified time.

3   Description of the Accomplished Projects

During a 12 year period of time, five projects in various scopes of medicine have been carried out. In each of these projects, attempts have been taken to construct data sets for the purpose of designing machine learning classifiers. The workflow and methods of construction of the datasets in these projects were modified iteratively among them. The IT-side team responsible for data acquisition was kept unchanged throughout the whole period. Responsible for the medical side, there were various medical teams composed of one to eight people. The number of medical facilities (hospitals) taking part in each project ranged from one to five. In four of the projects, the labeling process concerned computer tomography (CT), X-ray coronary angiography (XRA), and endoscopic images. In the remaining project, Impedance Cardiography Graphs (ICG, as raw data) and multimodal data were used. The number of patients from whom the data was collected was between 100 and 400. In the projects, both out-of-the-box (OOTB) and in-house developed software was used.

In the following paragraphs, a concise description of each project is presented. Information about the subject matter, number of labeled classes, members of the IT and medical team, planned and real (dedicated exclusively for data acquisition), data type (its format and whether the examinations were prospective or retrospective), used annotation tools, assumed and achieved number of examinations and general notes about the quality of annotations in all projects are emphasised.

ERS (Endoscopy Recommender System) [3]. In this project an image dataset of retrospective examinations of gastrointestinal tract endoscopy was provided. Classification of diseases was fully compliant with MST 3.0 standard. The Minimal Standard Terminology (MST) was developed to standardise endoscopic reporting and it defines "minimal" list of terms that could be included within a system used to record the results of a gastrointestinal endoscopic examination[1]. 104 classes were described, 70 of which were upper digestive tract and 34 colorectal. The data was roughly segmented and was made up of movie clips and single frames. 5051 annotations were composed (1779, 2199 and 1019 for gastric, intestinal, and health respectively). It was the first of the projects, only for scientific research, used subsequently as a baseline for other projects. Some preparations had been done beforehand. The medical imaging expert was aided by an IT team member, responsible for monitoring the quality of the prepared data. The primary goal of this project was not to create a large training dataset, but rather to achieve a wide spectrum of diseases and real-time data processing. A dedicated software for labeling was used (ERSCutter[2]). Project objectives have been achieved, although it consumed a significant amount of time (48 months).

IPMed [4]. Analysis of hospitalization reports from patients who suffered an ischaemic stroke that involved medical history, blood tests, impedance cardiography, and observations to form a treatment and medication regimen. The project was commercial. Examinations were accelerated by involving 5 medical facilities with 8 medical experts. A strict clinical study protocol and data profile was formulated before data collecting. A dedicated application for data entry and quality control was used. Six different medical devices were responsible for data acquisition which, connected to multi-modality of data (clinical data, raw measurements, and follow-up), lead to a wide variety of aggregated information. A 4-h theoretical group training was executed to demonstrate the methods of labeling, although it lacked practical exercises on the part of the annotation team. The examinations were all prospective. Out of 500 planned studies, only 275 were accepted, with the remaining 235 being rejected due to not fulfilling the quality requirements. Initially, 4 facilities were taken into account, however, during the implementation of the project an additional one was involved. The project took 30 months to finish.

3D-Liver. The project concerned annotating liver parenchyma, tumours, and hepatic vessels on retrospective CT image data with contrast (artery phase). The project was commercial and included 3 experts (radiologists) from one medical facility, each of whom went through individual training. In the initial stage of the project, guidelines concerning the labeling process were developed, then a validation and presentation of the results from 3 different devices was carried out. Subsequently, the results were validated after every stage of the project. ITK-Snap[3] was used as annotation software. The project was assumed to last 9 months and 130 segmentations were to be yielded (130, 80 and 40 of parenchyma, tumours, and hepatic vessels respectively). Both factors were not satisfied, as the project was prolonged to 12 months and 130 parenchyma, 70 vessel and 30 tumour data was segmented.

Endoleaks.AI [5]. Commercial project based on retrospective image data collected from 3 different medical devices. It touched the issue of detection and

---

[2] https://cvlab.eti.pg.gda.pl/projects/ers.
[3] http://www.itksnap.org/pmwiki/pmwiki.php.

classification of endoleaks in abdominal aorta region using CT scans with arterial phase contrast injection. The medical team consisted of 4 cardiologists (1 highly experienced, 1 semi-experienced, 2 resident doctors) from one medical facility. Every medical team member went through individual annotation training. Guidelines concerning proper labeling were developed, which was done using three different programs that enable manual markup creation (ITK-Snap, OsiriX[4], 3D Slicer[5]). Two experts (both from IT and the medical team) were picked to watch over the quality of segmentation. The planned time for project completion was 9 months and 180 segmentations of the aorta, abdominal aorta, abdominal aorta aneurysm and 20 segmentations of stent graft with endoleak were to be obtained. All requirements have been successfully fulfilled.

Angioscore. The last of the described projects concerned automatic detection coronary tree and lesions on XRA images. Retrospective examinations taken from four different medical devices formed the data set of this commercial project. Two experienced medical experts from the same medical department were involved and took part in online group training concerning proper annotating methods. Guidelines regarding coherent tagging were developed. Dedicated supportive software was created (AngioTagger[6]). Project progress, both from the technical and medical side, was monitored at monthly intervals. Inter-observer error using the Cohen's kappa coefficient was calculated to keep track of the cohesion of the tagged data. Time limit of 10 months was not exceeded and the whole assumed set of data with 150 labeled scans was successfully obtained.

4   Medical Datasets Creation for ML Applications

During the implementation of these 5 projects, numerous approaches were explored to build medical datasets for machine learning applications. The methods used were iteratively modified to produce better results and to avoid encountered errors and problems, and to meet deadlines/milestones. The major changes in successive iterations are summarized in Table 1, whereas their motivations, aims, and details are described in the following sections.

Despite the review of scientific literature, no papers presenting comparative metrics on similar topics were found. Our described projects have no common type of modality. Thus, it is not possible to define any metrics to quantitative assessment. Instead, we present direct positive and negative effects of proposed modifications.

Endoscopy Recommender System. This was the first of the realized projects. The goal was to collect the largest possible comprehensive endoscopic

---

[4] https://www.osirix-viewer.com/.

[5] https://download.slicer.org.

[6] https://www.medmetric.ai/.

Table 1. Major changes/improvements in subsequent iterations

	Tool	Anno-tators	Annot./ patients	Instructions	Quality assurance	Novelty
ERS	Custom	1+1	6000 1100	Simple instructions given verbally to the middleman	Live technical validation by the middleman	–
IPMed	Custom	8	275 275	Precise, strict and rigorous data gathering and annotating protocol	Strict automatic validation according to the predefined medical protocol	More annotators, rigorous protocol with automatic validation
3D-Liver	OOTB	2	130 130	Precise text instructions with simple examples	Regular technical review	OOTB software with CV algorithms, more permissive QA
Endoleaks	OOTB	4	200 180	Precise text instructions with simple examples, plus individual training	Dice coefficient between subsequent stages	Preliminary annotation training, methods being adjusted during the project
Angioscore	Custom	2	150 150	Precise text instructions with simple examples, plus group training	Dice coefficient between subsequent stages, continuous technical review	Return to custom software, group training, continuous QA

data set. The annotations were to fully reflect the medical practice in describing examinations and videos. For this reason, a dedicated tool was implemented during the project to browse the videos, select frames, mark the findings using a polygon and assign a label from a predefined list according to the MST3 medical standard. One physician from one medical center was involved in the project. To minimize the required physician involvement, the annotation was done in tandem with an IT team member (middleman) who possessed the configured software, was familiar with the technical requirements of the annotations, and ensured their quality. During the project, most of the stated goals were achieved, although at great expense. The resulting collection was the largest endoscopic dataset in existence at the time, and it is still one of the most versatile collections today. Thanks to the work of the middleman, there was no need to train physicians or create elaborate annotation instructions. In addition, the middleman successfully monitored the progress and technical quality of the annotation on an ongoing basis and had a positive impact on physician engagement. Communication between the IT team and the physician was thus easy to understand and effective for both parties.

Unfortunately, the project revealed many limitations of the established workflow. Collaboration with only one center and one physician resulted in low diversity of data obtained and difficulty in determining objective substantive quality of data due to lack of comparison to other physicians and lack of histopathological verification. While the middleman reduced effort from the physician's perspective, his presence significantly increased the cost of annotation. The difficulty in coordinating the time and location of annotation meetings significantly increased the time of data collection and the overall project length.

IPMed. Several changes have been applied in the IPMed project. Data type was changed to prospective examinations, aggregated from various health facilities and devices. The annotation team member count was notably increased. A detailed data profile was drawn up. Direct supervision of tagging was desisted from. High emphasis was put on the quality of the obtained data with an addition of its automatic control. Dedicated software was created. The aim of the project was concentrated on a specific application (hemodynamic profiles), instead of a profile of various diseases, although it covered the full diagnostic process, including the 6 months follow-up. One common, concise, and rigorous protocol of data aggregation was introduced at the beginning of the project, which based on the substantive requirements presented by experts. Each of the annotation team members went through a tagging training before starting their work.

The implementation of the changes resulted in many positive effects, mainly: project was successfully deployed in each of the medical facilities, automatic distribution of application updates was implemented, obtained data were ready to easily be used in medical analysis and research, physicians were able to adjust the annotation time to suit their own needs.

It is worth noticing some drawbacks in conducted project. First of all, high initial cost of application production and installation in all facilities. Next, there was no communication between the medical facilities and annotation team which underwent personnel changes in the middle of the project and consequently noted a low level of commitment (as it was a part-time job). The protocol established beforehand turned out to be far too restrictive. Due to the prospective character of examinations and automatic validation, high amount of data was rejected. Prospective type of data and exorbitant requirements for quality control significantly decreased the number of acceptable examinations. The limit of 500 scans was not reached despite prolonging project's deadline and involving an additional healthcare facility. The problem described above led to a change of the protocol, which further resulted in many examinations being incoherent.

3D-Liver. In the 3D-Liver project, the type of acquired examination was changed back to retrospective with an additional selection, which was assumed to lower the time of data collection. Size of the tagging team was also reduced and individual training was conducted to increase data quality. Validation using Dice coefficient was introduced in the first stage of the project, which aimed to determine tagging errors between the labeling team members. A document that included labeling procedure guidelines was created. An off-the-shelf labeling software (ITK-Snap) was used. It contained a supportive active contour and a growing region system which aided with 3D semi-automatic tagging capabilities. Annotations were shared within project members using a file sharing cloud service, which demanded an appropriate file naming convention.

The changes implemented had mostly positive effects on the achieved outcomes. Supporting annotation with semiautomated computer vision methods made the data collection process both faster and easier, and helped to maintain physicians' engagement by reducing repetitive tasks. Utilizing existing annotation software and popular cloud-based file-sharing solutions allowed labeling to begin almost immediately with virtually zero start-up costs. The overall project time was also drastically reduced due to no need to wait for the tagging tool to be built or to coordinate meetings with the 'middleman'.

However, additional problems were encountered during the project. Despite the use of retrospective studies, the process of selecting appropriate data for labeling took a long time due to the lack of connection between the HIS and PACS systems. The lack of understanding of the business needs by the medical staff resulted in their limited involvement, which was especially evident in their lack of attention on details when using semiautomated annotation supporting methods. The initial phase of the project underestimated the required labeling team size, which resulted in the need to hire additional annotators during the project, yet changing the average experience level of the team. In addition, the annotators had a varying understanding of the details of the labeling instructions and objectives. The use of the cloud as a collaboration tool was successful, however, the requirement for a strict file naming convention resulted in many errors that were difficult to correct (integration with HIS would have helped here as well).

Endoleaks.AI. Endoleaks project was characterized by a very consistent team of annotators, whose members were employed in the same department that implicated fast and efficient communication. Various labelling software were being used throughout the project (3D Slicer, OsiriX, ITK-Snap). After each stage of the project, cross-member validation was executed using Dice metrics (4 times in total). Finally, all data was successfully obtained within the initially set time limit. Unfortunately, there was also an inter-stage delay related to tagging occurred in the third stage of the project. The main reason was no precise boundaries of an aneurysm had been written down, thus a high variance of segmentation was noted. A switching between different segmentation aiding tools caused a decline of interest in incorrect segmentation rectification. Little trouble with various saved data formats was also experienced.

Angioscore. In the Angioscore project, the number of medical devices used to obtain clinical data was increased to 4, which resulted in aggregation of different data profiles. A dedicated segmentation tool (AngioTagger) was used to standardize data conversion formats which also enabled to support the selection of sophisticated tasks (e.g. automated division of the coronary tree into segments or areas of calcification). This tool was also gradually expanded with new aiding ML algorithms without significant changes for annotators. Inter-rater reliability tests were conducted at monthly intervals. Regular meetings were taking place in order to note the data quality problems. Project team succeeded in collection intended, high-quality data within the project time limit although problem concerning irrelevant vessels was discovered.

5   The Proposition of Guidelines

On the basis of our experiences of successive workflow improvement iterations, a set of recommendations and good practices that positively impact the construction of medical datasets was created and provided in the following guidelines. The guidelines we suggest are an extension of the tips and checklist published by [11]. The recommendations are organized in the order of their appearance in the projects and their importance.

Project Preparation Stage. These are all the activities that need to be done before labeling begins, especially the design of labeling requirements and procedures.

1. **Circumstantial protocol for data collection and annotation.** The protocol should consider data profile and quality requirements. It should specify an acceptable error rate - excessively restrictive requirements result in the rejection of large amounts of data and discourage annotators.
2. **Selection of the appropriate labeling tool.** It should be kept unchanged during the project. If possible, it is worth customizing the tool to the specific requirements of the project in order to simplify the interface and for quality control. Consider the HITL model, if it is possible. Strictly technical requirements (e.g. file formats) should be enforced automatically.
3. **Annotators should be named and known from the very beginning of the labeling.** Keeping annotators actively involved and strong motivated in the process is very crucial - it is worthwhile to take motivational actions.
4. **A sufficiently large variety of data, employing no more than necessary collaborating institutions.** Having too many of them creates significant administrative overhead. On the other side, an appropriate number of devices should be available to ensure sufficient diversity.

Preliminary Annotation Stage. It includes annotator training and performing a trial labeling to eliminate initial errors and to equalize the levels of annotators.

5. **Consistent understanding of requirements and instructions by all of the selected annotators and that all achieve a comparable quality level.** Uninvolved and insufficiently experienced annotators generate a lot of erroneous data.
6. **Group training of annotators,** including requirements overview and demonstration of a few representative annotation processes
7. **Several trial annotations performed by every annotator, followed by inter-annotator validation** to determine annotation consistency and clarity of instructions, as well as edge cases. If necessary, modification and clarification of the annotation instructions/protocol.
8. **Data selection for annotation performed using HIS.** Manual data selection is subject to bias and causes technical errors.

The Main Annotation Process

9. **Maintaining the motivation of annotators in the project.** Showing progress and partial results frequently is a good way to keep motivation in a team (alike stand-ups in Agile methodology).
10. **Periodic validation of the quality and consistency of all the annotators' work.** Common discussing discrepancy in annotations by whole teams come to decrease inter-observer error.
11. **Ensuring systematic manner in which annotations are made.** Cyclic sharing data to ML team enables to discover errors in early stage and could prevent re-annotation on whole data set
12. **Special care on quality when using semi-automated annotation assistance algorithms.** Poorly motivated annotators tend to pay low attention on details and small annotations.
13. **Annotation in pairs** (medical expert and IT supervisor) **is excessively costly and time-consuming,** albeit helps to maintain high quality and facilitates communication.

6 Conclusions and Future Work

The process of data collection and annotation is critical in the entire workflow of developing machine learning-based systems. Building medical machine learning datasets is a non-trivial, multidisciplinary, time-consuming and expensive process. Therefore, it is essential to take measures that will produce datasets of the highest possible usefulness and quality, while keeping the cost and duration of the project under control. While conducting medical-IT projects, we have repeatedly revised and improved the data collection and annotation procedure. Five of the most significant completed projects are presented in this publication.

In successive iterations of the workflow adjustment, we identified the issues and modified the procedure to eliminate the errors or minimize their effects in order to establish acceptable datasets. In some cases, it caused other, new problems, which were dealt with in the next iterations - however, the total number of problems has been systematically reduced and the revealed defects were decreasing. We summarized the developed effective practices in short guidelines, which are successfully applied in current projects led by the team.

In the future, it is planned to extend proposed workflow to include collection of multimodal data, obtained from multiple imaging sources (including multiple imaging modalities and multiple examinations of a given patient performed at different times) or linked to clinical data. Due to the heterogeneity of this type of data, achieving an appropriate level of quality and consistency will require special care and further changes to the developed workflow and guidelines.

References

1. Aabakken, L., et al.: Minimal standard terminology for gastrointestinal endoscopy - MST 3.0. Endoscopy **41**(8), 727–728 (2009). https://doi.org/10.1055/s-0029-1214949
2. Albarqouni, S., Baur, C., Achilles, F., Belagiannis, V., Demirci, S., Navab, N.: AggNet: deep learning from crowds for mitosis detection in breast cancer histology images. IEEE Trans. Med. Imaging **35**(5), 1313–1321 (2016). https://doi.org/10.1109/TMI.2016.2528120
3. Blokus, A., Brzeski, A., Cychnerski, J., Dziubich, T., Jędrzejewski, M.: Real-time gastrointestinal tract video analysis on a cluster supercomputer. In: Zamojski, W., Mazurkiewicz, J., Sugier, J., Walkowiak, T., Kacprzyk, J. (eds.) Dependability and Complex Systems, vol. 170, pp. 55–68. Springer, Heidelberg (2012). https://doi.org/10.1007/978-3-642-30662-4_4
4. Dorożyński, P., Brzeski, A., Cychnerski, J., Dziubich, T.: Towards healthcare cloud computing. Adv. Intell. Syst. Comput. **431**, 87–97 (2016). https://doi.org/10.1007/978-3-319-28564-1_8
5. Dziubich, T., Białas, P., Znaniecki, Ł, Halman, J., Brzeziński, J.: Abdominal aortic aneurysm segmentation from contrast-enhanced computed tomography angiography using deep convolutional networks. In: Bellatreche, L., et al. (eds.) TPDL/ADBIS -2020. CCIS, vol. 1260, pp. 158–168. Springer, Cham (2020). https://doi.org/10.1007/978-3-030-55814-7_13
6. Glegoła, W., Karpus, A., Przybyłek, A.: MobileNet family tailored for Raspberry Pi. In: 25th International Conference on Knowledge-Based and Intelligent Information & Engineering Systems (KES) (2021)
7. Hanbury, A., Langs, G.: Cloud-Based Benchmarking of Medical Image Analysis. Springer, Heidelberg (2017). https://doi.org/10.1007/978-3-319-49644-3
8. Herrman, J.P.R., Azar, A., Umans, V.A., Boersma, E., Es, G.A.V., Serruys, P.W.: Inter- and intra-observer variability in the qualitative categorization of coronary angiograms. Int. J. Cardiac Imaging **12**(1), 21–30 (1996). https://doi.org/10.1007/BF01798114
9. Joskowicz, L., Cohen, D., Caplan, N., Sosna, J.: Inter-observer variability of manual contour delineation of structures in CT. Eur. Radiol. **29**(3), 1391–1399 (2019). https://doi.org/10.1007/s00330-018-5695-5

10. Kohli, M.D., Summers, R.M., Geis, J.R.: Medical image data and datasets in the era of machine learning-whitepaper from the 2016 C-MIMI meeting dataset session. J. Digit. Imaging **30**(4), 392–399 (2017). https://doi.org/10.1007/s10278-017-9976-3

11. Luo, W., et al.: Guidelines for developing and reporting machine learning predictive models in biomedical research: a multidisciplinary view. J. Med. Internet Res. **18**(12), e323 (2016). https://doi.org/10.2196/jmir.5870. http://www.jmir.org/2016/12/e323/. ISSN 1438-8871

12. Lutnick, B., et al.: An integrated iterative annotation technique for easing neural network training in medical image analysis. Nat. Mach. Intell. **1**(2), 112–119 (2020). https://doi.org/10.1038/s42256-019-0018-3.An

13. Montagnon, E., et al.: Deep learning workflow in radiology (2020). https://doi.org/10.1186/s13244-019-0832-5

14. Vinod, S.K., Min, M., Jameson, M.G., Holloway, L.C.: A review of interventions to reduce inter-observer variability in volume delineation in radiation oncology. J. Med. Imaging Radiat. Oncol. **60**(3), 393–406 (2016). https://doi.org/10.1111/1754-9485.12462

15. Willemink, M.J., et al.: Preparing medical imaging data for machine learning. Radiology **295**(1), 4–15 (2020). https://doi.org/10.1148/radiol.2020192224

ADBIS 2021 Workshop: Advances in Data Systems Management, Engineering, and Analytics – MegaData

MegaData: Advances in Data Systems Management, Engineering, and Analytics

Yaser Jararweh[1], Tomás F. Pena[2], and Feras M. Awaysheh[3]

[1] Duquesne University, USA
[2] University of Santiago de Compostela, Spain
[3] University of Tartu, Estonia

Description. The MegaData workshop aims to bring together researchers and practitioners from the data science and data engineering communities to share their latest studies in Big Data practices. It is an opportunity for novice and experienced Big Data users to learn, get help, and have exchanges with the system administrators, programmers, and others interested in sharing and presenting their perspectives on Big Data systems. The workshop focuses on effective data systems management, engineering, and analytics of different Big Data models ranging from databases to file systems and from data streaming to batch processing. MegaData offers an opportunity to showcase the latest advances in this area and to discuss and identify future directions and challenges in all aspects of Big Data systems' management and engineering.

Organization

Chairs

Yaser Jararweh	Duquesne University, USA
Tomás F. Pena	University of Santiago de Compostela, Spain
Feras M. Awaysheh	University of Tartu, Estonia

Program Committee

Ahmad Aburomman	Universidade da Coruña, Spain
Said Alawadi	Uppsala University, Sweden
Sattam Almatarneh	Middle East University, Jordan
Syed Attique Shah	University of Tartu, Estonia
James Benson	University of Texas at San Antonio, USA
Pablo Caderno	Universidade de Santiago de Compostela, Spain
Rosa Filgueira	University of Edinburgh, UK
Mehdi Gheisari	Guangzhou University, China
Arturo Gonzalez-Escribano	Universidad de Valladolid, Spain
Maanak Gupta	Tennessee Technological University, USA
Victor M. Muñoz	Universitat Oberta de Catalunya, Spain
Houshyar Honar Pajooh	Masey University, New Zealand
Xoan C. Pardo	Universidade da Coruña, Spain
Mohamed Ragab	University of Tartu, Estonia
Pablo Rodríguez-Mier	INRAE, France
Imed Romdhani	Edinburgh Napier University, UK
Manisha Sirsat	INESC, Portugal
Jose R. R. Viqueira	Universidade de Santiago de Compostela, Spain

A Federated Interactive Learning IoT-Based Health Monitoring Platform

Sadi Alawadi[1]([✉]), Victor R. Kebande[2], Yuji Dong[3], Joseph Bugeja[4],
Jan A. Persson[4], and Carl Magnus Olsson[4]

[1] Department of Information Technology, Uppsala University, Uppsala, Sweden
`sadi.alawadi@it.uu.se`
[2] Department of Computer Science, Electrical and Space Engineering,
Luleå University of Technology, Luleå, Sweden
`victor.kebande@ltu.se`
[3] School of Internet of Things, Xi'an Jiaotong-Liverpool University, Suzhou, China
`yuji.dong02@xjtlu.edu.cn`
[4] Department of Computer Science, Malmö University, Malmö, Sweden
`{joseph.bugeja,jan.a.persson,carl.magnus.olsson}@mau.se`

Abstract. Remote health monitoring is a trend for better health management which necessitates the need for secure monitoring and privacy-preservation of patient data. Moreover, accurate and continuous monitoring of personal health status may require expert validation in an active learning strategy. As a result, this paper proposes a Federated Interactive Learning IoT-based Health Monitoring Platform (FIL-IoT-HMP) which incorporates multi-expert feedback as 'Human-in-the-loop' in an active learning strategy in order to improve the clients' Machine Learning (ML) models. The authors have proposed an architecture and conducted an experiment as a proof of concept. Federated learning approach has been preferred in this context given that it strengthens privacy by allowing the global model to be trained while sensitive data is retained at the local edge nodes. Also, each model's accuracy is improved while privacy and security of data has been upheld.

Keywords: IoT · Healthcare · Federated · Machine learning

1 Introduction

Continuous advancement of the Internet of Things (IoT) healthcare systems has been experienced as a result of the sporadic technological changes, particularly in IoT device proliferation, and the need to manage the ever-rising quantity of patient data. Notably, these proliferation have allowed the usage of several health devices like wearable sensors that are able to measure and monitor several personal health parameters, which in some situations are able to create a trigger mechanism in case of a potential health incident. In order to make an accurate

© Springer Nature Switzerland AG 2021
L. Bellatreche et al. (Eds.): ADBIS 2021 Short Papers, Workshops and Doctoral Consortium,
CCIS 1450, pp. 235–246, 2021.
https://doi.org/10.1007/978-3-030-85082-1_21

prognosis using the data from IoT devices, most related healthcare systems currently leverages machine learning approaches for purposes of making decisions automatically. While this has seen an improved diagnosis and efficient detection of diseases [1], there have been several limitations such as lack of annotated data used to train the ML models, which in this context makes IoT-health systems unreliable and ineffective. Also, ineffective and malevolent coordination of ML model may lead to potential attacks and data leakage. In particular cases, this may lead to privacy infringement of patient data, which on similar situations puts the security of data at risk, hence creating mistrust among different parties. Indeed, in a recent study, Ponemon Institute identified that health data is the most targeted by cybercriminals [2] and that attacks on IoT devices were reported to be increasing by three-fold in 2018 [3].

Therefore, to improve both the ML model performance and accuracy in order to make IoT-health systems reliable and effective, there is need to interactively incorporate the expert's domain knowledge as 'human-in-the loop' to help in providing heuristic-based knowledge of the system while the IoT health system learns, and the need to preserve privacy, security and trust. Hence, Federated Machine Learning (FML) [4,5] will preserve data privacy by training the ML model over the user data locally without moving the data. In this context, ML model still can be adapted or contextualized locally which is more effective as opposed to leveraging a single trained model. Moreover, all the edge nodes will participate in training the ML model collaboratively using their data. Based on that fact, all ML models will share their learned knowledge among all participant nodes. To bring out the problem that is being addressed in this paper, we consider the following scenario:

The number of elderly people is on the rise and quite a good number of the them prefer to live in their homes (houses/apartments) devoid of privacy violations. However, in some cases, older people with chronic diseases are susceptible to other diseases like heart attacks or accidental falls etc. Monitoring their health remotely without human intervention using an IoT-based devices offers a suitable solution in this case. That notwithstanding, as the sensing data is massive and continuously generated, it may be impossible for humans to continually and accurately monitor and explore this data. A suitable solution is to use ML approaches to classify the sensed data into different events and let the domain experts only to validate those data deemed to have important events, for example, de-identification etc. Also, this kind of approach faces formidable challenges and issues. For example, the ML model's quality is paramount when it is required to give accurate classifications. A wrong classification or misdiagnosis could lead to serious consequences. Simultaneously, a good ML model needs massive data and many medical experts, which is probably impossible or too costly for a company. Additionally, enforcing the privacy and security of the data should be a priority in this context.

The authors take a step in addressing the aforementioned scenario by applying an IoT-based health monitoring platform with federated learning and interactive learning strategy that allows the knowledge from ML models that is triggered by the domain experts to be shared, at the same time the clients are able to reap the benefits of such domain knowledge given the accuracy of these ML models.

The remainder of this paper is structured as follows: Related work is presented in Sect. 2, while the system architecture is presented in Sect. 3. This is followed by experiments in Sect. 4 and a conclusion in Sect. 5.

2   Related Work

A thought that a human may instinctively outperform a machine learning algorithm has been explored based on existing evidence on the diagnostic radiologic image. This is represented as a suitable approach that solves the expert-in the loop technique [6]. However, while it looks relevant, its effectiveness is rarely investigated when it is mapped to the patients' privacy. Also, an architecture that acts as a remote human-in-the-loop named SENS-U allows health monitoring for Wireless Body Sensor Network (WBSN) for patients. It is able to monitor terminals of medical centers via four body vital signs for personal healthcare [7]. A cost-optimal multi-expert (Co-MEAL) approach that has machine learning adaptability allows the machine learning model to be able to learn from a variety of experts, e.g., the human oracle or a digital device. The advantage of this process is that it reduces the cost of labeling data while it capitalizes on the collaboration among experts with the main aim of enriching the knowledge [8]. Based on the expert selection module of the Co-MEAL, a collaborative-multi-expert architecture by [9] has been designed to be able to manage knowledge from heterogeneous sources by incorporating a technique that allows experts to collaborate in order to increase their knowledge. Additionally, this work, was able to propose an expert selection algorithm that could be applied in a real world scenario by utilising active and transfer learning strategy, while the expert selection is executed as an expert unit in the Co-MEAL architecture. Also, a FedHealth framework that utilises transfer learning has been able to build personalized models through activity recognition experiments. Based on this study, accurate healthcare is achieved by FedHealth while at the same time privacy and security is upheld. When federated learning is used an accuracy of 99.4% is achieved as opposed to 94.1% when it is not used [10].

While most of the aforementioned researches have a close inclination to the research proposed in this paper, key aspects like privacy preservation of patient data, security of data, use of active learning as human-in the loop is hardly explored, however, they provide useful insights that are used to build our suggested approaches.

3   System Architecture

A description of the proposed architecture is given in this section, where the concentration is on the architecture components and the mode of operation.

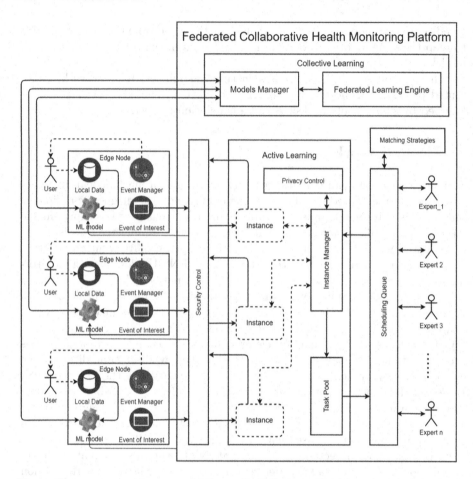

Fig. 1. The system architecture of federated collaborative health monitoring platform

3.1 Architecture Components

The proposed system architecture that is shown in Fig. 1 is composed of five different components that work in a coordinated mechanism in order to achieve the common objective and the role of each entity are shown as follows:

– **Collective Learning (Global Model):** The main role of the global model is to facilitate the aggregation of knowledge that originates from the edge nodes. For privacy concern, this platform will share only the learned knowledge that is coordinated by the collective learning component, while the personal or sensitive data are retained at the edge nodes. Since the nodes may have different types of data and ML models, the *Models Manager* is responsible for managing and coordinating the different models. The global model can be updated by way of synchronization between the edge nodes and the global model, which is managed by the *Federated Learning Engine*.

- **Edge nodes (Edge model):** Each received ML model from the collective learning component will be retrained over the personal local data. Then the model performs an incremental training approach in order to capture any new patterns or behavior. Moreover, the retrained edge models will be aggregated by the *collective learning* to re-update the global model's knowledge and redistribute it again to the edge nodes for further retraining. The edge nodes act as the clients of this platform who may belong to different stakeholders. For example, one node could be a hospital, which uses IoT devices to collect data from the patients and the hospital has its own local data and ML model to monitor the patients. When the hospital utilises this platform, it can increase the accuracy of its ML model by sharing learned knowledge with other stakeholders, and get emergent notification based on the opinions of the experts in the platform.
- **Security control:** Security control plays two major roles as follows: (1) Prevents, adversarial attacks, particularly, poisoning attacks, by creating a cryptographic hash during incremental training to retain the training data in its original form, and (2) Making it hard for a potential attacker to decipher the data contents as well as preventing malicious adversaries from accessing data when it is transmitted over the network to the experts during the Active learning process.
- **Active learning:** The edge nodes have their own strategies to send related data to the platform to get diagnosis from the experts. The strategies are managed by the *Event Manager* and the related data are sent via `Event of interest`. The validation responses from the experts will not only be used to give notifications to the related patients, but also gets annotated for the purpose of incremental learning. The instance in the active learning module that originates from any client is responsible for the related specific tasks from the client. The *Instance Manager* and *Privacy Control* need to pre-process the data, for example, by de-identification, before giving them to the experts. All the pre-processed data will be formatted and transferred to the *Task Pool*, which will assign the tasks to the experts. The expert validation responses, which prior also does checks with the client will be sent back to the *Instance Manager* where it can give the processed responses to the related clients.
- **Scheduling Queue:** Since there are many different types of experts with different levels, the tasks in the *Task Pool* are assigned to the appropriate expert based on the *Matching Strategies*. For example, the matching strategies could be based on the experts' reputation like specialty, experience of years and feedback in the past.

3.2   Modus Operandi

The suggested federated interactive IoT-based health monitoring platform that utilises domain experts (based on Fig. 1), is aimed at providing diagnose services to a variety of stakeholders, who own devices to monitor the users' health status. Consequently, each stakeholder can federate its ML model with the global model and create an instance in the active learning module in the platform to get feedback from many experts provided by the platform. Each created instance in the *Active Learning* module is able to generate related tasks that are inclined to the corresponding expert based on the requirements. All the tasks are pre-processed to protect the users' privacy and pushed to the task pool, so they can be ready to be annotated by the experts. After the experts finish the tasks assigned to them, the results will be sent back to the *Instance Manager*, and the related instances will notify the respective stakeholders. Then the stakeholders are poised to take actions and update their ML models based on the feedback from the experts. Finally, the updated ML models (local nodes) in all the stakeholders can improve the global model via a federated learning architecture by way of transferring the extracted knowledge. Eventually, when the global model is updated, it can also synchronize with all the related stakeholders' ML models to improve their models' accuracy.

4   Experiment

This section details the experiment setting and performance analysis which is aimed at providing proof of concept of the proposition that has been mentioned in this paper. Furthermore, it is worth noting that the experiment's focus is to leverage federated learning to ensure data privacy is upheld while the ML models' accuracy is improved at run-time through continuous learning by relying on the multi-expert validations.

4.1   Experimental Setting

The study employs Continuous Ambient Sensors Dataset (CASA) human activity recognition dataset[1]. The contents of CASA were collected from 30 different houses by using both ambient and PIR sensors. As listed in Table 1 each collected pattern comprises 37 features linked to different sensors distributed in those houses to monitor the user's daily activities. However, data linked to four houses were selected to conduct the experiment in order to evaluate the proposed architecture's behavior in terms of data privacy preservation by moving the ML model to the data location, and the improvement of the ML models accuracy's through the continuous 'human-in-the loop' learning.

To achieve this, Random-Forest classifier has been trained and tested over the selected data in both global and edge nodes. Hence, the selected houses data (csh105, csh108, csh111, and cthe sh123) have been associated to the global model, node1, node2 and node3 respectively [11,12]. At this stage, multiple local nodes (edge nodes) are trained over the local dataset's n number of iterations and the new learned knowledge is sent to the global node for aggregation.

---

[1] http://archive.ics.uci.edu/ml/datasets/Human+Activity+Recognition+from+Continuous+Ambient+Sensor+Data.

Table 1. CASA dataset features characteristics

Index	Features	Types	Index	Features	Types
1	lastSensorEventHours	Discrete	20	areaTransitions	Discrete
2	lastSensorEventSeconds	Continuous	21	numDistinctSensors	Discrete
3	lastSensorDayOfWeek	Discrete	22	sensorCount-Bathroom	Continuous
4	windowDuration	Continuous	23	sensorCount-Bedroom	Continuous
5	timeSinceLastSensorEvent	Continuous	24	sensorCount-Chair	Continuous
6	prevDominantSensor1	Discrete	25	sensorCount-DiningRoom	Continuous
7	prevDominantSensor2	Discrete	26	sensorElTime-Ignore	Continuous
8	lastSensorID	Discrete	27	sensorCount-Hall	Continuous
9	lastSensorLocation	Discrete	28	sensorElTime-Kitchen	Continuous
10	lastMotionLocation	Discrete	29	sensorElTime-LivingRoom	Continuous
11	complexity	Continuous	30	sensorElTime-Office	Continuous
12	activityChange	Continuous	31	sensorElTime-OutsideDoor	Continuous
13	sensorElTime-WorkArea	Continuous	32	sensorElTime-Hall	Continuous
14	sensorCount-Ignore	Continuous	33	sensorCount-Kitchen	Continuous
15	sensorCount-LivingRoom	Continuous	34	sensorCount-Office	Continuous
16	sensorCount-OutsideDoor	Continuous	35	sensorCount-WorkArea	Continuous
17	sensorElTime-Bathroom	Continuous	36	sensorElTime-Bedroom	Continuous
18	sensorElTime-Chair	Continuous	37	activity	Text (class label)
19	sensorElTime-DiningRoom	Continuous			

4.2 Performance Analysis

In our performance analysis, we evaluate the ultimate accuracy of the global model after aggregating the knowledge from the local edge nodes (see Fig. 2a and Fig. 2b). Observations from this analysis are presented as follows:

- In the first iteration the global and the edge nodes exhibited low accuracy.
- After aggregating the trained edge (distributed) models by taking the average of the learned knowledge, the model's learning behavior improved tremendously.
- The learned knowledge has been used in the derivation of the new (current) global model, which is then redistributed to the linked edge nodes for further training.
- Interactive learning has successfully aided the edge ML model's incremental learning by utilising the new annotated data. This data has played a significant role in the improvement of the model's learning process.

From the Figs. 2a and 2b, we notice that the performance of the global model is basically influenced by the availability of the linked edge nodes and the user-feedback from the experts which has shown an improvement on the model accuracy and performance.

4.3 Security and Privacy Analysis

In our security and privacy analysis, we hypothesise the threat model, a variety of possible attacks on the proposed IoT-Health platform and then an analysis is given on the same.

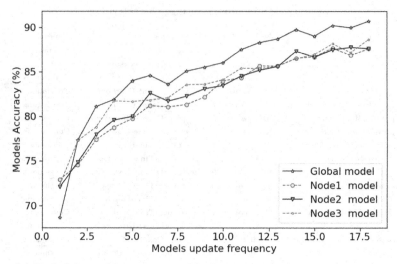

(a) Both federated learning model and participants nodes models accuracy

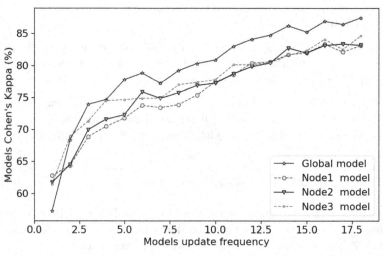

(b) Both federated learning model and participants nodes models Cohen's kappa

Fig. 2. Experiments depicting the federated global model and the participants of local nodes

Threat Model. Our threat model makes the following assumptions on adversarial perspectives: In the context of IoT-Health platform, there may exist malicious content that may hinder the global and local models accuracy, and this content may be channelled to the training data in an adversarial training attack at the local nodes during incremental learning. Also, we assume that an attacker may actively defeat the security of the platform by challenging the ciphertext in order to eavesdrop all communication when data is sent over the network.

Attack Analysis. Based on the assumptions of the threat model (Sect. 4.3), we analyse the security attacks as follows:

- Poisoning attack: An adversary may subject the training data to malicious content, which may end up affecting the knowledge that is extracted from the local nodes, thus affecting the accuracy of the global model: We have suggested the use of cryptographic hashes to maintain the integrity of training data at the local nodes.
- Ciphertext attack: Based on the security control technique in Fig. 1, we assume the role of an adversary is to corrupt sensitive patient data. Therefore, an adversary may obtain encrypted data or the secret keys to have a direct access to the data. We suggest the use of strong encryption approaches like homomorphic encryption and differential privacy.
- Eavesdropping attack: An adversary can attempt to eavesdrop on data that is sent between the local node and the global model, which ultimately has an impact on data privacy. We suggest maintaining strong privacy techniques which are discussed next.

Privacy Analysis. In our approach towards privacy-preserving technique, we have proposed an approach that utilises, federated learning and concepts of interactive learning that collaboratively are able to build a global model without sharing data whatsoever. Data is retained at the local nodes, where each node is able to maintain its data (Fig. 1). From this, the new global model is only able to learn from the knowledge from the local nodes which ensured that privacy is preserved locally, owing to the fact that only knowledge is transferred to the global model. Nonetheless, personal data of users for which the system was trained on might still be revealed indirectly through privacy attacks on the machine learning model. In particular, we identify the following attacks:

- Model inversion attack: An adversary having access to some data belonging to specific patients included in the training data, can infer further data about those same individuals by observing inputs/outputs of the machine learning model. For example, given some demographic information an adversary could infer genetic markers from the model despite having only partial access to the underlying training data [13]. A common mitigation against model inversion attacks is differential privacy.

Table 2. Evaluation of identified security and privacy issues

Security & privacy issue	Overview	Mitigation approaches
Poisoning attack	Contaminating training data with malicious content. The ultimate knowledge from the training is falsified. Affects the accuracy of the model	To retain the integrity, cryptographic hashes are preferred given that they are deterministic, where same input guarantees same output the fact that they are irreversible
Ciphertext attack	It is possible to tamper with encrypted data at the local nodes, by obtaining the public key used to encrypt the data from a source	Not only using strong encryption but employing digital certificate during data transmission
Eavesdropping attack	An adversary or a malevolent data labeler can listen to or gain access to data being transmitted with an elevated privilege	During incremental training and during the provision of learning model updates to use blockchain for it guarantees secure data transmission
Model inversion attack	An adversary making an inference about the data in possession	Use of differential privacy
Membership inference attack	Trained models can be used to leak information about a patient's record	Use of regularization
Model stealing attack	Constructing surrogate models from extracted model parameters	Use of information laundering

- Membership inference attack: An adversary may deduce whether a given patient is present in the training data of a machine learning model. For instance, if hospital records are used to train a model which predicts when a patient will be discharged from the hospital, adversaries could use that model with other data to reveal whether an individual had visited one of the hospitals that generated the training data during the period the data was collected. The use of regularization, e.g., through L_2 regularization, is identified as a technique for reducing membership inference attacks.
- Model stealing attack: Adversaries may extract parameters from a target model allowing them to reconstruct a surrogate model with similar performance as the target model. While this attack is harder to conduct in a federated learning setup as is the case for FIL-IoT-HMP, where multiple decentralized edge nodes are involved, in theory this attack may still be possible. Model stealing may indirectly compromise privacy, but more so, the confidentiality of the health platform users. Information laundering is a technique that can be used to mitigate against model stealing attacks.

Based on how the FIL-IoT-HMP model has been positioned, we argue that blockchain technology has been presented as a more suitable technique for enhancing secure data sharing at the local nodes by providing tamper-free adversarial attacks during incremental training [4]. Normally, adversarial attacks during active learning are common occurrences based on existing learning threat landscape like targeted attacks, unusual propagation attacks, malicious logic insertion and overall system manipulation [14].

Given that federated learning model is shared across the multiple nodes, we also argue that the following aspects transpire as a result, however a summary is given in Table 2:

- The data from the IoT health platform is regarded to be sensitive, as a result privacy preservation is a key aspect of consideration in this context
- At the edge, blockchain integration gives an assurance of the following: Secure data sharing during incremental learning process, resource location where smart contracts can be used as a way of access control and management
- From a security perspective, existing vulnerabilities arising from the learning model, especially during data transmission may enable an attacker to launch specific attacks that can lead to leakage of sensitive information. In this perspective federated learning guarantees privacy protection and verification through periodic updates during the transmission of learning models.

5   Conclusion

We have proposed a federated interactive IoT-based health monitoring platform that utilizes (active, interactive and human-in-the loop) This platform has a strong privacy-preserving feature and also its able to counter adversarial attacks during incremental learning. The problem of data leakage has been analyzed correctly by allowing the global model to only share the knowledge from local nodes while the data is retained at the local nodes. For future work, we aim to extend this work to incorporate multiple machine learning algorithms using different datasets in order to study the effect of the expert validation.

References

1. Kumar, P.M., Gandhi, U.D.: A novel three-tier internet of things architecture with machine learning algorithm for early detection of heart diseases. Comput. Electr. Eng. **65**, 222–235 (2018)
2. Ponemon Institute. 2020 State of Password and Authentication Security Behaviors Report. https://pages.yubico.com/2020-password-and-authentication-report/
3. Montalbano, E.: Kaspersky: Attacks on Smart Devices Rise Threefold in 2018. https://securityledger.com/2018/09/kaspersky-attacks-on-smart-devices-rise-threefold-2018/. Accessed 29 Mar 2021
4. Kebande, V.R., Alawadi, S., Bugeja, J., Persson, J.A., Olsson, C.M.: Leveraging federated learning & blockchain to counter adversarial attacks in incremental learning. In: 10th International Conference on the Internet of Things Companion, pp. 1–5 (2020)
5. Yang, Q., Liu, Y., Chen, T., Tong, Y.: Federated machine learning: concept and applications. ACM Trans. Intell. Syst. Technol. (TIST) **10**(2), 1–19 (2019)
6. Holzinger, A.: Interactive machine learning for health informatics: when do we need the human-in-the-loop? Brain Inf. **3**(2), 119–131 (2016)

7. Yuan, A., Yan, L., Cai-Wen, M., Li-Min, S., Zhi-Feng, X.: SENS-U: remote human in loop health-monitoring system at home. In: 2008 IEEE Pacific-Asia Workshop on Computational Intelligence and Industrial Application, vol. 1, pp. 441–445. IEEE (2008)

8. Saeedi, R., Sasani, K., Gebremedhin, A.H.: Co-meal: cost-optimal multi-expert active learning architecture for mobile health monitoring. In: Proceedings of the 8th ACM International Conference on Bioinformatics, Computational Biology, and Health Informatics, pp. 432–441 (2017)

9. Saeedi, R., Sasani, K., Gebremedhin, A.H.: Collaborative multi-expert active learning for mobile health monitoring: architecture, algorithms, and evaluation. Sensors **20**(7), 1932 (2020)

10. Chen, Y., Qin, X., Wang, J., Yu, C., Gao, W.: FedHealth: a federated transfer learning framework for wearable healthcare. IEEE Intell. Syst. (2020)

11. Alawadi, S., Delgado, M.F., Pérez, D.M.: Machine learning algorithms for pattern visualization in classification tasks and for automatic indoor temperature prediction. Ph.D. thesis, Universidade de Santiago de Compostela (2018)

12. Alkhabbas, F., Alawadi, S., Spalazzese, R., Davidsson, P.: Activity recognition and user preference learning for automated configuration of IoT environments. In: Proceedings of the 10th International Conference on the Internet of Things, pp. 1–8 (2020)

13. Veale, M., Binns, R., Edwards, L.: Algorithms that remember: model inversion attacks and data protection law. Philos. Trans. Roy. Soc. A: Math. Phys. Eng. Sci. **376**(2133), 20180083 (2018)

14. Kebande, V.R., Alawadi, S., Awaysheh, F.M., Persson, J.A.: Active machine learning adversarial attack detection in the user feedback process. IEEE Access **9**, 36908–36923 (2021)

Augmenting SQLite for Local-First Software

Iver Toft Tomter and Weihai Yu[✉]

UIT - The Arctic University of Norway, Tromsø, Norway
weihai.yu@uit.no

Abstract. Local-first software aims at both the ability to work offline on local data and the ability to collaborate across multiple devices. CRDTs (conflict-free replicated data types) are abstractions for offline and collaborative work that guarantees strong eventual consistency. RDB (relational database) is a mature and successful computer industry for management of data, and SQLite is an ideal RDB candidate for offline work on locally stored data. CRRs (conflict-free replicated relations) apply CRDTs to RDB data. This paper presents our work in progress that augments SQLite databases with CRR for local-first software. No modification or extra software is needed for existing SQLite applications to continue working with the augmented databases.

Keywords: Data replication · Eventual consistency · Conflict-free replicated data types · Conflict-free replicated relations · Edge computing

1 Introduction

Local-first software suggests a set of principles for software that enables both collaboration and ownership for users. Local-first ideals include the ability to work offline and collaborate across multiple devices [7].

RDB (relational database) is a mature and successful computer industry for management of data, and SQLite is an open source RDB engine that is an ideal candidate for local-first software, because its operation does not rely on network connectivity. Citing its homepage[1]: "SQLite is the most used database engine in the world. SQLite is built into all mobile phones and most computers and comes bundled inside countless other applications that people use every day".

One of the main challenges of supporting local-first software is the general limitation of a networked system, as stated in the CAP theorem [3,5]: it is impossible to simultaneously ensure all three desirable properties, namely consistency equivalent to a single up-to-date copy of data, availability of the data for update and tolerance to network partition.

---

[1] https://sqlite.org.

© Springer Nature Switzerland AG 2021
L. Bellatreche et al. (Eds.): ADBIS 2021 Short Papers, Workshops and Doctoral Consortium,
CCIS 1450, pp. 247–257, 2021.
https://doi.org/10.1007/978-3-030-85082-1_22

CRDTs (conflict-free replicated data types) [10] emerged to address the CAP challenges. With CRDT, a site updates its local replica without coordination with other sites. The states of replicas converge when they have applied the same set of updates (referred to as *strong eventual consistency* in [10]).

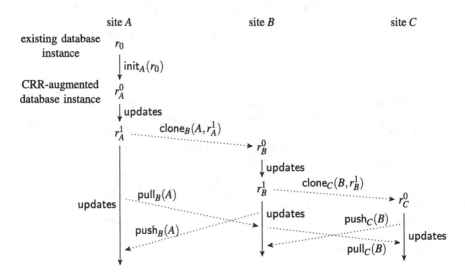

Fig. 1. A scenario of asynchronous database updates

CRRs (conflict-free replicated relations) apply CRDTs to RDBs [12]. In [12], we reported a CRR prototype that was built on top of an ORM (object-relation mapping) called Ecto[2]. Applications are therefore limited to those using the particular ORM. Unfortunately, Ecto does not support SQLite in the latest versions.

In this paper, we report our work-in-progress implementation that augments SQLite databases with CRR support. With a single command, we augment an existing SQLite database instance with CRR. All applications using the database, including the `sqlite3` shell[3], continue to work without any modification. Later, we can clone the augmented database to different sites. We can query and update the database instances at different sites independently. We can synchronize the instances when the sites are connected. We are not locked in, though. We can easily drop the CRR-augmentation on any of the database instances without losing any of the original database features.

Figure 1 shows a scenario of using our software. Initially we have a SQLite database instance r_0 at site A. We run $\text{init}(r_0)$ to augment r_0 to r_A^0 with CRR support. We then apply some updates that lead to instance r_A^1. Now at site B we run $\text{clone}(r_A^1)$ to get a clone of the database instance. Independently, we make

---

[2] https://github.com/elixir-ecto/ecto.
[3] https://sqlite.org/cli.html.

updates on the local instances at sites A and B. Later, we make yet another clone from site B to site C. From now on, we make local updates at all three sites and occasionally push our local updates to remote sites and pull remote updates to local instances.

2 Requirements

A primary requirement for local-first software is that a site should be able to independently perform queries and updates on the local database instances.

When two sites are connected, one site should be able to merge the updates performed at the other site without coordination. In particular, the site should be able to resolve conflicts without collecting votes from other sites.

The instances at different sites should be eventually consistent, or convergent [11]. That is, when they have applied the same set of updates, they should have the same state.

Database integrity constraints should be enforced. In particular, a merge of concurrent updates may cause the violation of an integrity constraint, though none of the updates violated any constraint locally at the sites. When this happens, one of the offending updates should be undone. It is important that the sites independently undo the same offending update.

Finally, existing applications should continue to work without any modification. In particular, performing queries and updates on local instances should not depend on the augmentation or any additional third-party software.

3 Technical Background

In this section, we review the necessary background information about CRDT and CRR.

3.1 CRDT

A CRDT is a data abstraction specifically designed for data replicated at different sites. A site queries and updates its local replica without coordination with other sites. The data is always available for update, but the data states at different sites may diverge. From time to time, the sites send their updates asynchronously to other sites with an anti-entropy protocol. To apply the updates made at the other sites, a site merges the received updates with its local replica. A CRDT has the property that when all sites have applied the same set of updates, the replicas converge.

There are two families of CRDT approaches, namely operation-based and state-based [10]. Our work is based on state-based CRDTs, where a message for updates consists of the data state of a replica in its entirety. A site applies the updates by merging its local state with the state in the received message. The possible states of a state-based CRDT must form a join-semilattice [4], which implies convergence. Briefly, the states form a *join-semilattice* if they are partially ordered with $\sqsubseteq$ and a join $\sqcup$ of any two states (that gives the least upper bound of the two states) always exists. State updates must be inflationary. That is, the new state supersedes the old one in $\sqsubseteq$. The merge of two states is the result of a join.

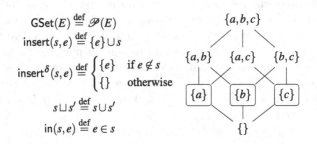

$$\mathsf{GSet}(E) \overset{\mathrm{def}}{=} \mathscr{P}(E)$$
$$\mathsf{insert}(s,e) \overset{\mathrm{def}}{=} \{e\} \cup s$$
$$\mathsf{insert}^\delta(s,e) \overset{\mathrm{def}}{=} \begin{cases} \{e\} & \text{if } e \notin s \\ \{\} & \text{otherwise} \end{cases}$$
$$s \sqcup s' \overset{\mathrm{def}}{=} s \cup s'$$
$$\mathsf{in}(s,e) \overset{\mathrm{def}}{=} e \in s$$

Fig. 2. GSet CRDT and Hasse diagram of states

Figure 2 (left) shows GSet, a state-based CRDT for grow-only sets [10], where E is a set of possible elements, $\sqsubseteq \overset{\mathrm{def}}{=} \subseteq$, $\sqcup \overset{\mathrm{def}}{=} \cup$, insert is a mutator (update operation) and in is a query. Obviously, an update through insert(s,e) is an inflation, because $s \subseteq \{e\} \cup s$. Figure 2 (right) shows the Hasse diagram of the states in a GSet. A Hasse diagram shows only the "direct links" between states.

Using state-based CRDTs, as originally presented [10], is costly in practice, because states in their entirety are sent as messages. Delta-state CRDTs address this issue by only sending join-irreducible states [1,2]. Basically, *join-irreducible* states are elementary states: every state in the join-semilattice can be represented as a join of some join-irreducible state(s). In Fig. 2, insert$^\delta$ is a delta-mutator that returns join-irreducible states which are singleton sets (boxed in the Hasse diagram).

Since a relation instance is a set of tuples, the basic building block of CRR is a general-purpose set CRDT ("general-purpose" in the sense that it allows both insertion and deletion of elements), or more specifically, a delta-state set CRDT.

We use CLSet (causal-length set, [12,13]), a general-purpose set CRDT, where each element is associated with a *causal length*. Intuitively, insertion and deletion are inverse operations of one another. They always occur in turn. When an element is first inserted into a set, its causal length is 1. When the element is deleted, its causal length becomes 2. Thereby the causal length of an element increments on each update that reverses the effect of a previous one.

$$\mathsf{CLSet}(E) \overset{\mathsf{def}}{=} E \hookrightarrow \mathbb{N}$$

$$\mathsf{insert}(s,e) \overset{\mathsf{def}}{=} \begin{cases} s\{e \mapsto s(e)+1\} & \text{if even}(s(e)) \\ s & \text{if odd}(s(e)) \end{cases}$$

$$\mathsf{insert}^{\delta}(s,e) \overset{\mathsf{def}}{=} \begin{cases} \{e \mapsto s(e)+1\} & \text{if even}(s(e)) \\ \{\} & \text{if odd}(s(e)) \end{cases}$$

$$\mathsf{delete}(s,e) \overset{\mathsf{def}}{=} \begin{cases} s & \text{if even}(s(e)) \\ s\{e \mapsto s(e)+1\} & \text{if odd}(s(e)) \end{cases}$$

$$\mathsf{delete}^{\delta}(s,e) \overset{\mathsf{def}}{=} \begin{cases} \{\} & \text{if even}(s(e)) \\ \{e \mapsto s(e)+1\} & \text{if odd}(s(e)) \end{cases}$$

$$(s \sqcup s')(e) \overset{\mathsf{def}}{=} \max(s(e), s'(e))$$

$$\mathsf{in}(s,e) \overset{\mathsf{def}}{=} \mathsf{odd}(s(e))$$

Fig. 3. CLSet CRDT [12]

As shown in Fig. 3, the states of a CLSet are a partial function $s\colon E \hookrightarrow \mathbb{N}$, meaning that when e is not in the domain of s, $s(e) = 0$ (0 is the bottom element of $\mathbb{N}$, i.e. $\perp_{\mathbb{N}} = 0$). Using partial function conveniently simplifies the specification of insert, $\sqcup$ and in. Without explicit initialization, the causal length of any unknown element is 0. In the figure, insert^{δ} and delete^{δ} are the delta-counterparts of insert and delete respectively.

An element e is regarded to be in the set when its causal length is an odd number. A local insertion has effect only when the element is not in the set. Similarly, a local deletion has effect only when the element is actually in the set. A local effective insertion or deletion simply increments the causal length of the element by one. For every element e in s and/or s', the new causal length of e after merging s and s' is the maximum of the causal lengths of e in s and s'.

3.2   CRR

The RDB supporting CRR consists of two layers: an Application Relation (AR) layer and a Conflict-free Replicated Relation (CRR) layer (see Fig. 4). The AR layer presents the same RDB schema and API as a conventional RDB system. Application programs interact with the database at the AR layer. The CRR layer supports conflict-free replication of relations.

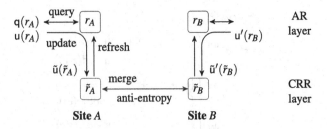

Fig. 4. A two-layer relational database system [12]

An AR-layer database schema R has an augmented CRR schema $\tilde{R}$. In Fig. 4, site A maintains both an instance r_A of R and an instance $\tilde{r}_A$ of $\tilde{R}$. A query q is performed on r_A without the involvement of $\tilde{r}_A$. An update u on r_A triggers an additional update $\tilde{u}$ on $\tilde{r}_A$. The update $\tilde{u}$ is later propagated to remote sites through an anti-entropy protocol. Merge with an incoming remote update $\tilde{u}'(\tilde{r}_B)$ results in an update $\tilde{u}'$ on $\tilde{r}_A$ as well as an update u' on r_A.

CRR has the property that when both sites A and B have applied the same set of updates, the relation instances at the two sites are equivalent, i.e. $r_A = r_B$ and $\tilde{r}_A = \tilde{r}_B$.

The two-layered system also maintains the integrity constraints defined at the AR layer. Any violation of integrity constraint is caught at the AR layer. A failed merge would cause some compensation updates.

We adopt several CRDTs for CRRs. Since a relation instance is a set of tuples or rows, we use the CLSet CRDT (Fig. 3) for relation instances. We use the LWW (last-write wins) register CRDT [6,9] for individual attributes in tuples.

4   Work-in-Progress Implementation

We implement all functionality required for local updates completely inside the SQLite database, so no modification to existing applications or extra software is required for the applications to be able to continue working with the database. We implement the command-line features in Python (that in turn calls SQL statements).

4.1   Command-Line Operations

For command-line operations, we adopt git[4] operation names.

- init augments an existing SQLite database instance with CRR support.
- clone copies a remote augmented database instance to a local location.
- pull merges remotely applied updates to the local instance.
- push merges locally applied updates to a remote instance.
- remote queries and configures the settings of remote instances.

---

[4] https://git-scm.com.

4.2   CRR-Augmented Database

For an AR-layer relation schema $R(A_1, A_2, \dots)$, we generate a new CRR-layer schema $\tilde{R}(K, L, T_1, T_2, \dots, A_1, A_2, \dots)$, ignoring all integrity constraints in R. K is the primary key of $\tilde{R}$. K values are globally unique. L is the causal-lengths (Fig. 3) of the tuples in $\tilde{R}$. T_i is the timestamp of the last update on attribute A_i. In other words, the (K, L) part represents the CLSet CRDT of tuples and the (A_i, T_i) parts represent the LWW register CRDT of the attributes.

In what follows, we write $t(K)$, $t(A_i)$ etc. for the K and A_i values of tuple t.

We use `randomblob(32)` of SQLite to generate K values. The chance that two tuples in the same relation have the same K value is extremely small.

For the AR-layer relations, we also generate triggers. We describe the triggers later in Sect. 4.3.

In addition to the augmentation of the AR-layer relations, we generate three more relations. A *Clock* relation implements a hybrid logical-physical clock [8] at a (virtual) nanosecond scale (Sect. 4.6). A *Site* relation maintains the information about the sites known at this instance. The information includes the hosts and paths of the remote instances, the last time this instance applied a push and a pull to the sites, etc. A *History* relation maintains a history of all the updates that have been applied.

4.3   Local Updates

The `init` operation generates triggers on relation R. Every insertion, deletion and update on an instance r of R triggers the corresponding update on the instance $\tilde{r}$ of $\tilde{R}$.

When inserting a new tuple t into r, we insert a new tuple $\tilde{t}$ into $\tilde{r}$, with the initial $\tilde{t}(L) = 1$. When deleting t from r, we increment $\tilde{t}(L)$ with 1, so that the new $\tilde{t}(L)$ becomes an even number. When inserting the deleted t back to r, we increment $\tilde{t}(L)$ with 1, so that the new $\tilde{t}(L)$ turns back to an odd number. When updating $t(A_i)$ in r, we update $\tilde{t}(A_i)$ and $\tilde{t}(T_i)$ in $\tilde{r}$.

Since no integrity constraint is defined in $\tilde{R}$, a successful local update on r will always lead to a successful update in $\tilde{r}$.

In addition to the triggers on R, the `init` operation also generates triggers on $\tilde{R}$. For every update on an instance $\tilde{r}$ of $\tilde{R}$, a trigger inserts a tuple in the *History* relation.

4.4   Merges

The pull of the concurrent updates from a remote site consists of the following steps: 1) generating the concurrent updates at the remote site; 2) transferring the generated updates to the local site (to be described in Sect. 4.5); 3) merging the received concurrent updates. A push is handled in a similar way.

As CRRs are based on delta-state CRDTs, the updates are join-irreducible states in a join-semilattice (Sect. 3.1). In our case, the updates are in fact the tuples in $\tilde{r}$ (Sect. 3.2). Using the *History* relation and the information of the last push and pull in relation *Site*, we can generate the updates since the last push or pull.

We generate the concurrent updates of the remote instance in a temporary database[5] and transfer it to the pulling site. This way, we avoid encoding individual tuples into an intermediate representation.

During the merge of received updates, we temporarily disable the generated triggers on AR-layer relation instances by setting a flag on the triggers.

An update on an relation instance $\tilde{r}'$ at a remote site is actually a tuple $\tilde{t}'$. If a tuple $\tilde{t}$ in the local instance $\tilde{r}$ exists such that $\tilde{t}(K) = \tilde{t}'(K)$, we update $\tilde{t}$ with $\tilde{t} \sqcup \tilde{t}'$ where the merge $\sqcup$ is the join operation of the join-semilattice (Sect. 3.1). Otherwise, we insert $\tilde{t}'$ into $\tilde{r}$. The merge $\tilde{t} \sqcup \tilde{t}'$ is defined as:

$$\tilde{t} \sqcup \tilde{t}' \stackrel{\text{def}}{=} \tilde{t}'', \text{ where } \tilde{t}''(L) = \mathsf{max}(\tilde{t}(L), \tilde{t}'(L)), \text{ and}$$

$$\tilde{t}''(A_i), \tilde{t}''(T_i) = \begin{cases} \tilde{t}'(A_i), \tilde{t}'(T_i) & \text{if } \tilde{t}'(T_i) > \tilde{t}(T_i) \\ \tilde{t}(A_i), \tilde{t}(T_i) & \text{otherwise} \end{cases}$$

After the update of $\tilde{r}$, we update r as the following. If $\tilde{t}(L)$ is an even number, we delete t (where $t(A_1) = \tilde{t}(A_1) \wedge t(A_2) = \tilde{t}(A_2) \wedge \ldots$) from r. Otherwise, we insert or update r with $\pi_{A_1, A_2, \ldots}(\tilde{t})$.

If the update on r violates an integrity constraint, we first roll back the updates on r and $\tilde{r}$ and then start an compensation update [12] (remaining to complete, see Sect. 4.7).

4.5   Network Connections

At present, we support access to remote database instances in two possible cases: 1) the remote database instance is located on the same host as the local instance; or 2) the remote instance is located on a host where we have ssh[6] access.

When performing a clone, we specify a database instance stored on a remote host as `ssh://user@host#port:path/to/db`.

Since a SQLite database instance is stored as a file, we may (accidentally) copy or move the file to a different location. Every time we open an instance for push or pull, we verify the location information of the local instance stored in the *Site* relation and make modifications accordingly. We run the `remote` command-line operation to explicitly set or modify the location information of remote instances.

4.6   Timestamp Values

The *Clock* relation, which the `init` operation creates, addresses two issues. The first issue is that the finest time resolution that SQLite provides is at a sub-millisecond level, so consecutive updates may have the same timestamp value.

---

[5] https://sqlite.org/lang_attach.html.
[6] https://www.ssh.com/.

Notice that a third-party library with higher time resolution does not help, since our goal is to implement all features related with local updates completely inside the SQLite database instance. The second issue is that the physical clock (or "wall" clock) values are not sufficient to represent the happen-before relationship between updates, so a concurrent update may mistakenly win a competition when the physical clocks at different sites are skewed.

To address the first issue, we use the $Clock(Ms, Ns)$ relation, where Ms is the physical clock value in milliseconds and Ns is the offset within a millisecond at nanosecond scale. The $Clock$ relation has only one tuple (τ_{ms}, τ_{ns}), which is the last timestamp value that has been generated or merged. The comparison of two timestamp values is defined as $(\tau'_{ms}, \tau'_{ns}) > (\tau_{ms}, \tau_{ns})$ iff $\tau'_{ms} > \tau_{ms}$ or $\tau'_{ms} = \tau_{ms} \wedge \tau'_{ns} > \tau_{ns}$.

To generate a new timestamp value, we first generate a new physical clock value τ'_{ms} (derived from the julianday function of SQLite) and a random number τ'_{ns} such that $0 \leq \tau'_{ns} < 10^6$. If the generated value (τ'_{ms}, τ'_{ns}) is greater than the old value (τ_{ms}, τ_{ns}), the new timestamp value is (τ'_{ms}, τ'_{ns}). Otherwise, we generate a new random number τ''_{ns} such that $\tau_{ns} < \tau''_{ns} < 10^6$ and the new timestamp value becomes (τ_{ms}, τ''_{ns}).

To address the second issue, we implement a hybrid logical-physical clock [8], which has the property that for two updates u_1 and u_2 with timestamp values τ_1 and τ_2, $\tau_1 < \tau_2$ iff u_1 happens before u_2 or u_1 and u_2 are concurrent. In other words, u_2 does not happen before u_1 when $\tau_1 < \tau_2$.

At a merge, if a timestamp value (τ'_{ms}, τ'_{ns}) of the incoming tuple is greater than the (τ_{ms}, τ_{ns}) tuple in the $Clock$ instance, we update the instance with (τ'_{ms}, τ'_{ns}).

4.7   Current Implementation Status

At the time of this writing, we have not finished all the features described in [12]. The remaining features include: enforcement of integrity constraints that are violated at the time of merge, and using a counter CRDT for lossless resolution of concurrent attribute updates with additive update semantic. In addition to the merge in batch mode (push and pull), we are going to implement a continuous synchronization mode, like in [12], so that the sites can frequently exchange latest updates without explicit command-line push and pull. We can use our earlier implementation on top of an ORM [12] as a guideline to implement the features that we have not implemented so far, in particular the features that do not need to be implemented fully in SQL.

We currently focus on making a working prototype and have not put much effort on performance issues. We expect performance penalties for database updates, since an update at the AR layer now involves multiple updates. The performance of queries should not be affected, since they do not involve the CRR-augmented parts.

5 Related Work

We limit the comparison to an alternative implementation reported in our earlier paper [12] and refer the interested reader to [12] for discussions on the other research work that are generally related to CRR.

Earlier, we implemented CRR on top of the Ecto ORM. One advantage of an implementation on top of an ORM is that it supports all RDBMSs (relational database management systems) that the ORM supports. It is even possible to synchronize between the instances running with different RDBMSs. There are some drawbacks, though. Only the applications using the ORM (and in the programming language of the ORM) can benefit from the CRR support. The supported RDBMSs are limited to those supported by the ORM. Unfortunately, the ORM of our choice, Ecto, does not support SQLite in the latest versions, and SQLite is an ideal candidate for local-first software (Sect. 1).

Implementing direct CRR support for SQLite addresses the above-mentioned drawbacks, at the cost of not benefiting from the advantages.

In [12], the implementation was mostly in the Elixir[7] programming language. Now, we try to keep the implementation as much in SQLite as possible. In particular, we aim at implementing all features related to local updates inside SQLite, so that existing applications continue to work without making any modification. We even restrict our implementation to the SQLite distribution that does not include any extension.

Since Elixir facilitates actor-based programs, data communication is built in. Little programming effort is needed for data communication. On the other hand, every database update is encoded and decoded between RDBMS and Elixir representations. This increases run-time overhead. Moreover, data security is not taken into account. Since we now transfer data through ssh connections, we do not have to worry about security and configuration issues.

There are some further differences in implementation details. In [12], a local update is first made in the CRR layer and then refreshed to the AR layer. Now the update first happens in the AR layer which then triggers updates in the CRR layer.

Currently, our implementation has not yet been as complete as the earlier implementation (Sect. 4.7).

6 Conclusion

We have presented a work-in-progress implementation of a software prototype that augments existing SQLite databases for local-first software. With a single command-line operation, we augment an existing database instance with CRR support. Existing applications using the existing database instance, without any modification, continue to work with the augmented instances. We can then maintain multiple instances of the same database at different devices and

---

[7] https://elixir-lang.org/.

independently query and update the different instances. We can synchronize the updates at different instances when the devices are connected. The instances are eventually consistent. That is, they will have the same state when they have applied the same set of updates. The implementation is still in its early stage and more features remain to be completed.

References

1. Almeida, P.S., Shoker, A., Baquero, C.: Delta state replicated data types. J. Parallel Distrib. Comput. **111**, 162–173 (2018)
2. Enes, V., Almeida, P.S., Baquero, C., Leitão, J.: Efficient synchronization of state-based CRDTs. In: IEEE 35th International Conference on Data Engineering (ICDE), April 2019
3. Fox, A., Brewer, E.A.: Harvest, yield and scalable tolerant systems. In: The Seventh Workshop on Hot Topics in Operating Systems, pp. 174–178 (1999)
4. Garg, V.K.: Introduction to Lattice Theory with Computer Science Applications. Wiley, Hoboken (2015)
5. Gilbert, S., Lynch, N.: Brewer's conjecture and the feasibility of consistent, available, partition-tolerant web services. SIGACT News **33**(2), 51–59 (2002)
6. Johnson, P., Thomas, R.: The maintamance of duplicated databases. Internet Request for Comments RFC 677, January 1976
7. Kleppmann, M., Wiggins, A., van Hardenberg, P., McGranaghan, M.: Local-first software: you own your data, in spite of the cloud. In: Proceedings of the 2019 ACM SIGPLAN International Symposium on New Ideas, New Paradigms, and Reflections on Programming and Software, (Onward! 2019), pp. 154–178 (2019)
8. Kulkarni, S.S., Demirbas, M., Madappa, D., Avva, B., Leone, M.: Logical physical clocks. In: Aguilera, M.K., Querzoni, L., Shapiro, M. (eds.) OPODIS 2014. LNCS, vol. 8878, pp. 17–32. Springer, Cham (2014). https://doi.org/10.1007/978-3-319-14472-6_2
9. Shapiro, M., Preguiça, N. M., Baquero, C., Zawirski, M.: A comprehensive study of convergent and commutative replicated data types. Rapport de recherche 7506, January 2011
10. Shapiro, M., Preguiça, N.M., Baquero, C., Zawirski, M.: Conflict-free replicated data types. In: 13th International Symposium on Stabilization, Safety, and Security of Distributed Systems, (SSS 2011), pp. 386–400 (2011)
11. Vogels, W.: Eventually consistent. Commun. ACM **52**(1), 40–44 (2009)
12. Yu, W., Ignat, C.-L.: Conflict-free replicated relations for multi-synchronous database management at edge. In: IEEE International Conference on Smart Data Services (SMDS), pp. 113–121 (2020)
13. Yu, W., Rostad, S.: A low-cost set CRDT based on causal lengths. In: Proceedings of the 7th Workshop on the Principles and Practice of Consistency for Distributed Data (PaPoC), pp. 5:1–5:6 (2020)

ADBIS 2021 Workshop: Computational Aspects of Network Science – CAoNS

CAoNS: Computational Aspects of Network Science

Dimitrios Katsaros[1] and Yannis Manolopoulos[2]

[1] University of Thessaly, Greece
[2] Open University of Cyprus, Cyprus, and Aristotle University of Thessaloniki, Greece

Description. The First International Workshop on Computational Aspects on Network Science (CAoNS 2021) aimed to bring together researchers and practitioners working on areas related to any kind of network, such as networks of devices; the World Wide Web; social, bibliographic, biological networks; complex systems; and so on, with the main focus on the management of networked data (retrieval, evolution) and on the use of analytics for predictive purposes and/or on the analysis of the networks themselves.

Selected Papers. The CAoNS 2021 workshop accepted for presentation three full papers. The program of the workshop also included one invited talk by Ginestra Bianconi (Queen Mary University of London, UK). The first paper entitled "Scalable and Explainable User Role Detection in Social Media" by Kastner and Fischer develops an automated and scalable detection of a wide range of roles – in particular bots or spammers – by clustering users hierarchically on salient, complementary features such as their actions, their ability to trigger reactions, and their network positions. The second paper entitled "Multi-dimensional Ranking via Majorization" by Stoupas and Sidiropoulos introduces the mathematical concept of majorization and exploits it in university ranking by developing the concept of "sets of equivalence", avoiding the pitfalls of present methodologies which include arbitrary choices about the key indicators, and their special weights. Finally, the third paper entitled "Inferring Missing Retweets in Twitter Information Cascades" by Neumann and Fischer analyzes the data set limitations and their impact concerning the identification of the paths on which information diffuses, and it also derives a damage model and a method that utilizes the social graph neighborhood to infer the missing messages.

Organization

Chairs

Dimitrios Katsaros	University of Thessaly, Greece
Yannis Manolopoulos	Open University of Cyprus, Cyprus, and Aristotle University of Thessaloniki, Greece

Program Committee

Jorge Bernardino	Polytechnic Institute of Coimbra, Portugal
Richard Chbeir	University of Pau and Pays de l'Adour, France
Paquale De Meo	University of Messina, Italy
Cicek Guven	Tilburg University, The Netherlands
Maria Halkidi	University of Piraeus, Greece
Andreas Kaltenbrunner	University of Pompeu Fabra, Spain
Evangelos Kanoulas	University of Amsterdam, The Netherlands
Fragkiskos Malliaros	University of Paris-Saclay, France
Fragkiskos Papadopoulos	Technical University of Cyprus, Cyprus
Dimitrios Papakostas	Hellenic Air Force, Greece
Evaggelia Pitoura	University of Ioannina, Greece
Dimitrios Rafailidis	University of Thessaly, Greece
Ryan Rossi	Adobe, USA
Andrea Tagarelli	University of Calabria, Italy
Alex Thomo	University of Victoria, Canada
Sherali Zeadally	University of Kentucky, USA

Scalable and Explainable User Role Detection in Social Media

Johannes Kastner and Peter M. Fischer[✉]

University of Augsburg, Universitätsstr. 6a, 86159 Augsburg, Germany
{johannes.kastner,peter.m.fischer}@uni-a.de

Abstract. While identifying specific user roles in social media -in particular bots or spammers- has seen significant progress, generic and all-encompassing user role classification remains elusive on the large data sets of today's social media. Yet, such broad classifications enable a deeper understanding of user interactions and pave the way for longitudinal studies capturing the evolution of users such as the rise of influencers.

We build on the fundamental role definitions of previous empirical studies and provide a *largely automated, scalable detection* of *fine-grained roles*. Our approach clusters users *hierarchically* and explains the salient features. To associate clusters with roles, we use *supervised classifiers*: trained by experts on completely new media, but transferable on related data. Furthermore, we employ the combination of samples in order to improve scalability and allow probabilistic assignments of user roles.

Our evaluation on Twitter indicates that a) *stable* and *reliable detection* of a wide range of *roles* is possible b) the labeling *transfers* well as long as the fundamental properties don't strongly change between datasets and c) the approaches *scale* well with little need for human intervention.

Keywords: Social media · User role detection · Classification

1 Introduction

Automatically identifying user roles in social media at scale and speed promises interesting insights on their prevalence and impact. Furthermore, a stable recognition explains how individual users and communities evolve over time.

We propose a method that combines *unsupervised learning* to discover *fine-grained* classes of users over a wide range of features with *supervised learning* - generalizing expert knowledge from manually labeled reference data to new datasets, mapping role candidates to well-known roles or identifying new roles.

The paper provides the following *contributions*:

© Springer Nature Switzerland AG 2021
L. Bellatreche et al. (Eds.): ADBIS 2021 Short Papers, Workshops and Doctoral Consortium,
CCIS 1450, pp. 263–275, 2021.
https://doi.org/10.1007/978-3-030-85082-1_23

- Learning the *structure* of *user groups* as well as assigning *suitable labels*.
- A study on *large, complementary datasets* shows that both *recognizing* and *transferring* roles is feasible over *longer time periods* or *topic variations*.
- The classification hierarchy with salient *feature detection* and the cluster metrics support reviewing, so that *identification* requires *little intervention*.
- *Sample combination* provides *scalability* and *probabilistic role assessment*.

The remainder of this paper is structured as follows: In Sect. 2 we discuss related work. We introduce our methodology in Sect. 3 and provide more details on structure discovery and labeling in Sects. 4 and 5, respectively. After an extensive evaluation (Sect. 6), we conclude the paper.

2   Related Work

Clearly, identifying user roles has been one of the textbook examples of classifier algorithms, yet the application to social networks has been limited to particular aspects. Often, the studies focus on detecting specific roles or describing only a small number of coarse-grained classes. Considering the negative dynamics of many social networks, most researchers focus on identifying malicious users like bots [2] and spammers [9] or aggressors in the context of cyber bullying [1,7]. In contrast, our goal is to comprehensively assign all users to roles. Multi-role approaches such as Varol et al. [11], Rocha et al. [4] and Lazaridou et al. [8] limit themselves to identify a small number (often 3–5) of major, course-grained groups, roughly corresponding the upper levels of our detection hierarchy. Du et al. [3] provide a somewhat higher number of rules (still lower than ours), but only give fairly generic descriptions. All of the previously mentioned methods are constrained on just detecting the structure by unsupervised learning: clustering via K-Means [8] or EM [4] or via topic models [3], leaving the analysis entirely to human experts. Varol et al. [11] fully rely on such human classification, using similarity matrices and handcrafted rules. In contrast, qualitative works like Tinati et al. [10] or Java et al. [6] provide a comprehensive overview on fine-grained roles and their semantics, but provide only general rules on how to detect them. An interesting, complementary direction is the work on content communities/web forum, often exploring complex temporal models, e.g., [5]. It should be noted that all of these works (with the exception of [3] (Weibo, 12K users), [7] (Instagram, 18K users), and [5] (Stack Overflow)) solely rely on Twitter due to the limited availability of data from other services.

3   Approach

In order to classify diverse user roles in large data, we phrase three questions:

1. To which extent can clusters of users be utilized to sensibly *detect* user roles in social media and build a *classifier* to *(semi-)automatically* label them?
2. Can this approach be applied individually over *a wide variety of data sets*, currently stemming from the same social media?
3. Can the *knowledge* on roles be *transferred* to *new data sets*?

As the related work only describes instances of user roles, but not the concept of a role itself, we use the following, *basic definition*: A **user role** is a *set of users* that *share similar feature values* and are *well separated* from other groups. The features *capture salient properties* of users and allow a *meaningful categorization*, typically capturing *behavior* and *position* in the network/media. Groups constitute roles if they are *present* in sufficient number within a data set and *re-occur* over multiple data sets.

Our approach can be applied in *two complementary scenarios*, requiring different quantities of human involvement given the amount reference data:

1) If only **unexplored datasets** are available, we *discover* groups of *similar users* and their *hierarchical relationship*, providing *candidates* for *user roles*. The analyst is then aided by *metrics* and *visualizations* in assigning *role labels*. In turn, these *labels* form the *input* for a *classifier* that captures this *knowledge*.
2) If a **reference dataset** with a *classifier* is available, the labeling process can be cut short by providing *candidate labels*. The user can *evaluate* these labels within the new dataset or *compare* roles across datasets. We explored causes of mislabelings and methods to adapt, yet a full exploration remains future work.

4   Feature Selection and Data Clustering

In this work, we aim to use *features* that cover *significant* and *complementary* aspects of *users* and are well established in the literature [1,4,8]. In addition, they should be feasible to compute in *large scale* so that data is commonly available and incur *moderate cost* to compute. Likewise, we want to avoid a large number of features, as this hurts both algorithm performance and explainability.

Figure 1 highlights our features: **static user properties** express (self-)description: most relevant is the *verified* status, traditionally reserved for VIPs. **User activity** is characterized by the number of original tweets of each user (observed and "offtopic"), the activities on other tweets such as retweets and replies within the topic as well as mentions of other users. Basic **network position** features like the number of *followers* and *followees* (aka degree centrality) of a user as well underpin the potential to exert influence. In turn, the user's ability to actually elicit **reactions from the network** is captured by the *ratio of tweets* to lead to *replies* and *retweets* as well as the frequency of *being mentioned*.

We investigated a wide variety of additional features from these classes, but dropped them as they were *correlated* or had *little discriminative power*. Furthermore, we did not include complex network metrics such as path-based centralities, spatio-temporal features [11] as well as explicit content analyses [1,7]. Even partial social graphs are exceedingly hard to get from any social media (including Twitter), while our *crawling strategy* already provides a *topic focus*.

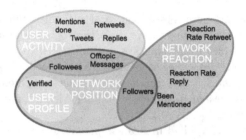

Fig. 1. User feature classification

As most features exhibit *significant skew* and *domain variation*, we *normalize* each dataset using *logarithmic transformation*, followed by a *Min-Max normalization*. This way, we capture the *relative distributions* and *feature drift*.

To *identify* the *structure* and *(sub)-groups* among the user data, we evaluated a broad range of *unsupervised learning* approaches based on centroids (e.g., K-Means), density (like DBScan) and probability distribution (e.g., EM). *Hierarchical clustering* with geometric linkage (Ward) was the best fit: a) it can *capture complex, irregular shapes* without requiring a fixed number of clusters and b) the *hierarchy* serves as an (yet unlabeled) *classification tree*.

To overcome the *limited scalability* ($O(n^2)$ scaling for both CPU and memory, single-threaded execution) and the lack of probabilistic approaches for hierarchical clustering, we chose a *sampling/ensemble-based approach*: Clustering a small number of samples allows us to *quickly discover* the *structure* while *drastically reducing* the *cost* compared to clustering the whole dataset. By incrementally drawing more samples, we see a *linear cost increase* (while allowing parallel execution) and provide a *faithful representation* of the data. With *overlapping cluster* results from several samples for the same user, we can choose to assign to a *majority role* or the probability for specific roles. Likewise, we can determine how *stable* the *role recognition* is. The number of samples becomes a tuneable, trading off the effort of computation (and labeling) with the *coverage of users* and the *amount of support* for the roles on each user or role. If all users need to be covered, we can also minimize the overlap or apply metric-based assignments.

5   User Role Identification

While the cluster structure identifies candidates, it does not provide the actual roles. We now describe the *assignment* and *transfer* of the roles to new datasets.

5.1   Manual Role Assignment

Considering the lack of precise, commonly established roles (see Sect. 2), we apply *complementary methods* to derive *candidates*: 1) *internal cluster quality metrics*, as well as *statistical metrics* and heuristics like the *elbow method* provide indications on the (approximate) number of clusters (typically around 20–30), but no meaning. 2) analyzing the *dendrogram* and *salient features* of *clusters splits* allows us to match clusters to *fine-grained* user roles. 3) *dimensionality reduction* such as *PCA* or *LDA* simplifies visual inspection, helps with *correlating user roles* across datasets and exposes the *drift/evolution*.

Certain heuristics support this process: 1) Role generalizations tend to occur further up the hierarchy, creating *subtrees* that could be *refined*. 2) some very *distinct roles* tend to show up in most datasets, providing an *"anchor"* for labeling. The process is stopped once we do not gain well-discernible clusters. In some cases, it may be useful to coarsen the roles or combine clusters into a single role.

We match these aspects to well-known role descriptions when possible (e.g., *Star*), but also observed stable, recurring clusters that did not align well with the known role descriptions. In our dataset, among of them were *Rising Stars* (gaining followers via activity, receiving significant retweet reactions, not yet star or semi-stars) and *Loners* (low activity and weak connections). Figure 3 provides an overview in terms of groups, features and frequency.

The same process can be applied *across datasets*, so that we can *track* the user roles and *evaluate concept shifts* and *drifts*, such as their *frequency/probability* of roles or their feature distributions. Typically, we observed around 10–15 class candidates that did show up in varying frequency, sometime disappearing entirely.

While we capture *domain expert knowledge* and produce *well-described role clusters*, we suffer from *limited scalability* and *reproducibility*.

5.2   Classifier

To overcome these issues, we trained an *n-class classifier* that on role samples of one dataset and determined the role labels on clusters in other datasets. While classifying individual users yielded lower runtimes (no clustering step needed), we observed lower classification quality due to the inherent "noise" shown by individual users. Clusters were represented by *aggregate feature values* of all members where *means* tended to provide better *separation* than median, while *pooled Cohens d* seemed to capture more *temporal evolution* than "pure" means. We took samples that showed the best cluster separation to minimize the noise. As initial experiments showed, the original number of dimensions in the data yielded better quality than reduced dimensionality.

As our (clustered) data sets are relatively small and skewed, yet we seek to express a large number of classes, we see little support for some classes. This more or less rules out deep learning. Instead, methods based on *ensembles of decision trees* (Gradient Boosted Decision Trees (GBM), Extremely Randomized Trees (ET)), *multi-class support-vector machines* (SVM) or *k-nearest-neighbor* (kNN) turned out to be most suitable. We utilized the Python implementations of scikit-learn for ET, SVM and kNN as well as XGBoost[1] for GBM.

The setup to build *training sets* utilized *repeated stratified cross validation* with three splits (leave-one out, due to the small amount of data) and three

---

[1] https://xgboost.readthedocs.io/en/latest/.

repetitions (with different permutations to cater for possibly missing groups). We used *F1-macro* as a metric to *compensate* for *class imbalance* and prevent focus on either precision or recall and applied grid search to tune parameters. All classifiers learn and generalize well, leading to 94–95% score in validation and training set with no obviously stronger or weaker candidates.

When *transferring* the *classification* to new datasets, we compensated for mislabelings by *varying training* and *prediction data* (e.g. cluster number) or choosing *more suitable training sets*. Explicitly including drift models and relevance feedback from the user remain future work.

6 Evaluation

After introducing our concepts, we provide an *evaluation* on *diverse data* from Twitter. We address the *three questions* outlined in Sect. 3, not only on the *technical aspects* but also on the *empirical observations*.

6.1 Datasets and Preparation

While our long-term goal is to *recognize* user roles over a variety of social media, we focused our *initial analysis* on *well-defined data* with a large number of users. As in most of the related work, we relied on Twitter, as it is one for the few social media services which fulfills these criteria.

In order to *transfer user role detection knowledge*, we are looking at several classes (Table 1): *major sports* tend to be *repetitive* and *predictable* with a very large number of messages and users, covering *significant periods of time*. Different types of sports provide a (albeit limited) thematic variance. These datasets are complemented by those of two *major disasters* which also tend to have a strong, yet very different *topic focus* and different *interaction patterns*.

Table 1. Overview on data sets.

Dataset	Messages	Users	Time period	Category
Olympic Games 2012	13.68M	2.27M	August 2012	Sport event
Olympic Games 2014	14.58M	1.96M	February 2014	Sport event
Olympic Games 2016	38.05M	4.76M	July/August 2016	Sport event
FIFA World cup 2014	109.00M	10.40M	June/July 2014	Sport event
2015 paris attacks	6.77M	0.74M	November 2015	Tragic incidence
NFL Superbowl LIV 2020	8.89M	0.89M	2. March 2020	Sport event
2016 Berlin truck attack	0.66M	0.15M	19. December 2016	Tragic incidence

Our datasets had each been *recorded* using the *Twitter Streams* and *Search API* using commonly proposed *hashtags*. We only considered users that were *active* at least twice to cater for aggregate metrics. Generally speaking, the *relative feature distributions* after *normalization* showed only minor changes from 2012 until today: The *verified* status is more prevalent. Overall activity increased moderately, while users tend to move into *"reactive"* behavior of *forwarding*.

6.2 Initial Dataset: 2012 Olympics

The first step focuses on a *single dataset* with *uniform feature usage* and *role stability* due to the relatively short period of time. The analyses provide insights to which extent such as clustering, user roles detection and automated labeling are feasible, as stated in Q1.

The *benefits* of sampling are shown in Table 2. The numbers were generated using `scipy.cluster` on an 8-core partition of an AMD Epyc 7401. A small dataset like Berlin 2015 may still be clustered completely, yet a sample can be generated almost instantly. For large datasets, *full clustering* is clearly impossible, while *samples* fit well. The cost can almost entirely be attributed to creating the *linkage matrix*, so *refinement/exploration steps* are interactive in all variants. Combining them (Fig. 2) shows how *coverage* and *certainty* of *roles* (number of role assignments per user) improve, while *cost scales linearly*. The decreasing "no majority" part gives insights on user that are not well identified - which is dataset-dependent, but often includes Spammer, Loners, etc.

Table 2. Runtime and memory of samples, full data sets and approximated (*).

	Oly12 5%	Oly12 10%	Oly12 100%	Berlin16 10%	Berlin16 100%
Runtime	19 min	136 min	226 h*	10 s	38 min
Memory	94 GB	375 GB	375 TB*	1.2 GB	184 GB

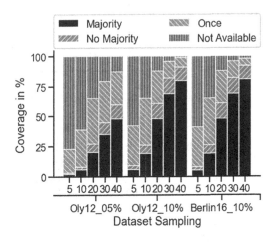

Fig. 2. Coverage and overlap of sampling.

After clustering, we manually labeled clusterings of the samples to get a ground truth as training and test data which can be done incrementally.

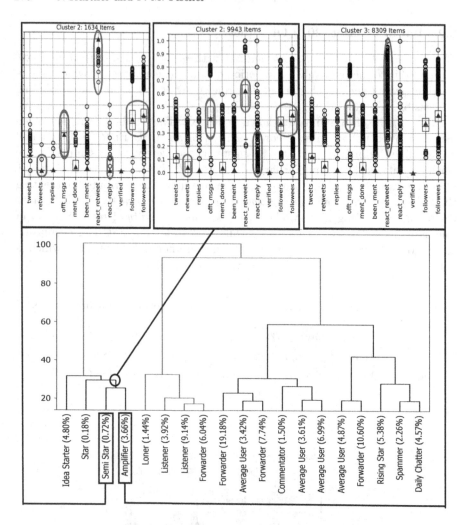

Fig. 3. Olympics 2012 10% sample Dendrogram with salient features.

In particular after applying PCA (see Fig. 4), we can *identify* a *number of well-separated clusters*. Despite some minor variances, the *dendrograms* (see Fig. 3) and the *features* over the set of samples exhibit a *high degree of similarity* that has become a part of our overall classification (Table 3, leftmost column): there are between 3 and 5 *subtrees* representing *major groups*: The first (green) shows users that are able to *trigger strong reactions* (*retweets, replies, being mentioned*), the second (red) shows *passive user* with fairly *weak positions in the network*, while the group(s) in between show various degree of *moderate activity and impact*. Even further down the tree, (as shown on the boxplots), we see a strong motivation for *fine-grained roles*. While the cluster sizes are often small, there are *salient feature differences* (which we can detect using *statistical tests* like *cohens d*) that explain the existence and semantics of this group. In the example one can see how *Semi Stars* and *Amplifiers* split on (among others) *retweet activities* and *reactions*. Overall, we determined 12 roles in the Olympics 2012 data set that are described in Table 3. Some *characteristics* are shown in the second column, in particular stronger *deviations* from the *average* as well as (broadly) similar user groups.

Table 3. User roles and their characterization: ↓/↑ denote feature deviation from whole dataset, ≈ closeness to other roles, ↘ / ↔ / ↗ changes over time

	Role	Characteristics	Freq./Trend
action triggering	Star	followers > followees, verified, ↓ activity, ↑ mentioned	0.2–0.8 ↗
	Semi Star	≈ Stars, ↓ followers, mentioned, ↑ react. (re)tweet, retweets, replies	0.2–1.4 ↘
	Idea Starter	≈Semi Star, ↓followers, ↑reactions	1–4 ↔
	Amplifier	≈ Idea Starters, Semi Stars, ↑ followers, followees	0.5–5 ↘
intermediates	Rising Star	≈ Semi Star, Idea Starter, Amplifier, ↑ followers, (re)tweets, replies	1.5–5.5 ↘
	Daily Chatter	≈ Average User, Spammer, ↓(re)tweets, offtopic	5–15 ↔
	Commentator	↑ replies, offtopic, reations	0.3–2 ↘
	Spammer	↑ (re)tweets, replies, offtopic ↓ followers, followees, reactions	1–7 ↔
passive	Average User	offtopic > tweets, retweets	8–30 ↓
	Forwarder	retweets > tweets, ↑ offtopic, followers, followees. ↓ reactions	25–65 ↑
	Listener	↓ (re)tweets, reactions	6–20 ↗
	Loner	↓↓ tweets, offtopic, followers	0–1.5 ↘

We evaluated the clustered and labeled samples (in total 507 clusters) with the *classifiers* mentioned in Sect. 5 and achieved nearly perfect results, as the leftmost data points in Fig. 5 show. There are only very few misclassifications between *Average User*, *Daily Chatter* and *Listener*, respectively - which are also *close in feature space* and *low in certainty*. The strong variance in the feature distribution (Fig. 3) also shows why training and classifying individual users instead of clusters yields inferior results.

Overall, the results show that both *clustering* and *classification* work well. *Expert knowledge* is needed to interpret the *dendrogram* and *assign roles*, but already within a single dataset, the *knowledge* can be transferred to additional samples and their clusters.

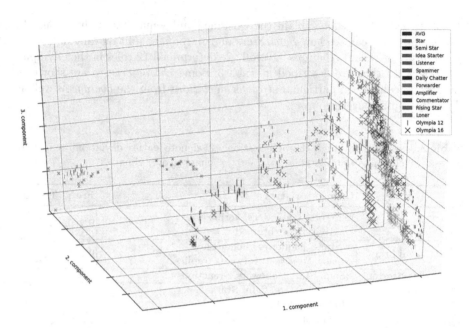

Fig. 4. PCA of clustered samples from Olympics 2012 vs 2016.

6.3 Multiple Individual Datasets

We now analyze several *datasets individually* to understand if the approach is more *widely applicable* (see Q2). Furthermore, this will show if the *same* or *similar roles* are *present* on how they *evolve* in *frequency* and *features*.

The 12 user roles identified on the *Olympics 2012* dataset are also *present* and *well-separated* in the other datasets, though -as the rightmost column of Table 3 shows- the frequency (in percent) varies over datasets (and over time):

In the *Olympics 2014* (278 clusters) and *FIFA World Cup 2014* (193 clusters) data very few changes can be observed: *Average User* and especially *Loner* occur less frequently, while *Forwarder* and *Listener* occur more frequently.

Significant changes occur in *Olympics 2016* (355 clusters). The PCA in Fig. 4 shows a *salient concept drift*, in particular for *Semi Stars* that tends to also cover a space much closer to *Stars*, as the *"verified"* status was more freely distributed by Twitter. The trends on the *Average Users /Loner* and *Forwarders* strengthens, and continues for the *Superbowl 2020* (345 clusters), which is otherwise (despite the difference in sports and time) similar to *Olympics 2016*.

The *2015 Paris Attacks* (160 clusters) covers a very different topic and distinct interactions (fewer offtopic messages, more retweets). Some user roles are not present (*Commentator, Loner*), yet most of the overall trends match the picture of the "sports events": *forwarding* instead of *content creation* becomes more dominant. In turn, *"influencer"* roles become pronounced, to the point where the *Semi Star* may have to split into two separate sub-roles.

The only exception where we could not apply our methodology was the random *Sample Stream*, as features based on topics lose their usefulness.

Overall, we see the *same features*, leading to *consistently recognizable user roles* that we can *correlate across data sets* to *trace shifting distributions*. Yet, at this step, labeling samples of each dataset manually is a limiting factor.

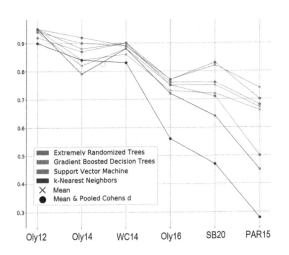

Fig. 5. Information retrieval measure F1 for classifiers.

6.4 Applying Models on New Datasets

In the third step, we *classify new datasets* with the *models* gathered from *reference data* to assess the *quality* and *effort* involved, answering Q3. We further study the *impact of variation* and *drift* to understand the limitations.

Figure 5 shows the *weighted F1 scores* when classifying the dataset based on the *Olympics 2012* as the reference, as it provides the longest prediction period. Overall, one can see a *gradual degradation* over time on the sport events, as the classification methods do not explicitly capture for the drifts, but still can generalize the roles over time. Still, the best methods achieve a 0.85 F1 score for "late" sport events. The *2015 Paris Attacks* data set sees the largest degradation, showing topic and interaction differences have a more profound impact than time. When comparing all these results to the slightly worse "macro" values, one can see that *small groups* are *captured well*, while larger clusters tend to be somewhat "blurry".

kNN and SVC keep up well for *short time intervals*, but tend to lose ground on longer distances. ET holds a small edge over GBM, while the latter stays still *competitive* and incurs much lower runtime cost. Both benefit from *enriching* the datasets with the *pooled Cohens d values*.

Roles that were either not well separated in the *Olympics 2012* data or drifted significantly are most affected. Yet, these *misclassifications* often leads to *adjacent roles*, e.g., *Average Users* as *Listener* and *Forwarder*, thus the F1-scores actually understate the result quality.

We added the data set of the *2016 Berlin Truck Attack* (Christmas market) that was not evaluated in the previous stages and provides *topic similarity* to *2015 Paris Attacks*, while being close to the *Olympics 2016* in time. This data set provides a good opportunity to assess the impact of *different training sets*: in addition to baseline of the *Olympics 2012* and close sets (*Olympics 2016, 2015 Paris Attacks*) and *Superbowl 2020* as a small, recent dataset, we tested two combinations: As Table 4 shows, these combined data sets provide the best results, *matching manual classification* or producing *misclassifications* to *close roles*. *2015 Paris Attacks* by itself seems to be too small to provide a sufficiently general model, but is able to *boost* the *full time range model*.

The experiments show that a *transfer* of *labeling knowledge* is effective with certain limitations: *large topic differences* or *very long time differentials* diminish the *usefulness*, yet a *good choice* of *reference data* can mitigate this effect.

Table 4. Classification of Berlin 2016 data set. Comb1: Oly12 & SB 20, Comb2: Oly12 & SB 20 & Par15.

Classifier	Oly12	Oly16	Par15	SB 20	Comb1	Comb2
XGB	0.58	0.59	0.51	0.70	0.78	0.92
ET	0.74	0.63	0.56	0.73	0.77	0.82

7   Conclusion and Future Work

In this paper we proposed a method to determine *user roles* in *large-scale social media data*. It combines *unsupervised learning* (i.e., hierarchical clustering) to *discover* and *explain* such roles over a *wide range of features* with *supervised learning* to generalize the *knowledge* from *manually labeled* smaller data.

Our analysis on a range of large data sets from Twitter shows that *well-separated roles* can consistently be *recognized* and *transferred*. The labeling achieves *high accuracy* not only within the same data set, but also on *new data sets* from *different event types* and/or *years* apart. *Scalability, incremental evaluation* and *probabilistic assignment* are achieved by *combining samples*.

For *future work*, we see a number of interesting directions: As the *quality* of *classification* begins to deteriorate over *longer time frames*, we plan to address *evolution*, considering both *temporal models* (for long-term studies of snapshots) and *stream clustering* (for short-term, continuous analyses). They may also pave the way for *longitudinal studies* of user groups and user mobility among groups. Likewise, adapting our model to cope with *topically non-related* or even *topically unconstrained data sets* poses a new set of challenges. Initial experiments show that the method should generally work, but *significant challenges* remain. In either case, testing our method on a *wider range of data sets* from Twitter or *other social networks* would be highly interesting.

References

1. Chatzakou, D., Kourtellis, N., Blackburn, J., Cristofaro, E.D., Stringhini, G., Vakali, A.: Mean birds: detecting aggression and bullying on Twitter. In: WebSci (2017)
2. Chu, Z., Gianvecchio, S., Wang, H., Jajodia, S.: Who is tweeting on Twitter: human, bot, or cyborg? In: ACSAC (2010)
3. Du, F., Liu, Y., Liu, X., Sun, J., Jiang, Y.: User role analysis in online social networks based on Dirichlet process mixture models. In: 2016 International Conference on Advanced Cloud and Big Data (CBD), pp. 172–177 (2016)
4. Edgar, R., Alexandre, P.F., Caladoa, P., Sofia-Pinto, H.: User profiling on Twitter. Semant. Web J. (2011)
5. Fu, C.: Tracking user-role evolution via topic modeling in community question answering. Inf. Process. Manage. **56**(6), 102075 (2019)
6. Java, A., Song, X., Finin, T., Tseng, B.: Why we Twitter: understanding microblogging usage and communities. In: WebKDD/SNA-KDD (2007)
7. Kao, H.T., Yan, S., Huang, D., Bartley, N., Hosseinmardi, H., Ferrara, E.: Understanding cyberbullying on instagram and Ask.Fm via social role detection. In: WWW 2019 Companion (2019)
8. Lazaridou, E., Ntalla, A., Novak, J.: Behavioural role analysis for multi-faceted communication campaigns in Twitter. In: WebSci (2016)
9. Li, H., et al.: Bimodal distribution and co-bursting in review spam detection. In: WWW (2017)
10. Tinati, R., Carr, L., Hall, W., Bentwood, J.: Identifying communicator roles in Twitter. In: WWW 2012, rel MSND Workshop (2012)
11. Varol, O., Ferrara, E., Ogan, C.L., Menczer, F., Flammini, A.: Evolution of online user behavior during a social upheaval. In: WebSci (2014)

Multi-dimensional Ranking
via Majorization

Georgios Stoupas[1] and Antonis Sidiropoulos[2(✉)]

[1] Aristotle University of Thessaloniki, Thessaloniki, Greece
grgstoupas@csd.auth.gr
[2] International Hellenic University, Thessaloniki, Greece
asidirop@imselab.iee.ihu.gr

Abstract. University ranking are ubiquitous as there are dozens of them. University ranking are not only used as tools in academia and funding agencies, but they are popular reading in newspapers and magazines as well. Also in academics is of major importance for decision-making. University ranking have been heavily criticized for a number of reasons, including that they are based on arbitrary choices about the key indicators, their special weights and so on. For these reasons, we argue that there is no meaning in ranking universities in an absolute ordered manner; instead universities should be ranked in "sets of equivalence". To this end, we introduce the mathematical concept of majorization and perform experiments on a dataset extracted from NTU Ranking. This approach is applicable to other entities, such as authors or other higher order multi-dimensional entities.

Keywords: Dominance · Skyline · Rainbow ranking · Majorization · University ranking

1 Introduction

University rankings are of major importance for decision-making by prospective students, academic staff, and funding agencies. The placement of universities in these lists is a crucial factor and academic institutions adapt their strategy according to the particular criteria of each evaluation system.

Probably, the most popular global rankings are: ARWU[1] (Academic Ranking of World Universities) of Shanghai, THE[2] (Times Higher Education) World University Rank. and QS[3](Quacquarelli Symonds) Rank. All these lists base their respective ranking on some set of performance indicators, which differ from one

---

[1] https://www.shanghairanking.com/.
[2] https://www.timeshighereducation.com/.
[3] https://www.topuniversities.com/.

© Springer Nature Switzerland AG 2021
L. Bellatreche et al. (Eds.): ADBIS 2021 Short Papers, Workshops and Doctoral Consortium,
CCIS 1450, pp. 276–286, 2021.
https://doi.org/10.1007/978-3-030-85082-1_24

organization to the other. The reader can retrieve these indicators from the respective sites.

Despite their popularity, these rankings are heavily criticized for their reproducibility, statistical soundness, etc. [1,6,7,15]. Here, we propose a new approach to evaluate and rank universities. We argue that there is no meaning in stating that the university at the, e.g., 100-th position, is superior in comparison to a university at the 110-th or 120-th position, since all the university rankings are prone to opacity and arbitrariness by mixing academic and non-academic issues, by relying on questionnaires, by assigning random special weights and so on.

Departing from this point, in this paper we introduce a new approach to this end, by adopting the mathematical concept of majorization and performing experiments on a dataset extracted from the National Taiwan University Ranking (NTU Rank)[4]. Majorization comes as another method to group universities in ranked "equivalence sets", next to the methods of Skylines [12] and Clustering [13]. This approach is applicable onto other entities, such as authors or other higher order multi-dimensional entities.

The next section builds on the notion of Skylines and Rainbow Ranking, a further method based on skylines. Section 3 introduces the mathematical concept of majorization and combines it with skylines to deliver new ranking approaches. Section 4 we deliver two experiments, one by using a dataset extracted from the NTU ranking, whereas the second one consists of a small set of Greek researches of Computer Science, in an effort to apply the proposed techniques in other multi-dimensional datasets. The results are indicative and justify the reasoning of ranking multi-dimensional entities in "equivalence sets" instead of arbitrarily enforcing a strict absolute order.

2 Skyline and Rainbow Ranking

The skyline operator [3] has been utilized in the Databases field of Computer Science for decades and dates back to the definition of the Pareto frontier in Economics [16]. However, the skyline approach does not refer to efficient resource allocation; rather it provides a multi-criteria selection of "objects". In particular, the skyline operator is used as a database query to filter only those objects that are not worse than any other (they are not dominated) [3]. The skyline operator takes advantage of the concept of "dominance", which is symbolized with $\prec$ and $\succ$. There are various definitions of the dominance relationship:

Definition 1 (Dominance [4]). *Point s dominates point t, denoted as $s \succ t$, iff $\forall i \in [1,d]$, $s[i] \geq t[i]$ and $\exists j \in [1,d]$, such that $s[i] > t[i]$. If neither s dominates t, nor t dominates s, then s and t are incomparable, denoted $s \prec\succ t$. The relation $s \succeq t$ denotes that either $s \succ t$ or $\forall i \in [1,d], s[i] = t[i]$.*

Definition 2 (Strict Dominance [4]). *Point p strictly dominates point q, denoted as $p \ggg q$, iff $\forall i \in [1,d], p[i] > q[i]$.*

---

[4] https://nturanking.csti.tw/.

Here, we note that there are two opposite views in the literature about dominance, depending on the application. In some applications we may need positive domination, in the sense that the best value is the greater one. Elsewhere, we may need negative domination, i.e. the best value is the smallest one. For example, an attribute as *h-index* needs positive domination but the attribute *rank position* needs negative domination.

Definition 3 (Skyline [14]). *Given a dominance relationship in a dataset, a skyline query returns the objects that cannot be dominated by any other object.*

$$SKY(S) = \{s \in S : \nexists t \in S,\ t \succ s\} \tag{1}$$

Apparently, we can apply the skyline concept in Scientometrics. For example, given a set of attributes that characterize scientific performance, the skyline operator outputs the entities (e.g. authors, institutions etc.) that cannot be surpassed by any other entity in the dataset [8].

An extension of the skyline operator, the Rainbow Ranking [11], applies iteratively the skyline operator until all entities (i.e., scientists) of a dataset have been classified into a skyline level. More specifically, given a set of scientists X_1, the first call of the skyline operator produces the first skyline level, which is denoted as S_1. Next, the skyline operator is applied on the dataset $X_1 - S_1$, to derive the second skyline layer, denoted as S_2. This process continues until all the scientists of the dataset have been assigned to a particular skyline level S_i.

To give more semantics to the method, a particular value should characterize the skyline levels. Should this value be the iteration number, then this would convey limited interpretability since the relativeness would be lost. It is crucial to designate the position of a scientist among their peers. Therefore, a normalization of this value is necessary.

Definition 4 (RainbowRanking [11]). *Given a dominance relationship in a dataset, RainbowRanking is the process of iteratively applying the skyline operator. Each skyline operation produces a skyline level (i.e. $1, 2, 3, \ldots$). Thus, the RR-index of an object a is defined as:*

$$RR(a) = 100 - 100 \times \frac{|A_{above}(a)| + |A_{tie}(a)|/2}{|A|} \tag{2}$$

where A is the set of objects, $A_{above}(a)$ is the number of objects at higher skyline levels than object a, and $A_{tie}(a)$ is the number of objects at the same skyline level with object a, excluding object a.

Apparently: $0 < RR(a) \leq 100$, and the best object has the greatest value.

3 Majorizarion

Definition 5 (Weak Majorization [2,5]). *For a vector $a \in R^d$, we denote by $a^{\downarrow} \in R^d$ the vector with the same components, but sorted descendingly. Given that $a, b \in R^d$, a weakly majorizes (or dominates) b, denoted as $a \succ_w b$, iff:*

$$\sum_{i=1}^{k} a_i^{\downarrow} \geq \sum_{i=1}^{k} b_i^{\downarrow} \quad \forall k \in [1, d] \tag{3}$$

Equivalently, we define that b is weakly majorized (or dominated) by a, denoted as $b \prec_w a$.

Definition 6 (Majorization [2,5]). *If $a \succ_w b$ and in addition:*

$$\sum_{i=1}^{d} a_i = \sum_{i=1}^{d} b_i \tag{4}$$

then a majorizes (or dominates) b, denoted as $a \succ b$. Equivalently, we say that b is majorized (or dominated) by a, denoted as $b \prec a$.

As an example of the above definitions, the following expression holds: $(1,2,3)$ $\prec_w (1,3,3) \prec_w (1,3,4)$ according to the weak majorization scheme, whereas the expression $(1,2,3) \prec (0,3,3) \prec (0,0,6)$ holds. In the sequel, based on the majorization notion, we define a new method of dominance.

In the literature, the skyline operator by default uses the dominance relationship as mentioned in Definition 3. However, the skyline operator could use any dominance relationship of the above, each one giving a different set of results. The restriction set by the majorization dominance relationship is that all dimensions should be comparable. This is necessary because the attribute values will be sorted with-in the vectors. Thus, after sorting, an attribute at position i may be placed at position j and may be compared with an attribute than was generated at position k. Therefore, entity attributes (or vectors' entries) must be normalized.

To the best of our knowledge, the majorization dominance relationship (see Definition 5) have never been used in the framework of the skyline operator. Thus, we introduce the *Majorized Skyline operator* which is the skyline operator based on the weak majorization dominance relationship. Here, we notice that the (strict) majorization dominance relationship could not be applied because it presupposes that Eq. 4 holds.

Definition 7 (Majorized Skyline). *The Majorized Skyline operator is defined as the Skyline operator by using the weak majorization dominance relationship.*

Consequently, we define the Majorized Rainbow Ranking:

Definition 8 (Majorized Rainbow Ranking). *The Majorized Rainbow Ranking method is defined as the Rainbow Ranking method by using the Majorized Skyline operator.*

4   Experiments

For our experiments we extracted the top-500 universities from NTU Rank of year 2020, which uses eight attributes normalized into the range $[0, 100]$. These eight features are categorized into three categories, namely:

- *research productivity* comprised by two features: (i) number of articles in the last 11 years, and (ii) number of articles in the current year.
- *research impact* comprised by: (iii) the citations in the last 11 years, (iv) the number of citations in the last 2 years, and (v) the average number of citations in the last 11 years, and
- *research excellence* comprised by: (vi) the h-index of the last 2 years: (vii) the number of highly cited papers, and (vii) the number of articles in the current year in high-impact journals.

For these eight features, the special weights are 10%, 15%, 15%, 10%, 10%, 10%, 15% and 15%, respectively. Despite the fact that NTU is one of the very few rankings which consider only academic - and thus verifiable - performance indicators, it is also characterized by two drawbacks, as mentioned before:

- the special weights are arbitrary and, thus, different set of weights will produce different rankings.
- the absolute order of the universities does not have any substantial meaning.

Thus, we have proposed the use of layered groups for rank production [9]. The Skyline and Rainbow Ranking can overcome these drawbacks. We have applied the Rainbow Ranking based on each of the aforementioned types of domination.

Table 1 presents the first 9 layers as computed by using the Majorized Skyline method on the NTU dataset. The columns *NTU Score* and *NTU Order* present the original data as computed by NTU. The columns *Strict Skyline Layer* and *Strict RR* show the results as computed based on Definition 2. Notably, we get the same results for the first 5 skyline Layers by using the Weak Skyline definition (Definition 1), as well. Thus, we omit the inclusion of these results in this table.

The columns *Majorized Skyline Layer* and *Majorized RR* are computed by using the Definitions 6, 7 and 8. It can be seen that more layers are produced in comparison to the classical Skyline and Rainbow Ranking methods. This is also verified in Fig. 1, where we present the number of elements in each layer. It can be seen that there are about 20 layers for each skyline method, but 45 layers for the majorized version.

In Table 2 we present the Kendall τ, Spearman ρ and Pearson r coefficients [10]. It is noticeable that the Spearman coefficient values are equal to the Pearson coefficient ones for all pairs of the three Rainbow Ranking methods. For example, see values 0.90, 0.92 and 0.99 in the two tables for Spearman and

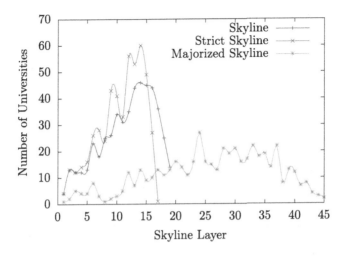

Fig. 1. Distribution of skyline sizes.

Pearson coefficients. This is a feature of the Rainbow Ranking Scoring which is the same with the transform applied during the Pearson coefficient computation. Thus, any two RR based rank lists will always produce the same Spearman and Pearson coefficients.

Next, we experiment with another dataset to further justify the concept of grouping multi-dimensional entities in "equivalence sets". To this end, we apply the above methods on a set of researchers. More specifically, in [12] a dataset of all the university staff members (539 persons) of the Greek CS faculties has been formed by extracting data from Google Scholar. Then, the Rainbow operator was applied on this dataset and resulted in 47 ranked groups. From this dataset, here we select the 22 individuals which comprise the top-3 skyline levels. For these persons, we kept only: (i) the number of citations (C), and (ii) their $h5$-index, for the simplicity of visualization purposes. Let's assume that we need to generate a unique ranking of the members of this dataset based on both attributes.

Table 3 summarizes the results of the experiment. The two basic features $(C$ and $h5)$ can be seen in the 2nd and 3rd columns. To compute the Majorized Skyline, a normalization step is previously applied. For the $h5$-index attribute we map $[0, 34] \rightarrow [0, 100]$, whereas for attribute C we set: $[0, 12568] \rightarrow [0, 100]$. The resulting values are shown in columns C_{nrm} and $h5_{nrm}$. The *Rank by C* and *Rank by h5* columns are self explained, whereas the last three columns depict the level of each individual according to the three methods: *Skyline*, *Strict Skyline* and *Majorized Skyline*, which produces more layers. Also, it is obvious that *Strict Skyline Layer* is less or equal to Skyline Layer.

Figure 2 shows the previous results of Table 3 in an illustrative manner. For example, Fig. 2a, Fig. 2b and Fig. 2c depict the skylines according to the definitions of Weak (Definition 1), Strict (Definition 2) and Majorized (Definition

Table 1. First 9 layers as of the Majorized Skyline method for the NTU dataset.

University	NTU score	NTU order	Strict skyline layer	Strict RR	Major. Skyl. layer	Major. RR
Harvard University	98.6	1	1	99.7	1	100.0
Stanford University	65.7	2	2	98.0	2	99.7
MIT	60.0	7	1	99.7	2	99.7
University of Toronto	62.2	3	2	98.0	3	99.0
University of London, UCL	61.2	5	2	98.0	3	99.0
University of CA, San Francisco	54.7	15	1	99.7	3	99.0
California Institute of Technology	44.9	65	2	98.0	3	99.0
University of CA, Santa Cruz	39.1	153	2	98.0	3	99.0
Johns Hopkins University	61.3	4	2	98.0	4	98.1
University of WA, Seattle	58.8	8	3	95.5	4	98.1
University of CA, Berkeley	54.4	16	1	99.7	4	98.1
Washington University in St. Louis	49.2	32	2	98.0	4	98.1
University of Oxford	60.9	6	2	98.0	5	97.3
University of Pennsylvania	56.8	10	3	95.5	5	97.3
Mayo Medical School	47.9	40	2	98.0	5	97.3
Univ. of TX, Anderson Cancer Center	45.5	58	2	98.0	5	97.3
University of MI, Ann Arbor	57.7	9	3	95.5	6	96.1
University of Cambridge	56.7	11	4	92.9	6	96.1
Tsinghua University	54.4	16	2	98.0	6	96.1
University of Chicago	51.2	22	3	95.5	6	96.1
Princeton University	43.2	86	3	95.5	6	96.1
University of CA, Santa Barbara	40.1	137	3	95.5	6	96.1
Weizmann Institute of Science	37.0	215	3	95.5	6	96.1
University of MA, Medical School	36.5	236	3	95.5	6	96.1
University of CA, Los Angeles	56.3	12	4	92.9	7	95.0
EPFL, Lausanne	41.4	109	4	92.9	7	95.0
Univ. of TX, Southwestern Medical Center	40.9	116	3	95.5	7	95.0
Columbia University	56.2	13	4	92.9	8	94.6
Imperial College London	55.2	14	4	92.9	9	94.3
University of CA, San Diego	54.0	18	3	95.5	9	94.3

5) dominance. In particular, in Fig. 2a we plot the results of column *Skyline Layer* of Table 3, and we show for each entity the skyline layer that it belongs. As seen on the plot, the skylines may tangent each other and also there are no vertical lines nor horizontal lines. In Fig. 2b we present the skylines based on strict dominance relationship according to the column *Strict Skyline layer* of Table 3. The skylines do not tangent each other and there are produced fewer layers than in the previous case. There exist vertical and horizontal lines. Finally, the *Majorized Skylines* are shown in Fig. 2c. The primary axes show the original values, whereas the secondary axes show the normalized ones. Although it seems that the lines follow the same notion as before, there are produced much more layers as well as the lines may not only be tangent but may intersect each other.

Table 2. Kendall τ, Spearman ρ, and Pearson r coefficients

	NTU	RR	Strict RR	Majorized RR
(a) Kendall τ				
NTU	–	0.79	0.80	0.71
RR	–	–	0.95	0.77
Strict RR	–	–	–	0.79
(b) Spearman ρ				
	NTU	RR	Strict RR	Majorized RR
NTU	–	0.91	0.92	0.86
RR	–	–	0.99	0.90
Strict RR	–	–	–	0.92
(c) Pearson r				
	NTU	RR	Strict RR	Majorized RR
NTU	–	0.78	0.78	0.75
RR	–	–	0.99	0.90
Strict RR	–	–	–	0.92

Table 3. A set of prominent professors of the Greek CS faculties.

Researcher	C	$h5$	C_{nrm}	$h5_{nrm}$	Rank order by C	Rank order by $h5$	Skyline layer	Strict skyline layer	Major. skyline layer
Ioannis Pitas	12568	25	100.0	73.5	1	3	1	1	1
G. Karagiannidis	8079	34	64.3	100.0	3	1	1	1	2
Nikos Hatziargyriou	11009	26	87.6	76.5	2	2	1	1	3
Petros Maragos	7499	17	59.7	50.0	4	12	2	2	4
John Psarras	3212	19	25.6	55.9	16	5	2	2	5
K.A. Antonopoulos	1744	20	13.9	58.8	22	4	2	2	5
Minos Garofalakis	4824	18	38.4	52.9	7	8	2	2	5
Y. Manolopoulos	5651	17	45.0	50.0	6	12	3	2	5
Ioannis Pratikakis	2729	19	21.7	55.9	17	5	3	2	6
Ioannis Vlahavas	3468	18	27.6	52.9	13	8	3	2	6
Aggelos Kiayias	5862	14	46.6	41.2	5	18	3	3	6
Aristidis Likas	4068	17	32.4	50.0	10	12	4	3	6
Konstantina Nikita	3405	18	27.1	52.9	14	8	4	2	7
St. Papathanassiou	2620	19	20.9	55.9	19	5	4	2	7
Yannis Theodoridis	3920	17	31.2	50.0	11	12	5	3	7
G. Tsoumakas	3383	18	26.9	52.9	15	8	5	2	8
Stefanos Kollias	4783	14	38.1	41.2	8	18	4	3	9
Anastasios Tefas	2675	17	21.3	50.0	18	12	6	3	9
Sergios Theodoridis	3535	16	28.1	47.1	12	17	6	4	9
Aggelos Bletsas	4285	13	34.1	38.2	9	21	5	4	10
Pavlos Georgilakis	2367	14	18.8	41.2	20	18	7	5	10
Sotiris Nikoletseas	2331	13	18.6	38.2	21	21	8	6	11

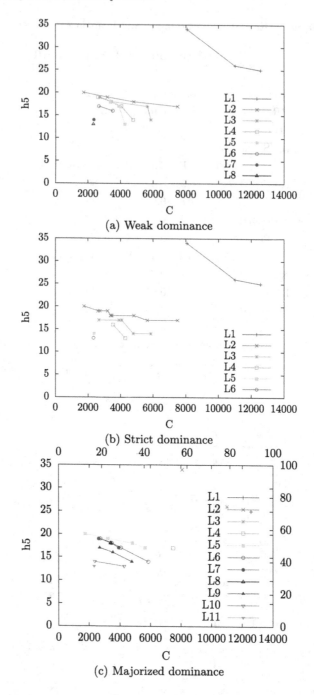

Fig. 2. Skylines under: (a) Weak, (b) Strict, (c) Majorized dominance.

5   Conclusion

In this paper we focus on university rankings and argue that listing universities in an absolute order does not have a substantial meaning. Instead, we propose to group universities in ordered equivalence sets. To this end, we apply the mathematical notion of majorization on the Skyline operator and the Rainbow Ranking method. We use a dataset extracted from the National Taiwan University (NTU) rank list, as well as on a dataset consisting of academic staff from Greek CS departments. The proposed method complements previous efforts to rank universities by using equivalence sets, which were based on the Skylines and Clustering. Further experimentation is necessary to compare these three approaches by exhaustively examining several university ranking, algorithms and datasets and provide a well-tuned method.

References

1. Angelis, L., Bassiliades, N., Manolopoulos, Y.: On the necessity of multiple university rankings. COLLNET J. Scientometr. Inf. Manage. **13**(1), 11–36 (2019)
2. Arnold, B.C., Marshall, A.W., Olkin, I.: Inequalities: theory of majorization and its applications. J. Am. Stat. Assoc. **76**(374), 492 (1981)
3. Borzsony, S., Kossmann, D., Stocker, K.: The skyline operator. In: Proceedings 17th International Conference on Data Engineering (ICDE), pp. 421–430 (2001)
4. Chester, S., Assent, I.: Explanations for skyline query results. In: Proceedings 18th International Conference on Extending Database Technology (EDBT). Brussels, Belgium (2015)
5. Hardy, G.H., Littlewood, J.E., Pólya, G.: Inequalities, 2nd edn. Cambridge University Press, London (1952)
6. Johnes, J.: University rankings: what do they really show? Scientometrics **115**(1), 585–606 (2018)
7. Manolopoulos, Y., Katsaros, D.: Metrics and rankings: myths and fallacies. In: Revised Selected Papers, 18th International Conference on Data Analytics & Management in Data Intensive Domains (DAMDID/RCDL), Moscow, pp. 265–280 (2017)
8. Sidiropoulos, A., Gogoglou, A., Katsaros, D., Manolopoulos, Y.: Gazing at the skyline for star scientists. J. Informet. **10**(3), 789–813 (2016)
9. Sidiropoulos, A., Stoupas, G., Katsaros, D., Manolopoulos, Y.: The rainbow over the Greek departments of computer science/engineering: a bibliometric study. In: Proceedings 21st Panhellenic Conference on Informatics (PCI), Larissa, Greece, pp. 1–6 (2017)
10. Sprent, P., Smeeton, N.C.: Applied Nonparametric Statistical Methods, 3rd edn. Chapman & Hall, Boca Raton (2001)
11. Stoupas, G., Sidiropoulos, A., Gogoglou, A., Katsaros, D., Manolopoulos, Y.: Rainbow ranking: an adaptable, multidimensional ranking method for publication sets. Scientometrics **116**(1), 147–160 (2018)
12. Stoupas, G., Sidiropoulos, A., Katsaros, D., Manolopoulos, Y.: Skyline-based university rankings. In: Proceedings of the ADBIS-TPDL-EDA Common Workshops and Doctoral Consortium, Lyon, France, pp. 347–352 (2020)

13. Stoupas, G., Sidiropoulos, A., Katsaros, D., Manolopoulos, Y.: Ranking universities via clustering. In: Proceedings of the 18th International Conference on Scientometrics & Informetrics (ISSI), Leuven, Belgium (2021)
14. Tiakas, E., Papadopoulos, A.N., Manolopoulos, Y.: Skyline queries: an introduction. In: Proceedings 6th International Conference on Information, Intelligence, Systems & Applications (IISA), Corfu, Greece (2016)
15. Van Raan, A.F.: Fatal attraction: conceptual and methodological problems in the ranking of universities by bibliometric methods. Scientometrics **62**(1), 133–143 (2005)
16. Voorneveld, M.: Characterization of Pareto dominance. Oper. Res. Lett. **31**(1), 7–11 (2003)

Inferring Missing Retweets in Twitter Information Cascades

Jennifer Neumann and Peter M. Fischer[(✉)]

University of Augsburg, 86159 Augsburg, Germany
{jennifer.neumann,peter.m.fischer}@uni-a.de

Abstract. Sharing (and thus amplifying) information on social media has an enormous impact, so there is a strong motivation to identify the paths on which information diffuses. Yet, in practice, there are many challenges which previous work address only to a limited degree. In our work, we systematically analyze the data set limitations and their impact, derive a damage model and present an method that utilizes the social graph neighbourhood to infer the missing messages. Initial results are promising, but there are still open questions on managing the huge search space to prioritize the most suitable results and reduce the cost, for which are investigating custom sorting strategies.

Keywords: Cascade reconstruction · Missing data · Twitter

1 Introduction

Social media have established themselves as one of the most important places for interaction and information exchange. Using means like sharing/retweeting, information can be spread to large audiences within short periods, amplifying and changing trends and opinions. Yet, as again observed in the ongoing pandemic, this also leads to unfounded rumours and outsize impact of questionable actors. Therefore, it is desirable to reliably and quickly capture the exact paths on which information spreads in order to identify sources, influential users or coordinated communities that push this information.

Studying such diffusion processes (also often called information cascades) has attracted significant interest (see e.g., the surveys in [1,7]), but so far the approaches tackling missing information such as restricted/deleted users, noisy data and API constraints have been limited in scope and effectiveness.

We currently focus our research on Twitter, as they are one of the few social media to provide access to a wide range of information types on a (somewhat) larger volume. The fidelity of their APIs has been studied [2,3], underpinning our own observations on the data quality.

The most prominent mean of sharing in Twitter are Retweets, which -on an API level- contain information about the user posting the original tweet but no information about the influence paths. Like other works, we rely on the message

© Springer Nature Switzerland AG 2021
L. Bellatreche et al. (Eds.): ADBIS 2021 Short Papers, Workshops and Doctoral Consortium,
CCIS 1450, pp. 287–292, 2021.
https://doi.org/10.1007/978-3-030-85082-1_25

sequence and the social graph to determine possible paths, deriving a diffusion graph. We observe that missing retweets are common and may cause altered or broken diffusion paths, thus changing the reconstructed cascades significantly.

The remainder of the paper is structured as followed: Sect. 2 provides more background information on relevant existing research. In turn, Sect. 3 describes briefly the algorithm for complete data (referred as one hop reconstruction). In Sect. 4 we describe the damage observed as well as as the model we derive on it, followed by our approach for incomplete data in Sect. 5 (referred as two hop reconstruction). The test setup and results are shown in Sect. 6. The paper concludes with research directions in Sect. 7.

2   Related Work

While studying information cascades has drawn considerable interest, the focus has been on conceptual models, offline or coarse-grained reconstruction and typically complete data, as shown in the extensive survey by Guille et al. [1]. Many researchers focus on cascade prediction instead (see survey by Zhou et al. [7]).

Missing data and information diffusion was the subject of only a few published papers so far. Most of them have other goals than reconstructing diffusion paths: [5] want to estimate size and diameter of complete cascades. [4] estimate the properties of an underlying social network given an incomplete set of nodes and information diffusion paths.

So far, only [8] reconstruct cascades with missing data, yet the work suffers from a number of limitations: Conceptually, it relies on the Independent Cascade (IC) model, which assumes that every active node n at a time t can only activate other notes at time $t+1$ and becomes completely inactive afterwards. This is not in line with observations on real-life cascades, since the original message (aka root) has typically significant impact and can trigger retweets for long periods, even after previous reactions have already become inactive.

Likewise, the authors assume that there is an upper bound for the diameter of the cascade, so they force paths to be below that length. Furthermore, they do not analyse whether their approach is suitable for real observed damage and do not show how a reasonable upper bound for the reconstruction tree can be found without knowing how many retweets will be added in the future.

Generally speaking, most approaches focus on coarse-grained cascades descriptions, may not be used for online reconstruction or make assumptions not in line with real, observed cascades. We aim to overcome these limitations.

3   Cascade Reconstruction for Complete Data

The approach introduced by Taxidou et al. [6] provides means to reconstruct fine-grained diffusion paths for retweets in an online fashion when data is mostly complete. Conceptually it relies on the assumption that information is spreading over network connections: Users incrementally forward the information, thus providing cascades that are temporal subgraphs of the social graph.

A newly received retweet T_{n+1} by a user U_{n+1} should be connected to the already given cascade C containing $\{T_1(U_1), T_2(U_2), ..., T_n(U_n)\}$ as soon as it is observed. For this purpose, we check if this user is connected to any of these users using the (directed) social graph, as information spreads to followers.

Due to its peculiar design, the Twitter API allows us to crawl those connections in both directions, so we can use follower lists (leading to Prefix iteration) or friend lists (leading to User iteration) (see Fig. 1). As the lists represent the same underlying structure, the sets of generated diffusion paths are the same. We are therefore free to choose the algorithm variant both on computational cost and data availability: If all lists are accessible, we select the shorter one, otherwise we can already compensate for some missing information.

As assessed in [6], this algorithm reliably gives plausible diffusion paths if the data is complete in terms of messages: In our data sets, we observe that above 90% of the retweets can be connected to the cascade if no messages are missing and the social graph is accessible. As long as popular users are well represented, gaps in the social graph had a minor effect. Missing messages, however, can have significant impact, breaking the cascade in fragments. To estimate how often this occurs we need to take a closer look at the observed damage on Twitter.

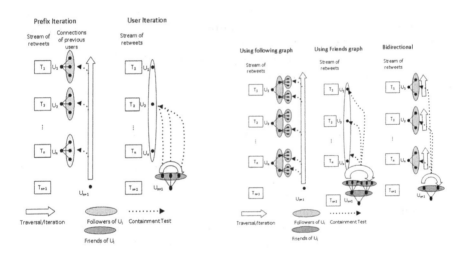

Fig. 1. Approaches for complete data.   **Fig. 2.** Approaches for incomplete data.

4 Observed Damage and Damage Model

As already pointed out, Twitter provides a rich set of APIs to collect data, but restricts the rates. This causes gaps, in particular in streaming data and popular hashtags. Further limitations arise due to privacy settings of users, deleted/banned users and "noise", i.e. smaller inconsistencies in the large volume of data. In practice, none of these can be fully avoided. Concerning the rate limits and noise [3] shows that using more than one crawler with the same filter does not lead to significant better data. While combining complementary hashtag filters may give better results [2], damaged cascades are common: When comparing the accessible messages with the retweet counter in different data sets (such about sports events like Olympics, disasters and scientific conferences), we obverse that for most cascades damage by noise or privacy settings amounts in 10 to 20% of the messages. In contrast, cascades affected by rate limiting are massively damaged, often lacking 50 or 60% of the messages. In such cases, only coarse-grained estimation of properties is feasible [5], while we aim for a fine-grained reconstruction of moderately damaged cascades.

Cascade reconstruction derives typically two types of nodes: Bridges connecting (larger) graph fragments and leaves not inducing new resharing activity. While leaves are more common, only missing bridges have a great impact on the cascade shape, connectedness and depth (see Fig. 3) - we therefore currently focus on inferring bridges. More precisely, our goal is to infer the missing bridge tweet and its edges, but avoiding false positives.

To evaluate our approach against a reference point, we damage existing, reconstructable cascades. Given a cascade with just one fragment (e.g. complete data given or all missing tweets are no bridges) our damage model removes one (or multiple) bridges. To detect the bridges we reconstruct the cascades with the one-hop approach in Sect. 3. Afterwards we delete a given number of bridge tweets randomly, matching noise or user damage.

When reconstructing such a damaged cascade with at least one tweet this one-hop reconstruction will fail and our two hop reconstruction (as described in the following section) will start.

5   Cascade Reconstruction for Incomplete Data

Our approach to extend [6] follows a seemingly simple intuition on the structure of diffusion processes and social graphs: Instead of just checking for a direct connection from the new tweet to the cascade we search for connections in the neighbourhood. We limit ourselves to two hops, matching the average path length between users. In multiple data sets we observe that between 75 and 90% of the gaps in the reconstruction can be filled using this idea, depending on the completeness of the neighbourhood in our social graph snapshot.

Like in the one-hop case, we may consider edges for followers, friends and -as a fallback- a bidirectional approach. All options lead to the same result, allowing use to choose on data availability and cost.

As shown in Fig. 2, we extend the one hop approach to be applied on each follower/friend of the original user, yielding all possible two hop connections between the cascade and the new retweet. Performing this idea naively, we may generate a large number of candidates and/or incur a high cost due to the size of such neighborhoods (for popular users or tightly connected clusters), while only a single node (or a small number) is/are correct. We therefore intend to limit the search space by sorting the follower and friends lists as well the prefix, thus enabling an early stop on the results. This way, we place the correct node early into the results and avoid the cost of irrelevant computations.

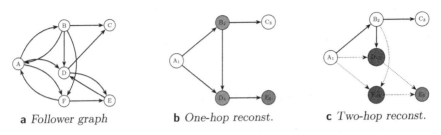

a *Follower graph*   **b** *One-hop reconst.*   **c** *Two-hop reconst.*

Fig. 3. One and two-hop reconstruction for the incomplete cascade after deleting D_4 for the given follower graph when adding E_5 to the cascade. The nodes are marked with the username and the timestamp of the tweet. All bridges nodes are green, the inferred nodes red, inferred edges dotted.

6   Proof of Concept Evaluation

In the first step want to show that relevant damage is present in real data. We use a data set about Olympics 2014, containing 34980 cascades with at least 10 retweets. Only 1.3% of those do not contain any damages, while for around 75% of the cascades less than 25% are missing according to the retweet count. Therefore approaches for incomplete data are necessary.

In the next step, we investigate our approach. To generate a sufficient number of test cases, we reconstruct the cascades using the one-hop method and keep only the fragment connected to the root. Now we modify these cascades using our damage model described in Sect. 4, removing one bridge per cascade, which yields around 10K cascades with at least 10 remaining retweets. Furthermore, we removed those cascades in which a significant part of two-hop social graph was missing, reducing the number to around 7.5K cascades.

In the first experiment, we estimate the consequences of a naive execution when all neighborhood candidates are generated. The correct node was always found, but the number of spurious candidates became rather high: For around 150 cascades at most 3 tweets are inferred, for 2.5K between 3 and 10, 2.4K between 10 and 25 and 2.45K above 25 (including 425 with at least 100 inferred nodes). To reduce the number of false positives we now limit the number of inferred tweets to 3. So for over 7K cascades sorting will influence the result.

Without any sorting of the prefix and the social graph we can find the correct node among the first 3 for 2K cascades (around 1.5K from between 3 and 10, 750 from between 10 and 25 and 100 above 25, also 3 with at least 100 inferred nodes). Even with this naive approach we are able to find the correct node for 27% of the cascades but -as expected- we are more successful for cascades with a small amount of possible inferred nodes.

After some experimentation, sorting the lists by friends count and the prefix by out-degree (both descending) provided the best results. Now we can reconstruct the correct node for around 3K cascades. The number of correctly reconstructed cascades increases especially for cascades with at most 10 reconstructed nodes. If we take a closer look at cascades in every category that are reconstructed correctly under specific sorting, this leads to the assumption that different cascade shapes solved best with different sorting.

7   Conclusion and Future Work

To reconstruct Twitter cascades an approach for incomplete data is needed because damaged data can not be avoided. Our main idea is using neighbourhood search and limiting the number of inferred nodes to suit the 'real' cascade. Our test shows that this approach yields to promising results, as the sorting of the friends and follower lists as well as the prefix can lead to better results. But especially for well connected nodes (which leads to a high amount of inferred nodes) more work has to be done. Currently we just tested simple sorting methods (sorting by user meta data like status count) for the social graph list or the prefix. In the future, we want to test advanced sorting approaches like pairwise influence or selecting the sorting method depending on the current cascade shape. Additionally we need to come up with a model to assess our solution for 'real' missing data and want to test our approach with different cascade models.

References

1. Guille, A., et al.: Information diffusion in online social networks: a survey. ACM Sigmod Rec. **42**(2), 17–28 (2013)
2. Ghosh, S., et al.: On sampling the wisdom of crowds: random vs. expert sampling of the twitter stream. In: Proceedings of the 22nd ACM International Conference on Information & Knowledge Management (2013)
3. Joseph, K., Landwehr, P.M., Carley, K.M.: Two 1%s don't make a whole: comparing simultaneous samples from Twitter's streaming API. In: Kennedy, W.G., Agarwal, N., Yang, S.J. (eds.) SBP 2014. LNCS, vol. 8393, pp. 75–83. Springer, Cham (2014). https://doi.org/10.1007/978-3-319-05579-4_10
4. Kossinets, G.: Effects of missing data in social networks. Soc. Netw. **28**(3), 247–268 (2006)
5. Sadikov, E., et al.: Correcting for missing data in information cascades. In: Proceedings of the 4th ACM International Conference on Web Search and Data Mining (2011)
6. Taxidou, I., Fischer, P.M.: Online analysis of information diffusion in Twitter. In: Proceedings of the 23rd International Conference on WWW (2014)
7. Zhou, F., et al.: A survey of information cascade analysis: models, predictions, and recent advances. ACM Comput. Surv. (CSUR) **54**(2), 1–36 (2021)
8. Zong, B., et al.: Inferring the underlying structure of information cascades. In: 2012 IEEE 12th International Conference on Data Mining. IEEE (2012)

ADBIS 2021 Doctoral Consortium

The ADBIS Doctoral Consortium

Mirjana Ivanović[1] and Olaf Hartig[2]

[1] University of Novi Sad, Serbia
[2] Linköping University, Sewden

The ADBIS 2021 doctoral consortium (DC) was a forum where PhD students working on ADBIS-related topics had a chance to present their research ideas and their current scientific achievements. They could gain inspiration and motivation, and receive useful feedback from their peers as well as from senior researchers. Moreover, the DC was also a forum for the PhD students to network and build cooperative relationships. The focus of the DC papers is on describing the current stages of PhD students' research and contributions in particular areas. The DC Committee accepted four papers for presentation at the conference and inclusion in the proceedings. The topics of these papers, discussed during the DC session, include semantically diverse constrained queries in location-based services, business-oriented smart contracts based on process mining, an architecture for a multi-stakeholder, interactive recommendation system, and the impact of dimensionality reduction on 3D feature descriptors.

Organization

Co-chairs

Mirjana Ivanović   University of Novi Sad, Serbia
Olaf Hartig   Linköping University, Sweden

Program Committee

Barbara Catania   University of Genoa, Italy
Bart Kuijpers   University of Hasselt, Belgium
Kamel Boukhalfa   USTHB, Algeria
Imen Megdiche   Institut de Recherche en Informatique de Toulouse, France
Jérôme Darmont   University of Lyon 2, France
Maja Zumer   University of Ljubljana, Slovenia
Michal Kompan   Kempelen Institute of Intelligent Technologies, Slovakia
Mikołaj Morzy   Poznan University of Technology, Poland
Oscar Romero   Universitat Politècnica de Catalunya, Spain
Peter Z. Revesz   University of Nebraska-Lincoln, USA
Sándor Laki   Eötvös Loránd University, Hungary
Theodoros Tzouramanis   University of Thessaly, Greece

Semantically Diverse Constrained Queries

Xu Teng$^{(\boxtimes)}$ and Goce Trajcevski

Department of Electrical and Computer Engineering,
Iowa State University, Ames, IA, USA
{xuteng,gocet25}@iastate.edu

Abstract. Location-Based Services are often used to find proximal *Points of Interest* (PoI) – e.g., nearby restaurants, museums, etc. – in a plethora of applications. However, one may also desire that the returned proximal objects exhibit (likely) maximal and fine-grained semantic diversity. For instance, rather than picking several close-by attractions with similar features – e.g., restaurants with similar menus; museums with similar art exhibitions – a tourist may be more interested in a result set that could potentially provide more diverse types of experiences, for as long as they are within an acceptable distance from a given (current) location. So far, we introduced a topic modeling approach based on Latent Dirichlet Allocation, a generative statistical model, to effectively model and exploit a fine-grained notion of diversity, based on sets of keywords and/or reviews instead of a coarser user-given category. More importantly, for efficiency, we devised two novel indexing structures – Diversity Map and Diversity Aggregated R-tree. In turn, each of these enabled us to develop efficient algorithms to generate the answer-set for two novel categories of queries. While both queries focus on determining the recommended locations among a set of given PoIs that will maximize the semantic diversity within distance limits along a given road network, they each tackle a different variant. The first type of query is kDRQ, which finds k such PoIs with respect to a given user's location. The second query kDPQ generates a path to be used to visit a sequence of k such locations (i.e., with max diversity), starting at the user's current location.

Keywords: Mobility · Diversity · Distance-bound

1 Introduction and Problem Statement

Since the late 1990s, many applications relying on *Location-Based Services* (LBS) have targeted the efficient search for *Points of Interest* (PoIs) – e.g., tourist attractions – relative to user's location. Since the cost (distance or travel-time) is an important factor when selecting PoIs, significant amount of research efforts have been invested into distance-oriented queries – e.g., range and *k-Nearest*

© Springer Nature Switzerland AG 2021
L. Bellatreche et al. (Eds.): ADBIS 2021 Short Papers, Workshops and Doctoral Consortium,
CCIS 1450, pp. 297–302, 2021.
https://doi.org/10.1007/978-3-030-85082-1_26

Neighbor (*k*NN) queries [5,6]. However, in addition to the proximity, the semantics of PoI is often an influential factor when planning one's motion and activities [23]. Modelling and querying of the semantic or activity trajectories have been a subject of intense research in the past decade [3,16,23] however, the semantic aspect was used to augment the traditional searches used in typical spatial and spatio-temporal queries.

In our work to date, we are tackling a novel variant of the problem – namely, coupling the *proximity* constraints (with respect to the user's location) with the *diversity* of the semantic descriptors of the PoIs.

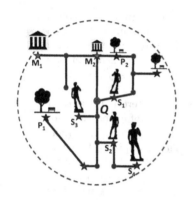

The scenario depicted in Fig. 1, illustrates a user at location Q, searching for $k = 3$ tourist attractions to visit. The user specifies a maximum distance that he/she is willing to travel. Processing this query could return the answer set $T_1 = \{S_1, S_2, S_3\}$, consisting of 3 nearest PoIs. However, one can readily see that in this case, all three returned PoIs are monuments/statues. If the user would like a more diverse experience, recommending these three sites would likely not be satisfactory. The concept of *diversity* in the spatial queries was introduced only recently [7,18].

Fig. 1. Example of diverse constrained query

*k***DRQ**: To cater to situations described above, we introduced a novel type of a query called *k-Diverse Range Query* (*k*DRQ), which aims at *maximizing the semantic diversity of the answer set of spatial queries within a bounded range*. Due to the hardness of the problem, we proposed approximate solutions (with slightly different variations of the constraint) in [18]. As a concrete illustration, in the context of Fig. 1, the user may have a preference for the answer set $T_2 = \{M_1, P_1, S_4\}$, which includes a statue (S_4), a museum (M_1) and a park (P_1) – within the desired distance bound.

The second part of our work is motivated by the observation that the existing approaches assume that the user will choose only one of the results, aiming at maximizing the diversity of the options of the user. However, there was no solution to the problem of determining whether there exists a path between all the PoIs that satisfies the range constraints as path. In this example, it is clear that, while any PoI in T_2 is within the spatial range, visiting all three of them relying on the existing road network (cf. Fig. 1) will exceed the distance limitation in terms of total travel.

*k***DPQ**: To address this limitation, we introduced another novel query, *k-Diverse Path Query* (*k*DPQ) in [17]. It finds a *path that maximizes diversity of PoIs along it, subject to the constraint that the length of the whole path is within user-specified limits*. Towards processing *k*DPQ, we propose two algorithms, along with an index structure, called *Diversity Aggregated R-tree* (DAR-tree), devised to improve the efficiency of the *k*DPQ processing. Note that the road network distance from user's location to any PoI is always $\geq$ to Euclidean distance (indicated by the blue dashed circle in Fig. 1), which is used for pruning, to speed up the processing of k**DPQ**.

In the rest of this paper, we survey related state-of-the-art in Sect. 2, and discuss the progress obtained so far as well as the plan for future work in Sect. 3.

2 State-of-the-art

Coupling motion and semantics have brought about the concepts of semantic and activity trajectories. Both the modelling aspects [3, 16] and the query processing aspects [10, 23], which combines spatial, temporal and descriptor contexts of the PoIs, along with transition mode (e.g., walk, drive), have been tackled. What separates our work from the aforementioned ones is that we are focusing on constructing a path that will be limited in its cost (travel-time or distance) along a road network, and will visit a collection of PoIs with highest diversity in terms of their semantic descriptors.

Finding the kNNs to a given query location such that the distance between any two points is greater than a predefined minimum diversity was introduced in [11], and selecting the most diverse set within a predefined radius in Hamming space is addressed in [1]. A k-similar diversification set which optimizes a linear functions combining the similarity (i.e., closeness) and diversity for a given trade-off between them has been studied in [19]. Monitoring the most diverse k-sized set over distributed sets was proposed in [4]. All these works have the goal of finding a k-cardinality subset of size k, among a set of candidates PoIs, that maximizes diversity. However, they do not consider the constrained travel along road networks and do not return any path for visiting the resulting PoIs.

Other recent results combining diversity and spatial contexts are presented in [7] in the context of NN queries, tackling the settings of optimizing the weighted sums of the constraints. The rationalé is to give users merely a set of diverse options, from which the user is expected to choose one. However, it is restricted within a path from a query location to a single PoI. The main difference with our works is that kDPQ generates a path that connects multiple PoIs while ensuring high diversity.

More distantly related approaches to spatial diversification include angular diversity [14] – which defines the nearest Surrounder Query to find the nearest objects from a query point from different angles; and the angular similarity – which have been used for diversified kNN problem in [13].

The idea of using keywords, i.e., a finer granularity in order to distinguish categories, to find diverse kNNs has been explored in [22]. The keywords are used for filtering data points, i.e., only points that contain *all* query keywords are considered. We use the concept of Latent Dirichlet Allocation as a more sophisticated notion of diversity based on the set of keywords that describe each object. Differently from the works above, we propose an indexing structure to speedup the processing of kDRQs.

3   Current Progress and Future Plan

We now present the progress to date and outline the plan for our future work.

Current Results: Three main results obtained so far are discussed below.

(1) Topic-based Diversity: Based on the attached descriptive items, the semantic description of each PoI p_i is illustrated by a vector of probability (topic) distribution θ_i whose length is the number of latent topics K. $\theta_{i,j}$ $(1 \leq j \leq K)$ represents the probability of p_i belonging to topic j. For a set of PoIs $P = \{p_1, ..., p_{|P|}\}$, we define a vector ProbDiv(P) that stores, for each topic j, the probability that it is covered by P as ProbDiv$(P)_j := 1 - \prod_{p_i \in P}(1 - \theta_{i,j})$, which is then aggregated into a diversity score, Topic-based Diversity, via expected number of topics covered: TBD$(P) = \sum_{j=1}^{K}$ ProbDiv$(P)_j$.

Intuitively, the probability $1 - \theta_{i,j}$ indicates that PoI p_i does not cover topic j. Exploiting that PoIs are stochastically independent, $\prod_{p_i \in P}(1 - \theta_{i,j})$ is the probability that none of the PoIs in P covers topic j. We define ProbDiv$(P)_j$ as the counter-probability, i.e., the probability of the complementary event that at least one PoI in P covers topic j. Finally, these probabilities are aggregated into the expected number of topics covered by P – i.e., TBD(P).

(2) k-diverse Range Query: Our query processing approach for kDRQ, presetns a novel index structure and algorithms (more details in [18]).

To efficiently support kDRQ on spatial networks, we propose a specialized index structure, *Diversity Map Index* (DivMap), which is inspired by the concept of distance signatures [9]. At each node v of the road network, the algorithm places a virtual "signpost". Pointing in each direction that can be taken from v, it gives a distance-approximated summary of the topics that can be found following that direction. In particular, this information includes the maximum topic values of sites found in each direction, and for each distance bucket (e.g., $0 \sim 500\,\mathrm{m}, 500\,\mathrm{m} \sim 1,000\,\mathrm{m}$, etc.), as defined by the distance signature. Intuitively, not only does it tell the algorithm which direction to follow to find interesting topics, but it also gives an approximation of how far it will have to travel to find these PoIs.

Instead of using a naive Dijkstra search to find all the PoIs inside the query range, our algorithm exploits DivMap index to greedily direct the search to sites that locally complement the diversity of the k-most diverse sites that were found so far. Similarly, we maintain a priority queue of "active" edges adjacent to vertices that have already been explored and processed. In each iteration, we greedily select an active edge which maximizes the utility given the currently best k-set of sites, based on topic-based diversity.

(3) k-diverse Path Query: To efficiently answer kDPQ, we adopt informed search with our proposed index structure (more details in [17]). The heuristic function we used during informed searching is: $h(v, \varepsilon) := $ ProbDiv$(P_E[v, \varepsilon])$, where $P_E[v, \varepsilon] = \{p \mid \|v, p\|_2 \leq \varepsilon\}$ denotes the set of all PoIs having Euclidean distance from node v at most ε. Let $P[v, \varepsilon]$ denote the set of all PoIs whose road-network distance from v at most ε. Note that Euclidean distance is always less or equal

to the network distance, thus the diversity from heuristic function must obtain an upper bound of $P[v, \varepsilon]$, which is utilized to prune unnecessary search space.

The main point of utilizing Euclidean distance in heuristic function is to leverage an R-Tree [8] to efficiently obtain the set $P_E(v, \varepsilon)$ while avoiding expensive network exploration to obtain the set $P(v, \varepsilon)$ by, for example, Dijkstra algorithm. We introduce a *Diversity Aggregated R-Tree* (DAR-Tree) to accelerate the computation of heuristic function, which stores the information related to probabilistic diversity of each *Minimum Bounding Rectangle* (MBR) in both leaf and inner nodes. Abstractly speaking, instead of always recursively iterating all the way down to the leaf node, we can terminate the search if an MBR of non-leaf node has already been fully contained by the searching region.

We proposed two variants of searching strategies, utilizing the heuristic function and DAR-Tree to estimate the expected gain from a potential direction [17]: (1) prune the search space by only considering simple paths, i.e., paths that do not visit the same node twice; (2) allow re-visiting nodes but a directed edge cannot be visited more than once. While the constraint variant (1) gains in efficiency, it must be noted that the most diverse path may very well be non-simple, for example if the path visits a PoI located in a dead end (a node of degree one).

Future Work: In the next stages of the PhD studies, we plan to extend the present results in several directions, each bringing theoretical (and even paradigm based) challenges, and addresses several problems of practical relevance. More specifically, our plan includes: (1) An immediate extension on how to tackle: – incorporating ranking to PoIs; – considering dynamic scenarios (i.e., ones involving updates in the semantic attributes such as menus in restaurants at a given time of day), as well as spatio-temporal attributes (e.g., the travel-time along a road segment). (2) To our knowledge, the problem of detecting a most suitable initial location (i.e., a hotel to be booked) for a given set of diversity and distance constraints, has not been addressed in the literature. While it may have some (limited) analogy with the (reverse) kNN query – the family of works addressing reverse kNN query and variants [21] have never considered any semantic diversity aspect. (3) Another challenge is in the realm of spatial crowdsourcing [15]. Despite the recent popularity of the topic, the large body of works have not addressed the issue of efficient and scalable solutions to the problem of re-calculating the assignments of workers to tasks when some of the spatio-temporal parameters used in the original assignments changes. This could happen, for example, when a worker takes longer than planned to complete a task in a particular job-site or when the traffic anomaly (e.g., and accident) causes a significant delay in the trip needed for a worker to get from one job site to another. (4) A broad category of challenges stems from the recent popularity of the learnt indexes [2]. Namely, most of the works in this recent topic are focusing on traditional DB learned indexes [12,20] – however, to our knowledge, the problem of combining numerical and categorical attributes in learned indexes (and their updates) has not been addressed.

Acknowledgment. Research supported by NSF SWIFT grant 2030249.

References

1. Abbar, S., et al.: Diverse near neighbor problem. In: SOCG, pp. 207–214 (2013)
2. Al-Mamun, A., Wu, H., Aref, W.G.: A tutorial on learned multi-dimensional indexes. In: ACM SIGSPATIAL (2020)
3. Alvares, L.O., Bogorny, V., Kuijpers, B., de Macedo, J.A.F., Moelans, B., Vaisman, A.: A model for enriching trajectories with semantic geographical information. In: ACM GIS, p. 22. ACM (2007)
4. Amagata, D., Hara, T.: Diversified set monitoring over distributed data streams. In: DEBS, pp. 1–12. ACM (2016)
5. Bao, J., Chow, C.Y., Mokbel, M.F., Ku, W.S.: Efficient evaluation of k-range nearest neighbor queries in road networks. In: MDM, pp. 115–124. IEEE (2010)
6. Benetis, R., Jensen, C.S., Karĉiauskas, G., Saltenis, S.: Nearest and reverse nearest neighbor queries for moving objects. VLDB J. **15**(3), 229–249 (2006)
7. Costa, C.F., Nascimento, M.A., Schubert, M.: Diverse nearest neighbors queries using linear skylines. GeoInformatica **22**(4), 815–844 (2018)
8. Guttman, A.: R-trees: a dynamic index structure for spatial searching. In: ACM SIGMOD (1984)
9. Hu, H., Lee, D.L., Lee, V.: Distance indexing on road networks. In: VLDB, pp. 894–905. VLDB Endowment (2006)
10. Issa, H., Damiani, M.L.: Efficient access to temporally overlaying spatial and textual trajectories. In: MDM, pp. 262–271 (2016)
11. Jain, A., Sarda, P., Haritsa, J.R.: Providing diversity in k-nearest neighbor query results. In: PAKDD, pp. 404–413 (2004)
12. Kraska, T., Beutel, A., Chi, E.H., Dean, J., Polyzotis, N.: The case for learned index structures. In: Proceedings of the 2018 International Conference on Management of Data, pp. 489–504 (2018)
13. Kucuktunc, O., Ferhatosmanoglu, H.: λ-diverse nearest neighbors browsing for multidimensional data. TKDE **25**, 481–493 (2013)
14. Lee, K.C.K., Lee, W.C., Leong, H.V.: Nearest surrounder queries. In: ICDE, p. 85 (2006)
15. Neto, F.R.A., Santos, C.A.: Understanding crowdsourcing projects: a systematic review of tendencies, workflow, and quality management. Inf. Proc. Manag. **54**, 490–506 (2018)
16. Parent, C., et al.: Semantic trajectories modeling and analysis. ACM Comput. Surv. (CSUR) **45**, 42 (2013)
17. Teng, X., Trajcevski, G., Kim, J., Züfle, A.: Semantically diverse path search. In: IEEE MDM (2020)
18. Teng, X., Yang, J., Kim, J.S., Trajcevski, G., Züfle, A., Nascimento, M.A.: Fine-grained diversification of proximity constrained queries on road networks. In: SSTD, pp. 51–60 (2019)
19. Vieira, M.R., et al.: On query result diversification. In: ICDE, pp. 1163–1174. IEEE (2011)
20. Wang, H., Fu, X., Xu, J., Lu, H.: Learned index for spatial queries. In: IEEE MDM (2019)
21. Yang, S., Cheema, M.A., Lin, X., Wang, W.: Reverse k nearest neighbors query processing: experiments and analysis. In: Proceedings of the VLDB Endow, pp. 605–616 (2015)
22. Zhang, C., et al.: Diversified spatial keyword search on road networks. In: EDBT, pp. 367–378 (2014)
23. Zheng, K., Shang, S., Yuan, N.J., Yang, Y.: Towards efficient search for activity trajectories. In: ICDE, pp. 230–241. IEEE (2013)

An Approach to Validation of Business-Oriented Smart Contracts Based on Process Mining

Vladimir Ivković[1]($\boxtimes$) and Ivan Luković[2]

[1] Faculty of Technical Sciences, University of Novi Sad, Novi Sad, Serbia
vladimir.ivkovic@uns.ac.rs
[2] Faculty of Organizational Sciences, University of Belgrade, Belgrade, Serbia
ivan.lukovic@fon.bg.ac.rs

Abstract. Enterprises are confronted with a highly competitive and fast-changing business environment resulting in an increasing level of cooperation between organizations. Cross-organizational processes (CBPs) aim to achieve an agreement between organizations at the process level in such an environment. An application of smart contracts and blockchain brings the unprecedented capability to execute CBPs in decentralized multi-peer environments and without central authorities to govern the workflow. However, the application of blockchain introduces various challenges due to the immutable nature of blockchain's transaction data and the specific development and deployment life-cycle of smart contracts. In this position paper, we propose a high-level platform-independent framework for monitoring and validation of smart contracts utilizing process mining techniques. The main purpose of such a framework will be to support process engineers in the detection of nonconforming behavior, resolving detected discrepancies, and improving smart contracts.

Keywords: Cross-organizational business processes · Smart contracts · Process mining · Blockchain technology

1 Introduction

In recent years, smart contracts, along with blockchain technology, are seen as an opportunity for implementing multiparty collaborations and an intense exchange of information along multiple channels in a business environment. However, the application of blockchain introduces various challenges including immutability of transaction data, specific development and deployment life-cycle of smart contracts, and diversity of available blockchain platforms with different features.

This paper has been supported by the Ministry of Education, Science and Technological Development through the project no. 451-03-68/2020-14/200156.

L. Bellatreche et al. (Eds.): ADBIS 2021 Short Papers, Workshops and Doctoral Consortium, CCIS 1450, pp. 303–309, 2021.
https://doi.org/10.1007/978-3-030-85082-1_27

1.1   Cross-Organizational Business Processes on Blockchain

Nowadays, enterprises are facing a highly competitive global business environment resulting in an increasing level of cooperation between organizations. That leads to the necessity of implementing interoperable software systems and efficient modeling of cross-organizational business processes (CBPs) [1]. Blockchain technology and smart contracts have the potential to improve CBPs since the applications, which needed centralized architectures or trusted third parties to verify them, can now operate in a decentralized way with the same level of certainty [2]. Some advantages of smart contracts compared with conventional contracts are: (i) reducing risks, (ii) cutting down administration and service costs, and (iii) improving the efficiency of business processes [3].

Suitable mechanisms need to exists in order to enable monitoring of CBP execution. Such mechanisms should provide sufficient feedback for various phases in a life-cycle of CBP, including design, development, simulation, execution, and validation. Having CBPs executed on the blockchain, transaction data represents a raw form of *event log* for process instances that are executed via smart contracts. Thus, monitoring and validation can be performed using *process mining* techniques [4] over blockchain transaction data.

1.2   Smart Contracts Validation

Smart contracts rely on a non-standard software life-cycle and delivered applications can hardly be updated by releasing a new version of the software [5]. This implies that validation and verification of smart contracts is a crucial step in the development process in order to minimize risk in the production phase. A distributed ledger provides a single source of truth about all activities contained within transactions and can serve as a source of event data. Therefore, an application of process mining can lead to a better understanding of the actual usage of smart contracts. Our goal is to make use of transaction data to establish an approach to the validation of smart contracts using process mining.

In this paper, we present a high-level and platform-independent approach to the validation of smart contracts based on process mining techniques assuming that a smart contract represents an implementation of a multi-party business process. Applying the proposed approach will help in (1) reducing the risk of deploying a faulty smart contract onto blockchain network and (2) analyzing existing smart contracts' behavior. The remainder of this paper is structured as follows. Section 2 contains present related works available in the literature. In Sect. 3, research challenges and potentials are listed and discussed. Section 4 describes proposed high-level approach. The paper concludes with Sect. 5 including implications for further research.

2   Related Work

In recent years, the first significant research efforts appeared addressing the possibility of the application of process mining on blockchain event data. We selected the three most relevant papers from a plethora of related work.

Mühlberger et al. [6] presented a proof-of-concept software artifact and its application on a data set produced by the smart contracts of a process execution engine stored on the public Ethereum blockchain network. Their approach requires manually injecting knowledge into the extraction algorithm and it is mainly restricted to the Ethereum blockchain.

Klinkmüller et al. [7] proposed the framework with three main parts: (i) a manifest specifying how data is logged into the blockchain, (ii) an extractor for retrieving data in XES format, and (iii) a generator that produces logging code to support smart contract developers. They focused on Ethereum and disregarded other blockchain systems which might use different logging mechanisms, potentially requiring a generalization of the proposed framework.

Duchmann et al. [8] presented an approach for the extraction of meaningful event logs from Hyperledger Fabric blockchain, as well as sample results of process mining on extracted logs. Their extraction approach is based on the comparison of block writes and transcribing each block as a trace of an event log in CSV format.

Most research efforts are related to blockchain systems that are not initially designed to support smart contracts that represent CBPs. The majority of papers barely discuss the process awareness of smart contracts and do not cover major challenges in event log extraction such as event and trace identification.

3 Research Challenges and Potentials

In this section, we discuss and analyze general challenges in adopting process mining in the context of blockchain. These challenges represent a result of preliminary research conducted to find associations between concepts used in blockchain systems and in process mining. Identified challenges range from extraction of event data to the application of process mining on the extracted event logs.

3.1 Event Log Extraction and Data Quality

One of the most important preconditions for the application of process mining is the existence of a well-structured and meaningful event log. In the blockchain context, each business process activity can be related to one or more smart contract functions. Minimal requirements for constructing an event log in a standard XES format [4] include the existence of events with activity names and timestamps. Events, related to the same process instance, are grouped in a trace that represents a sequence of events and is identified by a *caseId* [4].

Challenge 1.1 - *Extraction of event attributes.* All events derived from a particular transaction can utilize a transaction identifier (*txId*) and mandatory timestamp, as well as attributes from objects that the smart contract function reads from or write to the shared distributed ledger.

Challenge 1.2 - Trace discovery. As each event needs to have an activity name, one possible approach is to use the name of the smart contract's function. However, smart contracts are usually non-process-aware or do not leave the information on particular process instances that are executed.

Challenge 1.3 - Distinguishing multiple event logs. Sometimes, a group of dependent CBPs is implemented as a single smart contract. Therefore, criteria for distinguishing event logs for different CBPs should be defined with help of the domain experts. Otherwise, the application of process mining will result in discovering an extremely complex and unrealistic business process model.

Challenge 1.4 - Off-chain data. To get a larger picture of its processes, a collaborating party may want to include event logs from its internal information system which are related to CBP of interest. That will require unifying event logs from both sources. Eventually, process models discovered by process mining techniques will provide a comprehensive view of the particular CBP and other internal business processes that are related to it.

Challenge 1.5 - Event data quality. For example, transaction data contains only a timestamp representing a moment of appending the transaction to the shared ledger. Consequently, the actual timestamp when the invocation of the transaction happened remains missing.

3.2   Business Process Monitoring and Validation

The importance of validation of business-oriented smart contracts comes from its specific purpose in a distributed environment with multiple parties involved. The goal of the validation process is to find potential discrepancies between actual and expected behavior. Actual behavior is derived from the blockchain transactions and expected behavior is described by a business process model.

Challenge 2.1 - Process model discovery. A business process model can be utilized in a (semi-)automated generation of smart contract source code or for a better understanding of CBP. Sometimes, a smart contract exists without any related business process model, and process model discovery techniques can help to find the most appropriate process model. A real-time monitoring platform for blockchain-based CBPs should provide all kinds of insights about the execution of smart contracts and other events on the blockchain network.

Challenge 2.2 - Upgrading smart contracts. Offline process mining helps in detecting discrepancies that should be further analyzed in order to find out if there is nonconforming behavior. Resolving issues caused by nonconforming behavior mainly include implementational fixes and redeployment of smart contracts. Considering that Challenges 2.1 and 2.2 include potentially complex obstacles, these challenges should be further refined in the future work.

4   Process Mining-Based Approach

Our approach aims to address a majority of the challenges identified in the previous section in order to acquire high-quality event logs and enable an application of process mining techniques. To achieve that, we propose a framework with the following principal components (Fig. 1):

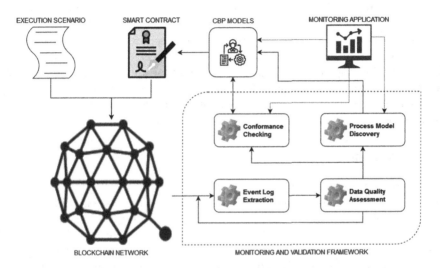

Fig. 1. The proposed framework for monitoring and validation of smart contracts.

- **Event log extraction from distributed ledger data.** Blockchain data, including blocks, transactions, and their attributes will be transformed into an event log in a standard XES format. That transformation will be conducted automatically by predefined rules for the discovery of events, traces, and their attributes. Furthermore, the proposed approach should support a semi-automatic event log extraction based on user-defined mappings. These mappings will enable extending or overriding predefined discovery rules, and they are supposed to be designed by a domain expert.
- **Event log data quality assessment.** Due to challenges 1.1-1.5, a transformation of object-centric data to event-oriented data needed in most process mining techniques can result in serious data quality issues. Thus, clear and comprehensive guidelines to assess the data quality of extracted event logs will be determined. These guidelines should include general rules for easier identification of most frequent data quality problems. Particular guidelines will depend on a chosen blockchain platform and the abilities of that platform to capture and represent events closer to the XES meta-model.
- **Business process model discovery using the extracted event log.** An application of the process discovery techniques will result in business process models usually represented as Petri Net or Business Process Model and Notation (BPMN).

- **Conformance checking of CBPs.** Conformance checking will point out parts of CBP that should be examined. Eventually, that will result in the modification of the smart contract's source code. Consequently, this implies informing all collaborating parties about necessary changes and deploying a new version of the smart contract.

Even though most segments of the validation process will be automated, the role of the process engineer is desirable. A process monitoring application will serve as a user interface designed to inform process engineers about all kinds of events happening throughout a system. As already discussed in *Challenge 2.2*, making changes in smart contracts can be an unexpectedly difficult task. Therefore, a systematic platform-independent method for applying such adjustments will be designed. It should provide clear mechanisms for fixing nonconforming behavior and redeployment of updated smart contracts through versioning of smart contracts.

5   Conclusion

In this paper, we presented a high-level approach to the validation of smart contracts using process mining methods. Various challenges regarding process-awareness of smart contracts, event log extraction, detection of misbehavior, and upgrading and redeployment of smart contracts are identified. The proposed framework for monitoring and validation of smart contracts can incorporate solutions for these challenges. Finally, process engineers and other people involved in managing CBPs over blockchain networks will benefit from such a framework. The evaluation process of the proposed framework will be conducted throughout several case studies involving multiple organizations with their CBPs.

The proposed approach will help in closing the gap between process design and analysis on one side, and blockchain-based process execution on the other side. Eventually, such a framework will enable better insight into how the process definition is deployed and executed on a target blockchain network with the capability of identifying and fixing non-conforming behavior.

References

1. Ziemann, J., Matheis, T., Freiheit, J.: Modelling of cross-organizational business processes. Enterp. Model. Inf. Syst. Archit. **2**(2), 9 (2007)
2. Di Ciccio, C., et al.: Blockchain-based traceability of inter-organisational business processes. In: Shishkov, B. (ed.) BMSD 2018. LNBIP, vol. 319, pp. 56–68. Springer, Cham (2018). https://doi.org/10.1007/978-3-319-94214-8_4
3. Zheng, Z., et al.: An overview on smart contracts: challenges, advances and platforms. Future Gener. Comput. Syst. **105**, 475–491 (2020). https://doi.org/10.1016/j.future.2019.12.019
4. van der Aalst, W.: Process Mining. Springer, Berlin, Heidelberg (2016)
5. Destefanis, G., et al.: Smart contracts vulnerabilities: a call for blockchain software engineering? In: 2018 IWBOSE, Campobasso, pp. 19–25 (March 2018). https://doi.org/10.1109/IWBOSE.2018.8327567

6. Mühlberger, R., et al.: Extracting event logs for process mining from data stored on the blockchain. In: Business Process Management Workshops, pp. 690–703 (2019). https://doi.org/10.1007/978-3-030-37453-2_55

7. Klinkmüller, C., et al.: Mining blockchain processes: extracting process mining data from blockchain applications. In: BPM: Blockchain and Central and Eastern Europe Forum, pp. 71–86 (2019). https://doi.org/10.1007/978-3-030-30429-4_6

8. Duchmann, F., Koschmider, A.: Validation of smart contracts using process mining. In: ZEUS. CEUR Workshop Proceedings, vol. 2339, pp. 13–16 (2019)

Challenges in Lifelong Pathways Recommendation

Nicolas Ringuet[1,2(✉)]

[1] University of Tours, Tours, France
`nicolas.ringuet@etu.univ-tours.fr`
[2] Neolink, Blois, France
`nicolas.r@neolink.link`

Abstract. Social services are notoriously complex to model and manage. It is essential to turn to mutualized systems that are tailored for the actors of these services. While examples of tools in the general public are legion today, there are still very few dedicated solutions in the social work and health sector, and even fewer offering long-term care for users. The objective of this thesis is to propose a multi-stakeholder and interactive recommendation system allowing to recommend social projects in the form of ordered set of actions, which we will qualify in this paper as lifelong pathways. This paper sets the basis of such a system and present preliminary results: (i) a tentative modeling of the system, by presenting the different actors and components, (ii) first experiments to show the complexity of this long-term composite items recommendation problem, (iii) a list of challenges that motivates the need for new solutions.

Keywords: Lifelong pathways · Recommender system · Chain composite items · Multi-stakeholders

1 Introduction

We consider in this work the problem of building a recommendation system to support social actors and beneficiary users in the interactive co-construction of a personal lifelong project, for example the assistance of job seekers, or elderly home support. This problem is illustrated by the French active solidarity income (RSA) that concerns 2.06 millions low resources people in 2020 in France. These beneficiaries are supervised by social actors in order to build their project to improve their social integration or their financial situation. Until now, these procedures and this support are carried out mostly manually and administered by each administrative region. As these procedures are complex, regions turn to alternatives allowing them to facilitate certain aspects of this system. In this context, the Neolink company proposes a tool dedicated to the social actors to support the beneficiaries of this social benefit in the definition of their socio-professional project including the search for a job. This multi-stakeholders system is currently designed to allow the different social actors to collaborate and

© Springer Nature Switzerland AG 2021
L. Bellatreche et al. (Eds.): ADBIS 2021 Short Papers, Workshops and Doctoral Consortium,
CCIS 1450, pp. 310–316, 2021.
https://doi.org/10.1007/978-3-030-85082-1_28

continue their work in a centralized manner. As such, the objectives are to enrich user experience and stakeholders interaction [8] by mutualizing the best practices among the social actors [3] and by helping them build a personal project for the beneficiaries from a very large set of possible actions related to health, social or professional aspects. We denote in this work, without loss of generality, the long-term sequence of actions, being it related to personal, medical, social or professional objectives, proposed to beneficiaries as their *lifelong pathways*.

Several research challenges are expected for devising such a complex recommendation system:

1. modeling the context of recommendation: what are the actors, their particularities and their interactions with one another [8] and how are they expected to interact with the recommender system?
2. dealing with the multi-stakeholders context: how to conciliate the objectives or constraints of each stakeholder among beneficiaries and the different social actors or services providers? This translate in the definition of the objective function to be optimized by our system to conciliate the different actors [4].
3. formalizing the recommender output: we consider that lifelong pathways are specific composite items that are built from atomic actions related to health, professional or personal objectives of a beneficiary. These combinations can be sets, sequences or more complex structures depending on the use case and will be denoted in this paper as stages. Several works already exists on the problem of building composites items for travel package planning [9], crowdsourcing tasks recommendations [7] or museum experiences [2]. Generating chain composite items [8] seems a relevant problem to our case.
4. providing a meaningful recommendation by determining the most adapted semantic and time granularity level for each action that is part of a composite item. Indeed, the expression of each action should be a balance between generality, for the ease of reuse, and the precision, to achieve meaningful recommendations for a specific beneficiary;
5. Ensuring the fairness of the recommendations and providing an explanation mechanism of the recommendations [5] to increase trust of stakeholders in the proposed composite items and to improve the acceptance of our solution as decisions taken impact directly the life of the beneficiaries and the workflow of social actors.
6. evaluating a multi-stakeholders, interactive lifelong pathways recommendation system, by determining the quality of the projects built by the system but also the quality of each of the recommended stage.

The objective of this paper is to propose a model involving all actors and their interactions with the envisioned recommender system (challenge 1) and to produce a preliminary classification model to predict meaningful actions for a user in the context of lifelong pathways recommendation. Noticeably, we rely on this prediction model to propose a first solution as a chain composite items recommender system (challenge 3) [1].

The rest of the document is organized as follows. Section 2 presents the modeling of the recommender system's context, while Sect. 3 announces the first results obtained on this recommendation task as a classification problem. Finally, Sect. 4 concludes and discusses how to deal with the aforementioned research challenges.

2   Recommender in Its Context

Figure 1 introduces an activity diagram of a preliminary model for lifelong pathways recommendation in the specific context of the RSA use case. This diagram represents a modeling of the real processes observed from a careful study of our RSA data:

1. `Initialization`: a *beneficiary* enters the system and defines mid to long-term objectives with the help of an *expert* that conducts a personal diagnosis ; this starts the lifelong pathways of the beneficiary.
2. `Stage`: a *supervisor* co-constructs with the beneficiary a combination of actions that is denoted as *stage*. These actions, after mutual agreement and validation, are then performed by the beneficiary.
3. `Report`: actions taken during the previous stages are evaluated and the beneficiary path is updated accordingly. If some of the long-term objectives are not met, the whole process is reproduced from the diagnosis step in the Initialisation layer.

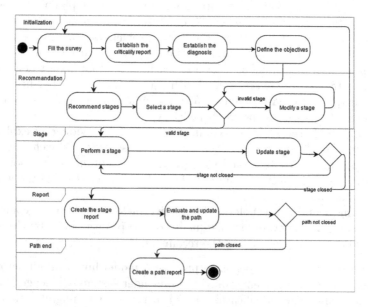

Fig. 1. Activity diagram of a generic lifelong pathways recommender system

Clearly, our main scientific challenges lie in the layer denoted `Stage` where a supervisor has to choose from a very large set of actions the ones to propose to the beneficiary. This step is the one for which we introduce the recommendation system that helps building stages as composite items or combination of actions. Interestingly, this recommender system has to take into account several stakeholders: the beneficiary, the supervisor and the expert as well as in the case of RSA, the funding entity (the region) and the preferences of actions providers (for example, a social service training people to write their CV).

This leads to the introduction in our model of a new `Recommendation` layer for the automatic proposal of several combinations of actions to the supervisor. If these combinations match the context hard constraints (related to the date, the availability or the cost of the actions), the supervisor can establish a tailored proposal for the beneficiary based on those recommendations. As a consequence, recommendations have to be explainable since the supervisor may have to explain to the beneficiary how they were devised. Another challenge pertains to the interactions of the supervisor with the recommended stages, when selecting the most appropriate actions combinations, and how to transfer back this knowledge to the recommender system so as to propose more personalized recommendations in the future.

Finally, it is important to notice from Fig. 1 that the objective of this research is not to replace any of the current social actors involved in the process but to facilitate their work in building better proposals with the beneficiary in mind. Final decisions are left to the supervisor and the beneficiary, the supervisor having both the roles of intermediate target for the recommendation and source of recommendation for the beneficiary, the latter being the end user.

3   Lifelong Pathways Actions Prediction Problem

These preliminary experiments introduce a method to predict if an action or a set of actions is meaningful and should be part of lifelong pathways.

Experimental protocol. We use datasets detailing for each beneficiary a diagnosis containing descriptive features and a project proposed by a supervisor as a set of recommended actions. From this set of manually suggested actions, it is possible to define 2 multiple binary classification problems: **1** a problem where we predict for each action if it should be part of the project for a specific user or, **2** a problem where each unique set of actions is considered as a class. Note that Problem **2** mimics the recommendation of composite items. Due to space limitations, this paper only reports results obtained with a Naive Bayes model. For each problem, each period and each class label (being an action or a set of actions) a 10-fold cross-validation is run and averaged depending on the classical evaluation metrics listed in Tables 1 and 2.

Datasets. We consider 2 "RSA" datasets D_1 and D_2, with respectively 419 and 2812 instances, D_1 being a subset of D_2 to test the sensitivity of the prediction to the number of available instances. Each dataset is composed of user diagnoses with 30 descriptive features that represent a set of context-aware information such as the need for housing assistance, the need for child care, health diagnosis, being a job seeker or not, etc. A data cleaning was conducted to replace null values with the most frequent ones. The number of unique actions varies from 46 in D_1 to 56 in D_2 when considering classification Problem 1. In the case of Problem 2, there are 129 unique combinations of actions in D_1 and 1108 in D_2. These problems translate into numerous binary classification problems that are sometimes represented by too few examples to allow good performance of classifiers. For this reason, we expect better performances on Problem 1 than on Problem 2.

Results. Our results confirm that predicting set of actions is a complex problem for several reasons. First, because of our problems formulation, class unbalance is very important as revealed by strong differences observed between the f-measure and the accuracy, in Table 1. Indeed, if we consider weighting each class by its cardinality (see Table 2), we observe much better results. Second, we observe a trade-off between the number of training instances with the advantage of accessing more user data, and the difficulty to handle the numerous (combinations of) actions that were recommended to them. Indeed, it can be observed that Problem 1 can be solved more efficiently on larger D_2 dataset with significantly better precision (0.861) and recall (0.686) than for dataset D_1 (both scores below 0.5). However, this result does not hold for Problem 2 because of the explosion in the number of classes. This observation stands even with weighted measures as in Table 2 which illustrates the difficulty of our problem. A general conclusion of these experiments is that it's possible to learn a recommender system for the most frequent (combinations of) actions as indicated by the weighted scores in Table 2. However, the task becomes complex when dealing with minority users, and is close to intractable for Problem 2 on both datasets, i.e. when recommending combinations of actions that correspond to the composite items that we envision, as observed in Table 1. As shown by the decrease in performance in this case when scaling from dataset D_1 to D_2, this solution is not feasible in the context of larger datasets.

Table 1. Prediction accuracy depending on the classification problem.

Problem	Dataset	Instances	Classes	Recall	Precision	F1	Accuracy
P1	D_1	419	46	0.298	0.479	0.351	0.994
P2	D_1	419	129	0.089	0.095	0.089	0.997
P1	D_2	2812	56	0.686	0.861	0.721	0.998
P2	D_2	2812	1108	0.036	0.019	0.022	0.999

Table 2. Average results based on the weight of each class in each problem.

Problem	Dataset	Weighted Recall	Weighted Precision	Weighted F1
P1	D_1	0.741	0.894	0.794
P2	D_1	0.663	0.62	0.637
P1	D_2	0.95	0.956	0.946
P2	D_2	0.50	0.223	0.296

4   Conclusion and Future Works

This paper presents a model and first experiments on real use case data, for a lifelong pathways recommender system. First, our model summarizes the context of the recommender system by representing all the related stakeholders and their interactions (among them and with the recommender system) and highlights the importance of selecting meaningful actions. Second, our first experiments, that consider the problem of predicting actions, clearly show that this is a difficult problem to recommend combination of actions aligned with real suggestions by a social actor to a beneficiary. We are conscious that addressing the 6 challenges mentioned in this paper is ambitious and to maximize the chance of success of this thesis[1] our priority will be: (i) to validate our context modeling (ch. 1), (ii) to extend our first recommendation model (ch. 2-3) [1], (iii) to propose a lifelong pathways model based on a careful study of life trajectories [6] literature (ch. 4), (iv) to ensure the fairness and the explainability of our system (ch. 5) and (v) to build a rigorous evaluation protocol (ch. 6).

References

1. Labroche, N., Marcel, P., Ringuet, N., T'Kindt, T., Chanson, A., Devogele, T.: A chain composite item recommender for lifelong pathways. In: DaWaK, LNCS (2021)
2. Benouaret, I., Lenne, D.: Personalizing the museum experience through context-aware recommendations. In: SMC, pp. 743–748 (2015)
3. Jameson, A., et al.: Human decision making and recommender systems. In: Recommender Systems Handbook, pp. 611–648 (2015)
4. Jameson, A., Smyth, B.: A collaborative filtering probabilistic approach for recommendation to large homogeneous and automatically detected groups. Int. J. Interact. Multimed. Artif. Intell. **6**(2), 90–100 (2020)
5. Aniek, F.M., et al.: The role of explainability in creating trustworthy artificial intelligence for health care: a comprehensive survey of the terminology, design choices, and evaluation strategies. J. Biomed. Inform. **113** (2021)
6. Noël, D., Villanova-Oliver, M., Gensel, J., Le Quéau, P.: Design patterns for modelling life trajectories in the semantic web. In: Brosset, D., Claramunt, C., Li, X., Wang, T. (eds.) W2GIS 2017. LNCS, vol. 10181, pp. 51–65. Springer, Cham (2017). https://doi.org/10.1007/978-3-319-55998-8_4

---

[1] This PhD thesis is funded by French ANRT CIFRE 2020/0731.

7. Leroy, V., Pilourdault, J., Borromeo, R.M., Toyama, M., Amer-Yahia, S., Gaussier, E.: Task composition in crowdsourcing. In: DSAA, pp. 194–203 (2016)
8. Amer-Yahia, S., Yu, C., Basu Roy, S., Das, G.: Interactive itinerary planning. In: ICDE, pp. 15–26 (2011)
9. Singh, M., Borromeo, R.M., Hosami, A., Amer-Yahia, S., Elbassuoni, S.: Customizing travel packages with interactive composite items. In: DSAA (2017)

The Descriptiveness of Feature Descriptors with Reduced Dimensionality

Dániel Varga$^{(\boxtimes)}$, János Márk Szalai-Gindl, and Sándor Laki

Eötvös Loránd University, Pázmány Péter sétány 1/C, Budapest 1117, Hungary

Abstract. Nowadays, depth data has an important role in many applications. The sensors which can capture depth data became essential parts of autonomous vehicles. These sensors record a huge amount of 3D data (point clouds with x, y, and z coordinates). Furthermore, for many point cloud processing applications, it is important to calculate feature vectors that aim at describing the neighborhood of each point. Usually, a feature vector has high dimensionality, and storing it in a database is a difficult task. One of the most common operations on feature descriptors is the nearest neighbor search. However, earlier works show that nearest neighbor search with spatial index structures in high dimensions could be outperformed by sequential scan. In this work, we investigate how dimensionality reduction on 3D feature descriptors affects the descriptiveness.

Keywords: Feature descriptor · PCA · Nearest neighbor search

1 Introduction

One important sub-area of point cloud processing is registration, where the task is to align two point clouds to each other, i.e., to make a transformation which moves one point cloud to another in such a way as to minimize the distance between identical parts of the surfaces defined by the clouds. For larger point clouds, feature-based registration methods are the most common [5]. These processes have the following steps: (1) detect the keypoints to select such points for which the neighborhood is distinguishable from the rest of the point cloud, (2) compute the feature vectors to create descriptors that represent the geometric properties of the neighborhoods, (3) determine the correspondences to match the key points of two point clouds based on the distance between their feature vectors, (4) compute the transformation based on the previous step. The feature descriptors mentioned in the second step are usually elements of a high-dimensional vector space which are needed to store in a database due to various data management and data analysis tasks.

Suppose that we examine sequences of data distributions where the dimensionality increases along the sequences. For example, such a sequence can consist of uniform distibutions of different dimensions. Beyer et al. showed in their classic

© Springer Nature Switzerland AG 2021
L. Bellatreche et al. (Eds.): ADBIS 2021 Short Papers, Workshops and Doctoral Consortium,
CCIS 1450, pp. 317–322, 2021.
https://doi.org/10.1007/978-3-030-85082-1_29

paper [1] that it is true for a wide range of the aforementioned sequences of data distributions that the distances of a query point from the nearest and the farthest neighbors tend to be equal as the number of dimensions increases. In these cases, the concept of the nearest neighbor becomes meaningless in a data distribution with sufficiently high-dimensionality. However, even if the dimensionality is not so high, it can still occur that the indexing techniques are outperformed by sequential scan. They also point out that there are situations in which high-dimensional data vectors spread in a subset of the space such that their intrinsic dimensionality is significantly lower than the original dimensionality of the data space. (Intrinsic dimensionality is the minimum number for attributes needed to describe data [2] at least with a good approximation.) In these cases, the nearest neighbor queries and indexing may make sense. Further works addressed different data distributions with different dimensions and analyzed the performance of indexing techniques such as R-trees for the nearest neighbor search (e.g. [7,15]). For example, if the data set has the self-similar property, then the performance depends on the intrinsic ("fractal") dimensionality of it [7]. As Samet writes in his book [13, Chapter 4], one of the reasons is the intrinsic dimensionality of a data set why dimension reduction techniques are used for it before storing in a database. Another reason mentioned in this book is that the possible number of child nodes decreases at a given page capacity as the dimensionality increases for some disk-based search trees (e.g. the R-trees) because, for example, the space requirement for the representation of a minimum bounding box also depends on the dimensionality. However, if the number of child nodes decreases, the height of the search tree increases, leading to a decrease in performance. Because of the above, on one hand, we apply Principal Component Analysis (PCA) on the feature descriptor data; on the other hand, we modify the internal parameters of the feature descriptors to reduce the dimensionality. We examine how their performance changes.

Based on Guo et al. [4], one of the most important attributes of a feature descriptor is descriptiveness, which shows the capability to encapsulate the information of the underlying surface. There are existing works in which the authors compared 3D point feature descriptors. Guo et al. compute precision and recall values based on point pair correspondences, obtained from nearest neighbor searches [4,14]. Usually, these papers compare feature descriptors using only their default number of dimensions. Prakhya [10] compares the descriptiveness of feature descriptors, using PCA to reduce the dimensionality of the descriptors. They use RRR metric to measure descriptiveness, which is slightly different from the previously mentioned methods [14].

2   Method

In this work we investigate three well-known 3D point feature descriptors: Point Feature Histogram (PFH) [11], Fast Point Feature Histogram (FPFH) [12] and Spin Image (SI) [6]. Their default dimensions are 125, 33 and 153, respectively. Each feature descriptor has a parameter that can define the resulting number of bins. In the case of PFH, it is the number (n) of subdivisions on each axis, and the resulting number of dimensions is n^3. For FPFH the number of subdivisions also defines the number of dimensions, but the bins on each axis are concatenated. The resulting number of dimensions is $n * 3$. For SI the parameter which defines the length of the descriptor is *image_width*. The resulting number of dimensions is $(image\_width + 1) * (2 * image\_width + 1)$. More information can be found in the cited articles. We reduce the dimensionality of feature descriptors in two ways: 1) modify their internal parameters, and 2) perform PCA on a computed feature with its original dimensionality.

Followed Guo et al. [4], to evaluate the descriptiveness of feature descriptors we use the Precision-recall curve (PRC). According to them, the PRC is more suitable for evaluating 3D point descriptors than widely used ROC. Creating a PRC consists of the following steps. We select *sample_num* points from both clouds as keypoints, with uniform sampling. It is important to note that there are many keypoint detectors, which are designed to detect keypoints that are highly distinctive (unique), repeatable, and stable [9]. With using keypoint detection algorithms instead of randomly selected keypoints, the descriptiveness evaluation of the feature descriptors results may vary. We think that by using randomly selected keypoints we avoid any bias. We compute the feature descriptors for the selected keypoints (when we use PCA, we compute the feature descriptor for all point in all point cloud, to obtain the dimensionality reduction transformation). To compute the feature descriptors we need to set a radius that defines the neighborhood of a point. As the radius increases the computational time increases too. We used the same radius with each feature descriptor. In the next step, we build a k-d tree to find the nearest neighbors of the features of the selected keypoints from the two clouds. We use the nearest neighbor distance ratio to decide if a point-pair considered is a match [8]. If the ratio between a feature's nearest and second nearest neighbor is less than a threshold τ, we consider the feature and its first neighbor a match. We use the dataset's ground truth transformations to decide if a match is really correct or not. If the two points are closer than a given ϵ threshold after the transformation, it is counted as a correct match. Finally, we compute the *precision* and *recall* of the matches. To generate the PCR we increase the threshold τ from 0.5 to 1 with 0.1 steps. It starts from 0.5 because there are only a small number of feature pairs that satisfy the condition that the difference between the point and its nearest neighbor and the point and its second neighbor is at least twice as much. In the case of real-world data, outlier points like this are usually considered as noise. The AUC denotes the area under the Precision-recall curve, which shows how descriptive is a feature descriptor, using different τ values to filter the matches, in a very compact way.

3 Evaluation

To evaluate the feature descriptors with different number of dimensions we used the 7-Scenes RGB-D dataset [3]. The dataset consists of point cloud fragments, with approx. 250000 points in each cloud. Before the evaluation, we performed the following preprocessing steps on the point clouds. 1) We downsampled the point cloud using a voxel grid filter with 0.01 voxel leaf size for faster processing

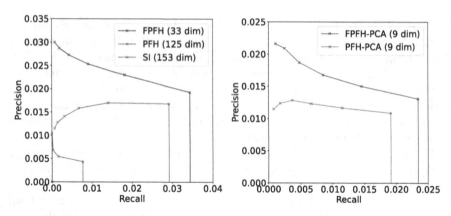

Fig. 1. Left: Precision-recall curve of 3 feature descriptor: FPFH, PFH, SI. Right: Precision-recall curve of FPFH and PFH. Both features were reduced to 9 dimensions.

time. The filter reduced the number of points in each cloud fragment to approx. 100000 points. 2) We filtered out the outlier points using a radius-based filtering method: a point was removed if it did not have 10 neighbors in a 0.04 radius. This step removed approx. 100 points from each fragment at most. 3) Finally, we estimated the normal vectors at each point, because it is needed to compute the feature descriptors. For normal estimation we used 0.04 as radius. The runtimes of computing feature descriptors are different from each other and it depends on the feature radius. In this work, we do not investigate computational times. For every feature descriptor, we used the same radius (0.06). For evaluation, we worked on 10 overlapping point cloud fragment pairs. The *sample_num* was 5000. Because of the randomness of keypoint selection, every point cloud fragment pair was evaluated 10 times. The PRC and AUC values on the figures show aggregated results.

First, we evaluated the feature descriptors with their original number of dimensions. Figure 1 (left) shows the PRC of three feature descriptor: FPFH, PFH, SI. Based on our evaluation the Spin Image descriptor is significantly worse than the other two. Note that SI has the most number of dimensions. The FPFH has a higher AUC than PFH, but the difference between the two descriptors is decreasing as the value of τ increases. Because the Spin Image descriptor has significantly smaller AUC value, we decided to investigate the dimensionality reduction on PFH and FPFH only.

Figure 2 shows the AUC values of FPFH and PFH descriptors with a varying number of dimensions. The left columns denote dimension reduction with parameters of the descriptors. The right columns denote the dimension reduction with PCA. We used PCA on the feature descriptor with its original dimension. Therefore the first column has only one value. In the case of FPFH we could compute the descriptor with many different dimensions: 33, 27, 15, 12, 9, and 6. Figure 2 (left) shows that from 15 dimensions FPFH-PCA gives better results. The difference between the AUC values becomes significant from 12 dimensions.

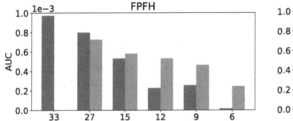 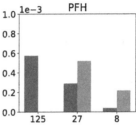

Fig. 2. Area Under Curve (AUC) of FPFH (left) and PFH (right) feature descriptors with different dimensions. The left columns denote the descriptors with reduced dimensionality using their internal parameters. The right columns denote dimensionality reduction using PCA.

If we need the use less than 15 dimensions for the FPFH, the PCA reduction is a better choice. Note that PCA computation adds time to the total runtime of computing and storing descriptors. Therefore, if the speed of runtime is critical, we should consider using FPFH without PCA with 15 dimensions despite FPFH-PCA is slightly better in this case also. With the parameters of PFH, we could only compute the feature descriptor with 3 different dimensions: 125, 27, and 8. In Fig. 2 (right), we want to highlight that the original descriptor with 125 dimensions has almost the same AUC value as the FPFH-PCA with 27 dimensions. The reason is, that for an ordinary point a PFH descriptor with 125 dimensions contains many zero values. In the case of PFH the PCA versions give significantly better results with 27 dimensions already. On Fig. 1 (right) we computed both descriptors with 9 dimensions. This figure is very similar to Fig. 1 (left). Based on our evaluation the FPFH-PCA gives better results than PFH-PCA with every number of dimensions.

4 Conclusion

Storing and indexing feature descriptors are difficult problems, especially when nearest neighbor searches are important. The biggest challenge is to handle descriptors with a high number of dimensions. In this work, we investigated how the descriptiveness of feature descriptors changes when we reduce their dimensionality. We reduced the dimensionality in two ways: modifying the descriptor's internal parameters and apply PCA dimension reduction technique. Based on our evaluation it is better to reduce the dimensionality with PCA, and FPFH is the more descriptive even with a low number of dimensions.

Acknowledgements. The authors thank the support of project "Application Domain Specific Highly Reliable IT Solutions" that has been implemented with the support provided from the National Research, Development and Innovation Fund of Hungary, financed under the Thematic Excellence Programme TKP2020-NKA-06 (National Challenges Subprogramme) funding scheme.

References

1. Beyer, K., Goldstein, J., Ramakrishnan, R., Shaft, U.: When is "nearest neighbor" meaningful? In: Beeri, C., Buneman, P. (eds.) ICDT 1999. LNCS, vol. 1540, pp. 217–235. Springer, Heidelberg (1999). https://doi.org/10.1007/3-540-49257-7_15
2. Fukunaga, K., Olsen, D.R.: An algorithm for finding intrinsic dimensionality of data. IEEE Trans. Comput. 100(2), 176–183 (1971)
3. Glocker, B., Izadi, S., Shotton, J., Criminisi, A.: Real-time RGB-D camera relocalization. In: 2013 IEEE International Symposium on Mixed and Augmented Reality (ISMAR), pp. 173–179 (2013)
4. Guo, Y., Bennamoun, M., Sohel, F., Lu, M., Wan, J., Kwok, N.: A comprehensive performance evaluation of 3d local feature descriptors. Int. J. Comput. Vis. 116, 66–89 (2015)
5. Holz, D., Ichim, A.E., Tombari, F., Rusu, R.B., Behnke, S.: Registration with the point cloud library: a modular framework for aligning in 3-d. IEEE Robot. Autom. Mag. 22(4), 110–124 (2015)
6. H'roura, J., Roy, M., Mansouri, A., Mammass, D., Juillion, P., Bouzit, A., Méniel, P.: Salient spin images: a descriptor for 3d object recognition. In: Mansouri, A., El Moataz, A., Nouboud, F., Mammass, D. (eds.) ICISP 2018. LNCS, vol. 10884, pp. 233–242. Springer, Cham (2018). https://doi.org/10.1007/978-3-319-94211-7_26
7. Korn, F., Pagel, B.U., Faloutsos, C.: On the" dimensionality curse" and the" self-similarity blessing". IEEE Trans. Knowl. Data Eng. 13(1), 96–111 (2001)
8. Lowe, D.: Distinctive image features from scale-invariant keypoints. Int. J. Comput. Vis. 60, 91–110 (2004)
9. Mian, A., Bennamoun, M., Owens, R.: On the repeatability and quality of keypoints for local feature-based 3d object retrieval from cluttered scenes. Int. J. Comput. Vis. 89, 348–361 (2010)
10. Prakhya, S., Liu, B., Lin, W., Li, K., Xiao, Y.: On creating low dimensional 3d feature descriptors with PCA. In: TENCON 2017–2017 IEEE Region 10 Conference, pp. 315–320 (November 2017)
11. Rusu, R., Marton, Z., Blodow, N., Beetz, M.: Persistent point feature histograms for 3d point clouds. In: Proceedings of the 10th International Conference on Intelligent Autonomous Systems (IAS-10), vol. 16 (January 2008)
12. Rusu, R.B., Blodow, N., Beetz, M.: Fast point feature histograms (fpfh) for 3d registration. In: 2009 IEEE International Conference on Robotics and Automation, pp. 3212–3217 (2009)
13. Samet, H.: Foundations of Multidimensional and Metric Data Structures. Morgan Kaufmann, Burlington (2006)
14. Spezialetti, R., Salti, S., Di Stefano, L.: Performance evaluation of learned 3d features. In: Ricci, E., Rota Bulò, S., Snoek, C., Lanz, O., Messelodi, S., Sebe, N. (eds.) ICIAP 2019. LNCS, vol. 11751, pp. 519–531. Springer, Cham (2019). https://doi.org/10.1007/978-3-030-30642-7_47
15. Weber, R., Schek, H.J., Blott, S.: A quantitative analysis and performance study for similarity-search methods in high-dimensional spaces. VLDB 98, 194–205 (1998)

Author Index

Akbar, Monika   133
Alawadi, Sadi   235
Aslam, Sidra   193

Babayev, Rufat   146
Barauskas, Andrius   61
Berti, Alessandro   169
Brilingaitė, Agnė   61
Bugeja, Joseph   235
Bukauskas, Linas   61

Čeikutė, Vaida   61
Čivilis, Alminas   61
Cordeiro, Robson L. F.   38
Costa, Umberto   97
Cychnerski, Jan   205, 217

Dahanayake, Ajantha   82
Deßloch, Stefan   27
Dong, Wenqin   50
Dong, Yuji   235
Doniparthi, Gajendra   27
Dziubich, Tomasz   205, 217

Espinosa-Oviedo, Javier A.   121

Farokhnejad, Mehrdad   72
Fischer, Peter M.   263, 287
Floriano, Luis   109

Ghahfarokhi, Anahita Farhang   169

Halfeld-Ferrari, Mirian   139
Hara, Carmem   109
Hertzberg, Vicki Stover   50
Ho, Joyce C.   50

Ivković, Vladimir   303

Jablonski, Stefan   14
Junior, Sylvio Barbon   157

Käppel, Martin   14
Kastner, Johannes   263
Kebande, Victor R.   235

Kossmann, Jan   3
Kozievitch, Nádia P.   133

Laki, Sándor   317
Lautert, Tatiane   133
Lee, Eric W.   50
Lescano, Alexis I. Aspauza   38
Lindner, Daniel   3
Löser, Alexander   3
Luković, Ivan   303

Machado, Sidnei   109
Magalhães, Dimmy   109
Medeiros, Ciro   97, 139
Mrissa, Michael   193
Mühlhaus, Timo   27
Musicante, Martin   139
Musicante, Martin   97

Neumann, Jennifer   287
Novitzká, Valerie   181

Olsson, Carl Magnus   235

Park, Gyunam   169
Persson, Jan A.   235
Pozo, Aurora   109
Pranesh, Raj Ratn   72

Ringuet, Nicolas   310

Šaltenis, Simonas   61
Schönig, Stefan   14
Shekhar, Ambesh   72
Sidiropoulos, Antonis   276
Simpson, Roy L.   50
Sodré, Ana   109
Steingartner, William   181
Stoupas, Georgios   276
Szalai-Gindl, János Márk   317

Tavares, Gabriel Marques   157
Teng, Xu   297
Thalheim, Bernhard   82

Tomter, Iver Toft   247
Trajcevski, Goce   297
Triapitcin, Ilia   82

van der Aalst, Wil M. P.   169
Varga, Dániel   317
Vargas-Solar, Genoveva   72, 121

Vilches-Blázquez, Luis M.   121
Villanueva-Miranda, Ismael   133

Wiese, Lena   146

Yu, Weihai   247

Zechinelli-Martini, José-Luis   121

Printed in the United States
by Baker & Taylor Publisher Services